PURE AND APPLIED MATHEMATICS

A Series of Texts and Monographs

Edited by: R. COURANT · L. BERS · J. J. STOKER

Vol. I: **Supersonic Flow and Shock Waves**
By R. Courant and K. O. Friedrichs

Vol. II: **Nonlinear Vibrations in Mechanical and Electrical Systems**
By J. J. Stoker

Vol. III: **Dirichlet's Principle, Conformal Mapping, and Minimal Surfaces**
By R. Courant

Vol. IV: **Water Waves**
By J. J. Stoker

Vol. V: **Integral Equations**
By F. G. Tricomi

Vol. VI: **Differential Equations: Geometric Theory**
By Solomon Lefschetz

Vol. VII: **Linear Operators—Parts I and II**
By Nelson Dunford and Jacob T. Schwartz

Vol. VIII: **Modern Geometrical Optics**
By Max Herzberger

Vol. IX **Orthogonal Functions**
By G. Sansone

Vol. X: **Lectures on Differential and Integral Equations**
By K. Yosida

Vol. XI: **Representation Theory of Finite Groups and Associative Algebras**
By C. W. Curtis and I. Reiner

Vol XII: **Electromagnetic Theory and Geometrical Optics**
By Morris Kline and Irvin W. Kay

Vol XIII: **Combinatorial Group Theory**
By W. Magnus, A. Karrass, and D. Solitar

Vol. XIV: **Asymptotic Expansions for Ordinary Differential Equations**
By Wolfgang Wasow

Vol. XV: **Tchebycheff Systems: With Applications in Analysis and Statistics**
By Samuel Karlin and William J. Studden

Vol. XVI: **Convex Polytopes**
By Branko Grünbaum

Vol. XVII: **Fourier Analysis in Several Complex Variables**
By Leon Ehrenpreis (*in preparation*)

Vol. XVIII: **Generalized Integral Transformations with Applications**
By Armen H. Zemanian

Vol. XIX: **Introduction to the Theory of Categories and Functors**
By Ion Bucur and Aristide Deleanu

Vol. XX: **Differential Geometry**
By J. J. Stoker

Vol. XXI: **Ordinary Differential Equations**
By Jack K. Hale

Vol. XXII: **Introduction to Potential Theory**
By L. L. Helms

Additional volumes in preparation

PURE AND APPLIED MATHEMATICS

A Series of Texts and Monographs

Edited by: R. COURANT · L. BERS · J. J. STOKER

VOLUME XXII

INTRODUCTION TO POTENTIAL THEORY

L. L. HELMS

Department of Mathematics
University of Illinois, Urbana

WILEY-INTERSCIENCE

a Division of John Wiley & Sons
New York · London · Sydney · Toronto

10 9 8 7 6 5 4 3 2 1

Library of Congress Catalog Card Number: 70-88312

SBN 471 36931 4

Printed in the United States of America

Preface

During the last decade there has been an upsurge in interest in potential theory which has created a need for a thorough exposition of classical potential theory. One cannot fully appreciate Markov processes or martingales in probability theory without knowing potential theory. The emergence of Brelot's axiomatic potential theory has opened the door to applications of potential theoretic methods in many areas of analysis. Choquet's theory of capacities has not only had an impact on potential theory and the theory of Markov processes but it promises much more. These new developments have their roots in classical potential theory and require a knowledge of the classical theory.

In writing *Introduction to Potential Theory*, I have tried to make the material accessible to the student who has just completed basic courses in measure theory and real variable theory. Much detail has been incorporated for this purpose. It is hoped that this book will provide the background necessary to understand what is now happening in probability theory and axiomatic potential theory.

Historical references are collected in "Notes and Comments." A glance at the bibliography will reveal an almost total absence of references to books or papers on complex variable type potential theory. This book is primarily about potential theory for higher dimensional spaces, but the two-dimensional case is included for the sake of completeness. An extensive search of the literature has not been made. I have, as far as possible, tried to identify the sources of some of the more important concepts. I hope I have not slighted anyone and apologize in advance if such is the case. There are very few references to contemporary potential theory, for I have limited the exposition to works that existed before 1958. References to papers published after 1958 were added only as a guide to the student who may wish to study contemporary potential theory.

I also wish to acknowledge several debts. Charles Lamb, a graduate student at the University of Illinois, read most of the manuscript and made many valuable suggestions; he has eliminated many errors and brought about the reorganization of some of the material to improve its readability. J. L. Doob furnished several references and insists on credit for the use herein of the phrase "the Green function" rather than "the Green's function." I appreciate the patience and understanding of E. H. Immergut of Wiley-Interscience during the prolonged period of manuscript preparation. Finally, my sincere thanks go to Mrs. Marilyn Cullen whose ability to place unerringly thousands of subscripts and superscripts has saved me many hours of toil.

Urbana, Illinois
May 1969

L. L. Helms

Contents

CHAPTER 0

Preliminaries

We shall be concerned primarily with a special class of functions defined on a part of the real n-dimensional Euclidean space E^n. Although the linear vector space properties of E^n are assumed to be known to the reader, we shall briefly review notation. Elements of E^n are denoted by $x = (x_1, \ldots, x_n)$. If $x = (x_1, \ldots, x_n)$ and $y = (y_1, \ldots, y_n)$ are in E^n, the inner product (x, y) is defined by $(x, y) = \sum_{j=1}^{n} x_j y_j$. The length of a vector $x \in E^n$ is defined to be the positive square root of (x, x) and is denoted by $\|x\|$. The distance between two vectors x and y is defined to be $\|x - y\|$. The angle between two nonzero vectors x and y is defined to be the angle ϕ such that $0 \leq \phi \leq \pi$ and

$$\cos \phi = \frac{(x, y)}{\|x\| \, \|y\|}.$$

The zero vector of E^n is denoted by $0 = (0, \ldots, 0)$. The ball $B_{x,\rho}$ with center x and radius ρ is defined by $B_{x,\rho} = \{y : \|x - y\| < \rho\}$. The closure of a set $F \subset E^n$ is denoted by $\bar{F}$, its boundary by ∂F, and its complement by $\sim F$.

Most functions are extended real-valued with domains in E^n. The restriction of a function f to a set E is denoted by $f|_E$. If u is an extended real-valued function with domain D, then u is locally bounded below (above) if $\inf_{y \in D \cap C} u(y) > -\infty$ $(\sup_{y \in D \cap C} u(y) < +\infty)$ for every compact set $C \subset E^n$. If u and v are extended real-valued functions with common domain, the maximum and minimum of u and v will be denoted by $u \vee v$ and $u \wedge v$, respectively; we shall let $u^+ = u \vee 0$ and $u^- = -(u \wedge 0)$, in which case $u = u^+ - u^-$ and $|u| = u^+ + u^-$. Continuity of such functions is relative to the metric topology on E^n unless otherwise specified. If u is an extended real-valued function with domain D, $x \in D$, and $u(x) = +\infty$ $(u(x) = -\infty)$, then u is continuous in the extended sense at x if $\lim_{y \to x} u(y) = +\infty$ $(\lim_{y \to x} u(y) = -\infty)$; u is continuous in the extended sense on D if either continuous or continuous in the extended sense at each point of D.

Let u be an extended real-valued function with domain $D \subset E^n$. For each $y \in E^n$ let $\mathcal{N}_y$ be the collection of neighborhoods of y. If $F \subset D$ and x is any point of $\bar{F}$, we define

$$\liminf_{\substack{y \to x \\ y \in F}} u(y) = \sup_{U \in \mathcal{N}_x} \left[\inf_{y \in U \cap F} u(y) \right]$$

$$\limsup_{\substack{y \to x \\ y \in F}} u(y) = \inf_{U \in \mathcal{N}_x} \left[\sup_{y \in U \cap F} u(y) \right];$$

if $F = D$ we simply write $\liminf_{y \to x} u(y)$ and $\limsup_{y \to x} u(y)$. The function u is lower semicontinuous (l.s.c.) at $x \in D$ if $u(x) = \liminf_{y \to x} u(y)$ and upper semicontinuous (u.s.c.) at $x \in D$ if $u(x) = \limsup_{y \to x} u(y)$. The function u is l.s.c. on D (u.s.c. on D) if l.s.c. at each point of D (u.s.c. at each point of D). If u is l.s.c. on D, then cu is l.s.c. or u.s.c. on D according as $c \geq 0$ or $c \leq 0$. In particular, $-u$ is u.s.c. on D if u is l.s.c. on D and the following statements for l.s.c. functions have u.s.c. duals. The function u is l.s.c. on D if and only if the set $\{x : u(x) > c\}$ is relatively open in D for every real number c. If $\{u_\alpha : \alpha \in A\}$ is a nonempty collection of l.s.c. functions with common domain D, then $u = \sup_{\alpha \in A} u_\alpha$ is l.s.c. on D. If u and v are l.s.c. functions with common domain such that $u + v$ is defined, then $u + v$ is l.s.c. If u and v are l.s.c. functions with common domain, then $u \wedge v$ is l.s.c. If u is any extended real-valued function on $D \subset E^n$, the lower regularization $\hat{u}$ of u defined for $x \in \bar{D}$ by

$$\hat{u}(x) = \liminf_{y \to x} u(y)$$

is l.s.c. on $\bar{D}$. If u is l.s.c. on the compact set D, then u attains its minimum on D. Lastly, if u is l.s.c. on $D \subset E^n$ and there is a continuous function f on E^n such that $u \geq f$ on D, then there is an increasing sequence of continuous functions $\{f_j\}$ on E^n such that $\lim_{j \to \infty} f_j = u$ on D.

If X is a locally compact Hausdorff space having a countable base for the topology, the class $\mathcal{B}$ of Borel subsets of X is the smallest σ-algebra of subsets of X containing the compact sets. A Borel measurable function is an extended real-valued function which is measurable relative to $\mathcal{B}$. A measure μ on $\mathcal{B}$ is a Borel measure if $\mu(C) < +\infty$ for every compact set $C \subset X$. **All measures will be Borel measures.** Each Borel measure μ on $\mathcal{B}$ is regular; that is, $\mu(E) = \sup\{\mu(C) : C \subset E,\ C \text{ compact}\}$ whenever $E \in \mathcal{B}$. In particular, any Borel measure on the Borel subsets of E^n is regular. A signed measure is an extended real-valued countably additive set function μ on $\mathcal{B}$ such that $\mu(\emptyset) = 0$ and such that μ takes on at most one of the two values $+\infty$ and $-\infty$. A signed measure μ has a decomposition $\mu = \mu^+ - \mu^-$ where μ^+ and μ^- are measures; the total variation $|\mu|$ of μ is the measure defined by $|\mu| = \mu^+ + \mu^-$; the norm $\|\mu\|$ of μ is defined by $\|\mu\| = \sup_{F \in \mathcal{B}} |\mu|(F)$. The signed measure μ is of bounded variation if $\|\mu\| < +\infty$. If μ is a measure on $\mathcal{B}$ and $E \in \mathcal{B}$, the restriction of μ to E is defined by $\mu|_E(F) = \mu(E \cap F)$, $F \in \mathcal{B}$, and vanishes on $\sim E$. The support of the measure μ is the complement of the largest open subset of X having μ-measure zero.

Under the same conditions on X, $C(X)$ will denote the collection of bounded continuous functions on X. $C(X)$ is a Banach space if the norm of an element $f \in C(X)$ is defined by $\|f\| = \sup_{x \in X} |f(x)|$. The set of functions in $C(X)$ having compact support (i.e., vanish on the complement of a compact set) will be denoted by $\mathscr{K}$. A sequence of signed measures $\{\mu_j\}$ on the Borel subsets of X converges in the w^*-topology to the signed measure μ if $\lim_{j \to \infty} \int f\, d\mu_j = \int f\, d\mu$ for all $f \in \mathscr{K}$; this convergence is denoted by $\mu_j \xrightarrow{w^*} \mu$.

The Lebesgue integral of a Lebesgue measurable function f defined on a Lebesgue measurable set $E \subset E^n$ is denoted by $\int_E f\, dz$; the function f is locally integrable if $\int_{E \cap C} |f|\, dz < +\infty$ for every compact set $C \subset E^n$.

Many of the functions that we shall deal with are functions only of the distance from the origin $0 \in E^n$. For such functions it is more convenient to use a spherical coordinate system rather than a rectangular coordinate system. The spherical coordinates of $x = (x_1, \ldots, x_n) \neq 0$ are defined as follows: if $r = \|x\|$, then

$$\theta = \left(\frac{x_1}{r}, \ldots, \frac{x_n}{r}\right)$$

is a point of $\partial B_{0,1}$, the unit sphere with center at 0. The pair (θ, r) uniquely determines x and are called the spherical coordinates of x. The spherical coordinates of 0 is the pair (0, 0). This transformation from rectangular coordinates to spherical coordinates is essentially the mapping $(x_1, \ldots, x_n) \rightsquigarrow (\theta_1, \ldots, \theta_{n-1}, r)$ where

$$\begin{aligned} \theta_1 &= \frac{x_1}{r} \\ \theta_2 &= \frac{x_2}{r} \\ &\vdots \\ \theta_{n-1} &= \frac{x_{n-1}}{r} \\ r &= (x_1^2 + \cdots + x_n^2)^{1/2}. \end{aligned}$$

We shall also let $\theta_n = x_n/r$. θ_n is the cosine of the angle between x and the vector $(0, 0, \ldots, 0, 1)$; that is, the angle between x and the "x_n-axis." The Jacobian of the mapping is easily calculated and its absolute value is given by

$$\left|\frac{\partial(x_1, \ldots, x_n)}{\partial(\theta_1, \ldots, \theta_{n-1}, r)}\right| = \frac{r^{n-1}}{(1 - \theta_1^2 - \cdots - \theta_{n-1}^2)^{1/2}} = \frac{r^{n-1}}{|\theta_n|}.$$

If $y = (y_1, \ldots, y_n) \in E^n$ and $\rho > 0$, then $\partial B_{y,\rho}$ is the surface defined by the equation

$$(x_1 - y_1)^2 + \cdots + (x_n - y_n)^2 = \rho^2.$$

Consider a Borel set $\Sigma \subset \partial B_{y,\rho} \cap \{(x_1, \ldots, x_n) : x_n - y_n \geq 0\}$. Let Σ_n denote the projection of Σ onto the subspace $\{(x_1, \ldots, x_n) : x_n = 0\}$; that is, $\Sigma_n = \{(x_1, \ldots, x_{n-1}, 0) : (x_1, \ldots, x_n) \in \Sigma\}$. For each $x \in \partial B_{y,\rho}$ let $\gamma = \gamma(x)$ be the angle between the "x_n-axis" and the outer normal to $\partial B_{y,\rho}$ at x. Then

$$\sec \gamma = \frac{1}{\cos \gamma} = \frac{\rho}{x_n - y_n}$$

and

$$\sigma(\Sigma) = \int \underset{\Sigma_n}{\cdots} \int \sec \gamma \, dx_1 \cdots dx_{n-1}$$

represents the surface area of Σ. If $\Sigma \subset \partial B_{y,\rho} \cap \{(x_1, \ldots, x_n) : x_n - y_n \leq 0\}$, the surface area of Σ is given by the same integral with $\sec \gamma = -\rho/(x_n - y_n)$. The integral of a Borel measurable function f defined on $\partial B_{y,\rho}$ relative to the surface area σ is denoted by

$$\int_{\partial B_{y,\rho}} f(x) \, d\,\sigma(x).$$

Consider an extended real-valued function f with domain in E^n. We shall take the following liberty with the functional notation. When f is considered as a function of the spherical coordinates (θ, r) of x, we shall denote the value of the composite function at (θ, r) by $f(\theta, r)$. Suppose f is integrable on $\bar{B}_{0,\rho}$. Then the integral of f over $\bar{B}_{0,\rho}$ can be evaluated using spherical coordinates as follows:

$$\int_{B_{0,\rho}} f(x) \, dx = \int_0^\rho \underset{\|\theta\|=1}{\int \cdots \int} f(\theta, r) \frac{r^{n-1}}{|\theta_n|} \, d\theta_1 \cdots d\theta_{n-1} \, dr$$

$$= \int_0^\rho r^{n-1} \left(\int_{\|\theta\|=1} f(\theta, r) \, d\,\sigma(\theta) \right) dr.$$

If $v_n(\rho)$ and $\sigma_n(\rho)$ denote the volume of a ball of radius ρ and the surface area of a sphere of radius ρ, respectively, then

$$v_n(\rho) = \underset{\|x\| \leq \rho}{\int \cdots \int} dx = \rho^n v_n(1)$$

and

$$\sigma_n(\rho) = 2 \underset{x_1^2 + \cdots + x_{n-1}^2 < \rho^2}{\int \cdots \int} \frac{\rho}{(\rho^2 - x_1^2 - \cdots - x_{n-1}^2)^{1/2}} \, dx_1 \cdots dx_{n-1}$$

$$= \rho^{n-1} \, \sigma_n(1),$$

the factor 2 being needed since the integral represents only the surface area of $\partial B_{0,\rho} \cap \{(x_1, \ldots, x_n) : x_n \geq 0\}$; but

$$v_n(1) = \int_0^1 r^{n-1}\left[\int_{\|\theta\|=1} d\sigma(\theta)\right] dr = \frac{\sigma_n(1)}{n}$$

and

$$\begin{aligned}
\sigma_n(1) &= 2 \underset{x_1^2+\cdots+x_{n-1}^2\leq 1}{\int\cdots\int} \frac{1}{(1 - x_1^2 - \cdots - x_{n-1}^2)^{1/2}}\, dx_1 \ldots dx_{n-1} \\
&= 2 \underset{x_1^2+\cdots+x_{n-2}^2\leq 1}{\int\cdots\int} \left[\int_{-(1-x_1^2-\cdots-x_{n-2}^2)^{1/2}}^{+(1-x_1^2-\cdots-x_{n-2}^2)^{1/2}} \frac{1}{(1 - x_1^2 - \cdots - x_{n-1}^2)^{1/2}}\, dx_{n-1}\right] \\
&\quad \times dx_1 \cdots dx_{n-2} \\
&= 2 \underset{x_1^2+\cdots+x_{n-2}^2\leq 1}{\int\cdots\int} \pi\, dx_1 \cdots dx_{n-2} \\
&= 2\pi v_{n-2}(1).
\end{aligned}$$

Using the equations $v_n(1) = \sigma_n(1)/n$ and $\sigma_n(1) = 2\pi v_{n-2}(1)$ and elementary facts about circles and spheres

$$\sigma_n(1) = \begin{cases} \dfrac{\pi^{n/2} n}{(n/2)!} & \text{if } n \text{ is even,} \\[2ex] \dfrac{2^{(n+1)/2}\, \pi^{(n-1)/2}}{1 \cdot 3 \cdot 5 \cdots (n-2)} & \text{if } n \text{ is odd and } > 1. \end{cases}$$

We use σ_n and v_n in place of $\sigma_n(1)$ and $v_n(1)$, respectively.

CHAPTER 1

Harmonic Functions

In this chapter we establish the basic tools of potential theory using the calculus. These tools include the Poisson integral formula by which a function harmonic on a neighborhood of a closed ball is represented as an integral over the boundary of the ball, the averaging principle for harmonic functions, and the maximum principle for harmonic functions. We also develop a preliminary version of the Riesz decomposition theorem according to which a superharmonic function can be expressed as the sum of a harmonic function and the potential of a mass distribution.

1. Green's Identity

Throughout this first chapter we deal only with real-valued functions defined on a neighborhood of the closure of an open set R and having continuous second partials thereon. Although the open set R can be rather general, we shall require that R has a smooth boundary. The particular cases of interest are those where R is a ball in E^n or the interior of the region between the surfaces of two balls of which one is contained in the other. We shall, in fact, limit R to one of these cases in this section.

Let $v = (v_1, \ldots, v_n)$ be a vector-valued function whose components v_j have continuous first partial derivatives on a neighborhood of $\bar{R} \subset E^n$. The divergence of v is defined by

$$\operatorname{div} v = \sum_{j=1}^{n} \frac{\partial v_j}{\partial x_j}.$$

We let $\mathbf{n}(x)$ denote the outer unit normal to the surface ∂R at the point $x \in \partial R$. We shall take as our starting point the following theorem which is adequately discussed in any advanced calculus textbook.

DIVERGENCE THEOREM

$$\int_R \operatorname{div} v \, dx = \int_{\partial R} (v, \mathbf{n}) \, d\sigma$$

We make the convention that whenever "**n**" appears in an integral over a smooth surface it is understood to be the outer unit normal to the surface.

Suppose now that u is a function defined on a neighborhood of $\bar{R}$ and has continuous second partial derivatives thereon. The Laplacian of u, Δu, is defined by

$$\Delta u = \sum_{j=1}^{n} \frac{\partial^2 u}{\partial x_j^2}$$

If u is a function of variables other than x and it is necessary to clarify the meaning of the Laplacian, we shall use $\Delta_{(x)}$ to signify that the Laplacian is relative to the coordinates of x. The gradient of u is defined by

$$\operatorname{grad} u = \left(\frac{\partial u}{\partial x_1}, \ldots, \frac{\partial u}{\partial x_n}\right).$$

Let v be a second such function. Then

$$u \operatorname{grad} v = \left(u \frac{\partial v}{\partial x_1}, \ldots, u \frac{\partial v}{\partial x_n}\right)$$

and

$$\operatorname{div}(u \operatorname{grad} v) = u\, \Delta v + (\operatorname{grad} u, \operatorname{grad} v).$$

It follows from the divergence theorem that

$$\begin{aligned} \int_R u\, \Delta v\, dx + \int_R (\operatorname{grad} u, \operatorname{grad} v)\, dx &= \int_{\partial R} (u \operatorname{grad} v, \mathbf{n})\, d\sigma \\ &= \int_{\partial R} u D_{\mathbf{n}} v\, d\sigma \end{aligned}$$

since $(u \operatorname{grad} v, \mathbf{n}) = u(\operatorname{grad} v, \mathbf{n})$ and the latter inner product is just the directional derivative $D_{\mathbf{n}} v$ of v in the direction **n**. By interchanging u and v and subtracting we obtain

GREEN'S IDENTITY

$$\int_R (u\, \Delta v - v\, \Delta u)\, dx = \int_{\partial R} (u D_{\mathbf{n}} v - v D_{\mathbf{n}} u)\, d\sigma$$

2. Fundamental Harmonic Functions

A real-valued function u defined on E^n and having continuous second partial derivatives thereon is called a **harmonic function** if $\Delta u = 0$ on E^n.

Numerous examples of harmonic functions can be constructed using the following theorem.

THEOREM **1.1** If f is an analytic function of the complex variable z, then the real and imaginary parts of f are harmonic functions.

Proof. Let $z = x + iy$ and $f(z) = u(x, y) + iv(x, y)$. Since f has derivatives of all orders, u and v have continuous second partial derivatives. Moreover, u and v satisfy the Cauchy-Riemann equations

$$\frac{\partial u}{\partial x} = \frac{\partial v}{\partial y}, \qquad \frac{\partial u}{\partial y} = -\frac{\partial v}{\partial x},$$

from which follows, for example,

$$\Delta u = \frac{\partial^2 u}{\partial x^2} + \frac{\partial^2 u}{\partial y^2} = \frac{\partial^2 v}{\partial y\,\partial x} - \frac{\partial^2 v}{\partial x\,\partial y} = 0.$$

EXAMPLE The function $u(x, y) = e^x \cos y$ is harmonic on E^2 since it is the real part of the analytic function e^z.

The above definition of harmonic function is relative to the region E^n. The function u is said to be **harmonic on the open set** $R \subset E^n$ if u has continuous second partial derivatives on R and $\Delta u(x) = 0$ for all $x \in R$. Suppose u is harmonic on the non-empty open set R and let y be a fixed point of R. Then $\Delta u(x)$ can be regarded as a function of the spherical coordinates (θ, r) of x relative to y for which $r = \|x - y\|$ and θ is point of intersection of the line segment joining x to y and a unit sphere with center at y. Suppose that u is a function of r alone. Then Δu, as a function of the spherical coordinates, is easily seen to be given by

$$\Delta u = \frac{d^2 u}{dr^2} + \frac{(n-1)}{r}\frac{du}{dr}, \qquad r \neq 0.$$

The only functions that are harmonic on $E^n \sim \{y\}$ and a function only of $r = \|x - y\|$ are those that satisfy the equation

$$\frac{d^2 u}{dr^2} + \frac{(n-1)}{r}\frac{du}{dr} = 0$$

on $E^n \sim \{y\}$. If $n = 2$, the general solution of this equation is $A \log r + B$ where A and B are arbitrary constants. The particular solution $-\log r$ is harmonic on $E^2 \sim \{y\}$ and is called the **fundamental harmonic function for** E^2 **with pole** y. If $n \geq 3$, the general solution of the above equation is $Ar^{-n+2} + B$. The particular solution r^{-n+2} is harmonic on $E^n \sim \{y\}$ and is called the

fundamental harmonic function for E^n, $n \geq 3$, **with pole** y. Since the fundamental harmonic functions occur repeatedly throughout this book, we shall introduce a notation for them and, at the same time, extend their domains so as to be defined on E^n by putting

$$u_y(x) = \begin{cases} +\infty & \text{if } \quad x = y, \\ -\log \|x - y\| & \text{if } \quad x \neq y, \end{cases}$$

in the $n = 2$ case and putting

$$u_y(x) = \begin{cases} +\infty & \text{if } \quad x = y, \\ \dfrac{1}{\|x - y\|^{n-2}} & \text{if } \quad x \neq y, \end{cases}$$

in the $n \geq 3$ case. The function u_y, so defined on E^n, will also be called the **fundamental harmonic function with pole** y.

The following theorem follows from the fact that the Laplacian is a linear differential operator.

THEOREM **1.2** The class of functions harmonic on an open set $R \subset E^n$ is a linear vector space over the reals which contains the constant functions.

3. The Averaging Principle

One of the principal problems of potential theory is that of extending a function on the boundary of an open set R to $\bar{R}$ in such a way that the extension is harmonic on R. This leads us to seek a representation of the harmonic function at interior points of R in terms of an integral over the boundary ∂R.

THEOREM **1.3** (*Gauss's Integral Theorem*). If h is harmonic on a neighborhood of the closure of a ball B, then

$$\int_{\partial B} D_{\mathbf{n}} h \, d\sigma = 0.$$

Proof. Put $v = 1$ in Green's identity.

In order to avoid needless duplication, the following theorem is stated more generally than is required at this point.

THEOREM **1.4** If u has continuous second partial derivatives on a neighborhood of the closure of a ball $B = B_{y,\rho}$, then (*i*) for $n = 2$ and $x \in B$

$$u(x) = \frac{1}{2\pi} \int_{\partial B} [(-\log r) D_{\mathbf{n}} u - u D_{\mathbf{n}}(-\log r)] \, d\sigma(z)$$
$$- \frac{1}{2\pi} \int_B \Delta u(-\log r) \, dz;$$

(*ii*) for $n \geq 3$ and $x \in B$

$$u(x) = \frac{1}{\sigma_n(n-2)} \int_{\partial B} (r^{-n+2} D_{\mathbf{n}} u - u D_{\mathbf{n}} r^{-n+2}) \, d\sigma(z)$$
$$- \frac{1}{\sigma_n(n-2)} \int_B \Delta u \cdot r^{-n+2} \, dz,$$

where $r = \|x - z\|$, $z \in \bar{B}$.

Proof of (i). Consider a fixed $x \in B$ and let $v(z) = -\log \|x - z\| = -\log r$ for $z \neq x$. Then v is harmonic on $E^2 \sim \{x\}$. Choose δ such that $\bar{B}_{x,\delta} \subset B_{y,\rho}$ and let R be the open set $B_{y,\rho} \sim \bar{B}_{x,\delta}$. By Green's identity

$$\int_R (u \, \Delta v - v \, \Delta u) \, dz = \int_{\partial R} (u D_{\mathbf{n}} v - v D_{\mathbf{n}} u) \, d\sigma(z).$$

Since v is harmonic on R and $\partial R = \partial B_{y,\rho} \cup \partial B_{x,\delta}$,

$$- \int_R v \, \Delta u \, dz = \int_{\partial B_{y,\rho}} (u D_{\mathbf{n}} v - v D_{\mathbf{n}} u) \, d\sigma(z)$$

(a)
$$- \int_{\partial B_{x,\delta}} (u D_{\mathbf{n}} v - v D_{\mathbf{n}} u) \, d\sigma(z).$$

The minus sign precedes the second integral on the right in view of the convention about the outer unit normal vector. The outer unit normal vector for ∂R at a point of $\partial B_{x,\delta}$ is the negative of the outer unit normal vector for $\partial B_{x,\delta}$. We will now let $\delta \to 0$ in equation (a). To show that

(b)
$$\lim_{\delta \to 0} \int_R v \, \Delta u \, dz = \int_{B_{y,\rho}} v \, \Delta u \, dz,$$

it suffices to prove that v is integrable over $B_{y,\rho}$ since Δu is bounded on $B_{y,\rho}$. By transforming to spherical coordinates relative to the pole x, for $\delta < 1$

$$\int_{B_{x,\delta}} |v| \, dz \leq 2\pi \int_0^\delta r \log \frac{1}{r} \, dr.$$

Since $\lim_{r \to 0} r \log(1/r) = 0$, the function $v \, \Delta u$ is integrable on $B_{x,\delta}$. Since $v \, \Delta u$ is integrable on R, it is integrable on $B_{y,\rho}$ and (b) holds. Consider the right side of (a). The first integral there does not depend upon δ and we need only consider the second integral. Since $|D_{\mathbf{n}} u| = |(\mathbf{n}, \text{grad } u)| \leq \|\mathbf{n}\| \, \|\text{grad } u\| = \|\text{grad } u\| = (\sum u_{x_i}^2)^{1/2}$ and the first partial derivatives u_{x_i} are bounded on $\bar{B}_{y,\rho}$, $D_{\mathbf{n}} u$ is bounded on $\partial B_{x,\delta}$ by some constant m. For small δ

$$\left| \int_{\partial B_{x,\delta}} v D_{\mathbf{n}} u \, d\sigma(z) \right| \leq m \int_{\partial B_{x,\delta}} \log \frac{1}{r} \, d\sigma(z) = 2\pi m \delta \log \frac{1}{\delta}.$$

Since

$$\delta \log \frac{1}{\delta} \to 0 \quad \text{as} \quad \delta \to 0,$$

$$\lim_{\delta \to 0} \int_{\partial B_{x,\delta}} v D_{\mathbf{n}} u \, d\sigma(z) = 0.$$

Now consider

$$\lim_{\delta \to 0} \int_{\partial B_{x,\delta}} u D_{\mathbf{n}} v \, d\sigma(z).$$

We can compute $D_{\mathbf{n}} v(z)$ at a point $z \in \partial B_{x,\delta}$ as follows. Since $v(z) = -\log r$ and the normal derivative of v is just the derivative with respect to r, $D_{\mathbf{n}} v(z) = -r^{-1}$. Therefore

$$\int_{\partial B_{x,\delta}} u D_{\mathbf{n}} v \, d\sigma(z) = -\frac{1}{\delta} \int_{\partial B_{x,\delta}} u \, d\sigma(z).$$

The integral on the right is just the average of the continuous function u over $\partial B_{x,\delta}$, expect for a factor of -2π, and has a limit $-2\pi \, u(x)$ as $\delta \to 0$. This shows that

$$\lim_{\delta \to 0} \int_{\partial B_{x,\delta}} u D_{\mathbf{n}} v \, d\sigma(z) = -2\pi u(x).$$

Taking the limit as $\delta \to 0$ in (a), we obtain

$$-\int_B v \, \Delta u \, dz = \int_{\partial B} (u D_{\mathbf{n}} v - v D_{\mathbf{n}} u) \, d\sigma(z) + 2\pi u(x).$$

The proof of (ii) is basically the same with $v(z)$ being replaced by $v(z) = \|z - x\|^{-n+2}$.

This brings us to the important averaging property of harmonic functions.

THEOREM **1.5** If h is harmonic on a neighborhood of the closure of the ball $B = B_{y,\rho}$, then the value of h at the center of the ball is equal to the average of h over the boundary of the ball.

Proof. Only the $n \geq 3$ case will be proved, since the $n = 2$ case is similar. By Theorem 1.4,

$$h(y) = \frac{1}{\sigma_n(n-2)} \int_{\partial B} [\|y - z\|^{-n+2} D_{\mathbf{n}} h - h D_{\mathbf{n}} \|y - z\|^{-n+2}] \, d\sigma(z),$$

since $\Delta h = 0$ on $\bar{B}$. For $z \in \partial B$, $\|y - z\|^{-n+2} = \rho^{-n+2}$ and $D_{\mathbf{n}} \|y - z\|^{-n+2} = D_r r^{-n+2}|_{r=\rho} = -(n-2)\rho^{-n+1}$. After substituting in the above integral

$$h(y) = \frac{\rho^{-n+2}}{\sigma_n(n-2)} \int_{\partial B} D_{\mathbf{n}} h \, d\sigma(z) + \frac{\rho^{-n+1}}{\sigma_n} \int_{\partial B} h \, d\sigma(z).$$

The first integral on the right is zero by Theorem 1.3 and the second integral is just the average of h over ∂B.

THEOREM **1.6** If h is harmonic on the ball $B = B_{y,\rho}$, the value of h at the center y of the ball is equal to the average of h on the ball itself.

Proof. By introducing spherical coordinates (θ, r) relative to the pole y

$$\frac{1}{v_n \rho^n} \int_B h \, dx = \frac{1}{v_n \rho^n} \int_0^\rho r^{n-1} \left(\int_{\|\theta\|=1} h(\theta, r) \, d\sigma(\theta) \right) dr.$$

If $\|\theta\| = 1$, then (θ, r) is a point of $\partial B_{y,r}$. The integral within parentheses is the integral over a sphere of radius r relative to a uniformly distributed measure of total mass σ_n and is equal to $\sigma_n h(y)$. Therefore

$$\text{average of } h \text{ over } B = \frac{1}{v_n \rho^n} \int_0^\rho r^{n-1} \sigma_n h(y) \, dr$$

$$= \frac{\sigma_n}{n v_n} h(y).$$

The theorem follows from the fact that $v_n = \sigma_n / n$.

The content of the two preceding theorems is summarized by saying that the harmonic functions are **mean valued** or satisfy the **averaging principle.** Averages occur so frequently that a notation will be introduced. If u is a function integrable relative to surface area on the boundary ∂B of $B = B_{y,\rho}$, define

$$L(u : y, \rho) = \frac{1}{\sigma_n \rho^{n-1}} \int_{\partial B} u \, d\sigma.$$

If u is integrable on B relative to Lebesgue measure, define

$$A(u : y, \rho) = \frac{1}{v_n \rho^n} \int_B u \, dx.$$

Using the notation, if h is harmonic on the open set R, then $h(y) = L(h : y, \rho) = A(h : y, \rho)$ whenever $\bar{B}_{y,\rho} \subset R$. A partial converse is proved later.

4. The Poisson Integral Formula

The results of the preceding section can be improved upon by showing that the value of a harmonic function h at an interior point of a ball is a certain weighted average of h over the boundary of the ball. In other words, the value of h at an interior point is completely determined by its values on the boundary. Let us re-examine the results of Theorem 1.4 for the $n = 2$ case. If h is harmonic on a neighborhood of the closure of $B = B_{y,\rho}$ and $x \in B$, then

$$\text{(c)} \qquad h(x) = \frac{1}{2\pi} \int_{\partial B} [(-\log \|z - x\|) D_{\mathbf{n}} h - h D_{\mathbf{n}}(-\log \|z - x\|)] \, d\sigma(z)$$

since $\Delta h = 0$ on B. Consider the integral on the right as a function of x. If $x^\star \notin \bar{B}$, then $-\log \|z - x^\star\|$ is harmonic on a neighborhood of $\bar{B}$ and by Green's identity

$$\text{(d)} \qquad 0 = \frac{1}{2\pi} \int_{\partial B} [(-\log \|z - x^\star\|) D_{\mathbf{n}} h - h D_{\mathbf{n}}(-\log \|z - x^\star\|)] \, d\sigma(z).$$

Although it is not essential for the following argument, we shall modify (d) to incorporate some constants for later use. By Theorem 1.3,

$$\text{(e)} \qquad \frac{1}{2\pi} \int_{\partial B} \left[-\log\left(\frac{\|y - x\|}{\rho} \right) \right] D_{\mathbf{n}} h \, d\sigma(z) = 0$$

since the factor in brackets is a constant. Moreover,

$$D_{\mathbf{n}}[-\log(\|y - x\| \, \|z - x^\star\|/\rho)] = D_{\mathbf{n}}[-\log\|z - x^\star\|],$$

since $\log(\|y - x\|/\rho)$ is a constant. It follows from (e), (d), and this remark that

$$\text{(f)} \quad 0 = \frac{1}{2\pi} \int_{\partial B} \left[-\log\left(\frac{\|y - x\| \, \|z - x^\star\|}{\rho} \right) D_{\mathbf{n}} h \right.$$
$$\left. - h D_{\mathbf{n}}\left(-\log \frac{\|y - x\| \, \|z - x^\star\|}{\rho} \right) \right] d\sigma(z).$$

It follows from (c) and (f) that

$$h(x) = \frac{1}{2\pi} \int_{\partial B} \left[\log\left(\frac{\|y - x\|}{\rho} \frac{\|z - x^\star\|}{\|z - x\|} \right) D_{\mathbf{n}} h \right.$$
$$\left. - h D_{\mathbf{n}}\left(\log \frac{\|y - x\|}{\rho} \frac{\|z - x^\star\|}{\|z - x\|} \right) \right] d\sigma(z)$$

whenever $x \in B$, $x^\star \notin \bar{B}$. If for each $x \in B$ we could choose $x^\star \notin \bar{B}$ as a function of x such that

$$\text{(g)} \qquad \frac{\|y - x\|}{\rho} \frac{\|z - x^\star\|}{\|z - x\|} = 1$$

independently of $z \in \partial B$, we would have the desired representation

$$\text{(h)} \qquad h(x) = -\frac{1}{2\pi} \int_{\partial B} h D_{\mathbf{n}} \left(\log \frac{\|y - x\|}{\rho} \frac{\|z - x^\star\|}{\|z - x\|} \right) d\sigma(z).$$

We shall now show how to choose $x^\star$ so that (g) holds independently of $z \in \partial B$. Consider the radial line joining y to x. For $x \neq y$, choose $x^\star$ on this radial line so that $\|x - y\| \, \|x^\star - y\| = \rho^2$. Then

$$\text{(j)} \qquad x^\star = y + \frac{\rho^2}{\|y - x\|^2} (x - y)$$

and is called the **inverse** of x relative to the sphere $\partial B_{y,\rho}$. Consider any $z \in \partial B$ and let ϕ be the angle between $z - y$ and $x - y$. Then

$$\text{(k)} \qquad \|z - x^\star\|^2 = \rho^2 + \|x^\star - y\|^2 - 2\rho \, \|x^\star - y\| \cos \phi,$$

$$\text{(l)} \qquad \|z - x\|^2 = \rho^2 + \|x - y\|^2 - 2\rho \, \|x - y\| \cos \phi.$$

By replacing $\|x^\star - y\|$ with $\rho^2/\|x - y\|$ in (k), solving (l) for $\cos \phi$, and substituting into (k), we obtain

$$\begin{aligned} \|z - x^\star\|^2 &= \rho^2 + \frac{\rho^4}{\|y - x\|^2} - \frac{2\rho^3}{\|y - x\|} \left(\frac{\rho^2 + \|y - x\|^2 - \|z - x\|^2}{2\rho \|y - x\|} \right) \\ &= \frac{\rho^2 \|z - x\|^2}{\|y - x\|^2}. \end{aligned}$$

The choice of $x^\star$ given by (j) thus establishes (g) independently of $z \in \partial B$ and we have proven the first part of the following theorem.

THEOREM **1.7** If h is harmonic on a neighborhood of the closure of the ball $B = B_{y,\rho}$, $x \in B$, and $x \neq y$, then

(i) for $n = 2$

$$h(x) = -\frac{1}{2\pi} \int_{\partial B} h D_{\mathbf{n}} \left(\log \frac{\|y - x\|}{\rho} \frac{\|z - x^\star\|}{\|z - x\|} \right) d\sigma(z);$$

(ii) for $n \geq 3$

$$h(x) = -\frac{1}{\sigma_n (n-2)} \int_{\partial B} h D_{\mathbf{n}} \left(\frac{1}{\|z - x\|^{n-2}} - \frac{\rho^{n-2}}{\|y - x\|^{n-2}} \frac{1}{\|z - x^\star\|^{n-2}} \right) d\sigma(z),$$

where $x^\star$ is the inverse of x relative to $\partial B_{y,\rho}$.

Proof of (ii). We follow the same steps in proving (i) with $-\log\|z-x\|$ replaced by $\|z-x\|^{-n+2}$ in (c) and with the appropriate constant before the integral. If $x^\star$ is defined as before, $-\log\|z-x^\star\|$ is also replaced by $\|z-x^\star\|^{-n+2}$ in (d), and the constant before the integral in (d) is adjusted, then upon multiplying both sides of the equation corresponding to (d) by α and subtracting the result from the equation corresponding to (c) we obtain

$$\text{(m)}\quad h(x)=\frac{1}{\sigma_n(n-2)}\int_{\partial B}\left[\left(\frac{1}{\|z-x\|^{n-2}}-\alpha\frac{1}{\|z-x^\star\|^{n-2}}\right)D_{\mathbf{n}}h\right.$$
$$\left.-hD_{\mathbf{n}}\left(\frac{1}{\|z-x\|^{n-2}}-\alpha\frac{1}{\|z-x^\star\|^{n-2}}\right)\right]d\sigma(z).$$

If we choose $\alpha=(\rho/\|y-x\|)^{n-2}$, then by (g)

$$\alpha=\frac{\rho^{n-2}}{\|y-x\|^{n-2}}=\frac{\|z-x^\star\|^{n-2}}{\|z-x\|^{n-2}}$$

for all $z\in\partial B$. As before, the quotient on the right is independent of $z\in\partial B$. With this choice of α, (m) reduces to the equation in (ii). The use of (g) in the $n\geq 3$ case is justified since the derivation of (g) in either case is strictly a two-dimensional vector subspace problem.

Although this theorem accomplishes the purpose we set out to achieve, we shall obtain another form of the integral representation by evaluating the normal derivatives. In doing this we shall take $y=0$, but this is not at all essential. Recall that $D_{\mathbf{n}}f=(\mathbf{n},\operatorname{grad}f)$. The outer unit normal to the surface $\partial B_{0,\rho}$ at $z\in\partial B_{0,\rho}$ is simply $z/\|z\|=z/\rho$. With $x\in B_{0,\rho}$ fixed, $x\neq 0$, and $x^\star$ the inverse of x,

$$\operatorname{grad}\log\|z-x\|=\frac{z-x}{\|z-x\|^2}$$

for $z\in\partial B_{0,\rho}$ and

$$\operatorname{grad}\left(\log\frac{\|x\|}{\rho}\frac{\|z-x^\star\|}{\|z-x\|}\right)$$
$$=\operatorname{grad}\log(\|x\|/\rho)+\operatorname{grad}\log\|z-x^\star\|-\operatorname{grad}\log\|z-x\|$$
$$=\frac{z-x^\star}{\|z-x^\star\|^2}-\frac{z-x}{\|z-x\|^2}.$$

Therefore at $z\in\partial B_{0,\rho}$

$$D_{\mathbf{n}}\left(\log\frac{\|x\|}{\rho}\frac{\|z-x^\star\|}{\|z-x\|}\right)=\left(\frac{z}{\rho},\frac{z-x^\star}{\|z-x^\star\|^2}-\frac{z-x}{\|z-x\|^2}\right).$$

If in (j) we put $y = 0$, substitute the resulting expression for $x^\star$ into the right side of this equation, and also use (g), we obtain

$$D_{\mathbf{n}}\left(\log \frac{\|x\|}{\rho} \frac{\|z - x^\star\|}{\|z - x\|}\right) = -\frac{1}{\rho} \frac{\rho^2 - \|x\|^2}{\|z - x\|^2}$$

for $z \in \partial B_{0,\rho}$. Applying this result to (i) of Theorem 1.7,

$$h(x) = \frac{1}{2\pi\rho} \int_{\partial B_{0,\rho}} h(z) \frac{\rho^2 - \|x\|^2}{\|z - x\|^2}\, d\sigma(z), \qquad x \in B_{0,\rho} \sim \{0\},$$

under the conditions of the theorem. The inverse of x does not appear in this representation. Note also that this equation holds when $x = 0$ since it reduces to $h(0) = L(h : 0, \rho)$. The preceding representation is known as the Poisson integral representation. Exactly the same procedure is used to establish the representation in the $n \geq 3$ case.

THEOREM **1.8** (*Poisson Integral Formula*) If h is harmonic on a neighborhood of the closure of a ball $B = B_{y,\rho}$ and $x \in B$, then

$$h(x) = \frac{1}{\sigma_n \rho} \int_{\partial B} \frac{\rho^2 - \|y - x\|^2}{\|z - x\|^n} h(z)\, d\sigma(z).$$

As it sometimes is necessary to compute a partial derivative of the Poisson integral with respect to one of the components of x, we shall justify the interchange of differentiation and integration once and for all.

LEMMA **1.9** Let U and C be open and compact subsets of E^n, respectively, let $u = u(x, y)$ be a real-valued function on $U \times C$ with the property that $(\partial u/\partial x_j)(x, y)$ is continuous on $U \times C$ for each j. Then

$$\frac{\partial}{\partial x_j} \int_C u(x, y)\, d\mu(y) = \int_C \frac{\partial u}{\partial x_j}(x, y)\, d\mu(y), \qquad x \in U, j = 1, \ldots, n,$$

for any measure μ on C.

Proof. Let η be a unit vector. We shall show that

$$D_\eta \int_C u(x, y)\, d\mu(y) = \int_C D_\eta u(x, y)\, d\mu(y).$$

Consider a fixed $x \in U$. The left side of this equation is just

$$\lim_{\lambda \downarrow 0} \int_C \frac{u(x + \lambda\eta, y) - u(x, y)}{\lambda}\, d\mu(y).$$

Let B be a ball with center x such that $\bar{B} \subset U$ and consider only those λ for which $x + \lambda\eta \in B$. By the mean value theorem of calculus for each such λ and each $y \in C$ there is a point $\xi_{\lambda, y}$ on the line segment joining x to $x + \lambda\eta$ such that

$$\int_C \frac{u(x + \lambda\eta, y) - u(x, y)}{\lambda}\, d\mu(y) = \int_C D_\eta u(\xi_{\lambda, y}, y)\, d\mu(y).$$

Since $D_\eta u$ is continuous on $\bar{B} \times C$, the integrand on the right is uniformly bounded for such λ and $y \in C$. We can therefore apply the Lebesgue dominated convergence theorem to obtain

$$\begin{aligned} D_\eta \int_C u(x, y)\, d\mu(y) &= \lim_{\lambda \downarrow 0} \int_C D_\eta u(\xi_{\lambda, y}, y)\, d\mu(y) \\ &= \int_C D_\eta u(x, y)\, d\mu(y). \end{aligned}$$

THEOREM **1.10** If h is harmonic on the open set R, then all partial derivatives of h are harmonic on R.

Proof. If $y \in R$, $\bar{B} = \bar{B}_{y, \rho} \subset R$, and $x \in B$, then

$$h(x) = \frac{1}{\sigma_n \rho} \int_{\partial B} \frac{\rho^2 - \|y - x\|^2}{\|z - x\|^n}\, h(z)\, d\sigma(z).$$

The preceding lemma may be used to show that

$$\frac{\partial h}{\partial x_j}(x) = \frac{1}{\sigma_n \rho} \int_{\partial B} \frac{\partial}{\partial x_j} \left(\frac{\rho^2 - \|y - x\|^2}{\|z - x\|^n} \right) h(z)\, d\sigma(z)$$

for each $j = 1, \ldots, n$. By computing the partial derivative under the integral and using the Lebesgue dominated convergence theorem, we can show that $\partial h/\partial x_j$ is continuous on B. In the same way, we can show that partial derivatives of all orders are continuous on B. Since B can be any ball with closure in R, the partial derivatives of h of all orders are continuous on R. Since

$$\sum_{i=1}^n \frac{\partial^2 h}{\partial x_i^2} = 0 \quad \text{on} \quad R,$$

$$\sum_{i=1}^n \frac{\partial}{\partial x_j} \frac{\partial^2 h}{\partial x_i^2} = 0$$

on R for each j. Using the continuity of the third partial derivatives, we can interchange the order of differentiation to obtain

$$\sum_{i=1}^n \frac{\partial^2}{\partial x_i^2} \frac{\partial h}{\partial x_j} = 0$$

for each j. Since $\partial h/\partial x_j$ has continuous second partial derivatives on R, $\partial h/\partial x_j$ is harmonic on R as well as all higher order partial derivatives.

According to the following theorem, there are no bounded harmonic functions on E^n except for the constant functions.

THEOREM **1.11** (*Picard*) If h is a harmonic function on E^n and either bounded above or bounded below, then h is a constant function.

Proof. Since $-h$ is harmonic if h is harmonic, we can assume that h is bounded below. Since the sum of a harmonic function and a constant is harmonic, we can assume that $h \geq 0$. Let x and y be distinct points. Consider a ball $B_{x,\delta}$ and a ball $B_{y,\varepsilon} \supset B_{x,\delta}$. By Theorem 1.6

$$v_n \delta^n h(x) = \int_{B_{x,\delta}} h(z)\, dz \leq \int_{B_{x,\varepsilon}} h(z)\, dz = v_n \varepsilon^n h(y)$$

and $h(x) \leq (\varepsilon/\delta)^n h(y)$. Now let $\varepsilon, \delta \to +\infty$ in such a way that $\varepsilon/\delta \to 1$ to obtain $h(x) \leq h(y)$. But, since x and y are arbitrary points, $h(y) \leq h(x)$ and h is a constant function.

5. The Maximum Principle

According to the Poisson integral formula, if a harmonic function is zero on the boundary of a ball then it is zero on the ball. This brings us to the important maximum principle of potential theory.

Definition. A function u defined on an open connected set R obeys the **maximum principle** if $\sup_{x \in R} u(x)$ is not attained on R unless u is a constant. There is a corresponding **minimum principle** if $\inf_{x \in R} u(x)$ is not attained unless u is a constant.

THEOREM **1.12** If u is continuous on the open connected set R and for each $x \in R$ there is a $\delta_x > 0$ such that $u(x) = L(u\colon x, \delta)$ whenever $\bar{B}_{x,\delta} \subset B_{x,\delta_x} \subset R$, then u obeys the maximum and minimum principles.

Proof. Let $M = \{x\colon u(x) = \sup_{y \in R} u(y), x \in R\}$. Since u is continuous, M is a relatively closed subset of R. We shall also show that M is open in R. If $x \in M$, then there is a δ_x such that $u(x) = L(u\colon x, \delta)$ whenever $\delta < \delta_x$. Consider any $y \in B_{x,\delta_x}$ and let $\delta_0 = \|y - x\| < \delta_x$. Since $u(x) = L(u\colon x, \delta_0)$, $L(u(x) - u\colon x, \delta_0) = 0$, but, since $x \in M$, $u(x) - u \geq 0$ on $\partial B_{x,\delta_0}$ and therefore $u(x) - u = 0$ a.e. (relative to surface area) on $\partial B_{x,\delta_0}$. By the continuity of u, $u = u(x)$ on $\partial B_{x,\delta_0}$ and, in particular, $u(y) = u(x)$. This shows that $y \in M$ and that $B_{x,\delta_x} \subset M$. We have shown that M is both relatively closed and open in R so that by the connectedness of R either $M = \varnothing$ or $M = R$. In the first case $\sup_{x \in R} u(x)$ is not attained on R and in the second case u is a constant function. This completes the proof that u satisfies the maximum principle. Now $-u$ also satisfies the hypotheses and therefore the maximum principle, but the maximum principle for $-u$ is the same as the minimum principle for u.

COROLLARY **1.13** If h is harmonic on the open connected set R, then h satisfies both the maximum and minimum principles.

6. Smoothing of Functions

We conclude this chapter with a study of certain smoothing operations. Let R be an open subset of E^n and let f be a function on R which is locally integrable. If $\delta > 0$, then R_δ will denote the set $\{x \in R : d(x, \sim R) > \delta\}$. Define a function f_δ on R_δ by $f_\delta(x) = A(f : x, \delta)$, $x \in R_\delta$.

THEOREM **1.14** If f is locally integrable on R, then f_δ is continuous on R_δ. If f is continuous on R, then f_δ has continuous first partial derivatives on R_δ. More generally, if f has continuous partial derivatives of order $k \geq 1$ on R, then f_δ has continuous partial derivatives of order $k + 1$ on R_δ.

Proof. Suppose $x, x_0 \in R_\delta$. Then

$$|f_\delta(x) - f_\delta(x_0)| = \left| \frac{1}{v_n \delta^n} \int_{B_{x,\delta}} f(z)\, dz - \frac{1}{v_n \delta^n} \int_{B_{x_0,\delta}} f(z)\, dz \right|$$
$$\leq \frac{1}{v_n \delta^n} \int_{B_{x,\delta} \Delta B_{x_0,\delta}} |f(z)|\, dz.$$

Since the Lebesgue measure of $B_{x,\delta} \Delta B_{x_0,\delta}$ tends to zero as x tends to x_0, the latter integral tends to zero as x tends to x_0 by the absolute continuity of the indefinite integral with respect to Lebesgue measure. It follows that

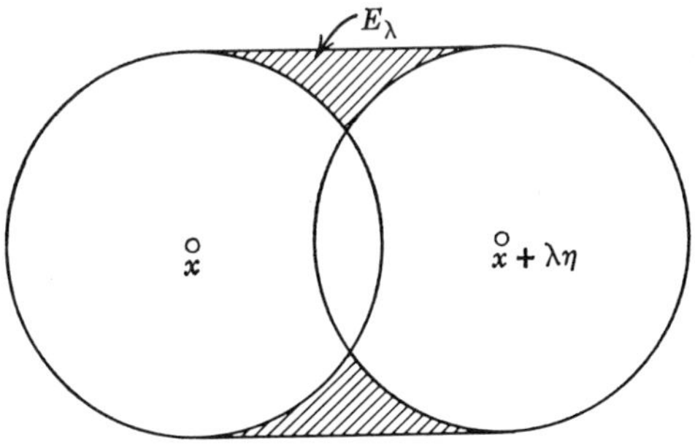

Figure 1

$\lim_{x \to x_0} f_\delta(x) = f_\delta(x_0)$ and that f_δ is continuous on R_δ. Suppose now that f is continuous on R. We shall show that f_δ has continuous first partials on R_δ. Let S denote the unit sphere in E^n; that is, $S = \{\theta : \|\theta\| = 1\}$. If η is a fixed element of S, let γ_θ denote the angle between θ and η and let $\lambda > 0$. Define $S^\eta = \{\theta \in S : \cos \gamma_\theta \geq 0\}$ and $S^{-\eta} = S \sim S^\eta$. Let E_λ be the set of points in $\bigcup_{0 \leq \rho \leq \lambda}\{\delta\theta + x + \rho\eta : \theta \in S^\eta\}$ but not in $B_{x,\delta} \cup B_{x+\lambda\eta,\delta}$. The former set is just the region swept out by translating a hemisphere in the direction η. Then

$$\int_{B_{x+\lambda\eta,\delta}} f(z)\,dz = \int_0^\lambda \int_{S^\eta} f(x + \rho\eta + \delta\theta)\cos\gamma_\theta\,d\sigma(\theta)\,d\rho - \int_{E_\lambda} f(z)\,dz + \int_{B_{x,\delta} \cap B_{x+\lambda\eta,\delta}} f(z)\,dz$$

Similarly,

$$\int_{B_{x,\delta}} f(z)\,dz = \int_0^\lambda \int_{S^{-\eta}} f(x + \rho\eta + \delta\theta)(-\cos\gamma_\theta)\,d\sigma(\theta)\,d\rho - \int_{E_\lambda} f(z)\,dz + \int_{B_{x,\delta} \cap B_{x+\lambda\eta,\delta}} f(z)\,dz$$

Therefore

$$\frac{f_\delta(x + \lambda\eta) - f_\delta(x)}{\lambda} = \frac{1}{v_n\delta^n}\frac{1}{\lambda}\int_0^\lambda \left[\int_S f(x + \rho\eta + \delta\theta)\cos\gamma_\theta\,d\sigma(\theta)\right]d\rho$$

Since f is continuous on R, the inner integral is continuous as a function of ρ in some small interval about $\rho = 0$. Since the limit of the left side of the last equation is just the directional derivative $D_\eta f_\delta(x)$,

$$D_\eta f_\delta(x) = \lim_{\lambda \downarrow 0}\frac{f_\delta(x + \lambda\eta) - f_\delta(x)}{\lambda} = \frac{1}{v_n\delta^n}\int f(x + \delta\theta)\cos\gamma_\theta\,d\sigma(\theta).$$

Since the integral on the right is a continuous function of x on R_δ, each directional derivative $D_\eta f_\delta$ is continuous on R_δ. As the last equation gives a representation of the first partials of f_δ, second partials can be computed by taking the derivative under the integral; this is permissible if f has continuous first partials on R. This procedure can be repeated as many times as the highest order of continuous partials of f.

CHAPTER 2

Functions Harmonic on a Ball

If R is a nonempty open subset of E^n with compact closure and f is a real-valued function defined on ∂R, the *classical Dirichlet problem* is that of finding a function h harmonic on R such that $\lim_{y \to x} h(y) = f(x)$ for all $x \in \partial R$. In this chapter we shall be concerned only with the case in which R is a ball. Generally speaking, the classical Dirichlet problem does not have a solution even when R is a ball. This occurs if f is not continuous on ∂R. We shall show that the problem is solvable if the boundary function is continuous. We shall also obtain an integral representation for functions which are harmonic on a ball B.

1. The Classical Dirichlet Problem

A typical problem in many applied mathematics textbooks is that of solving the classical Dirichlet problem for a semi-infinite slab where the boundary function, representing temperatures, is discontinuous. This classical Dirichlet problem, as we have defined it, is not solvable.

THEOREM **2.1** The classical Dirichlet problem is not solvable for discontinuous boundary functions.

Proof. Suppose the classical Dirichlet problem is solvable for a boundary function f defined on the boundary ∂R of a nonempty open set R. Let h be a solution. Consider any $x \in \partial R$ and $\varepsilon > 0$. Since $\lim_{y \to x} h(y) = f(x)$, there is a $B_{x,\delta}$ such that $|h(y) - f(x)| < \varepsilon/2$ for all $y \in B_{x,\delta} \cap R$. Suppose $x' \in B_{x,\delta/2} \cap \partial R$. Since $\lim_{y \to x'} h(y) = f(x')$, there is a ball $B_{x',\delta'}$ with $\delta' < \delta/2$ such that $|h(y) - f(x')| < \varepsilon/2$ for all $y \in B_{x',\delta'} \cap R$. Consider any $z \in B_{x',\delta'} \cap R \subset B_{x,\delta} \cap R$. For such a z

$$|f(x) - f(x')| \le |f(x) - h(z)| + |h(z) - f(x')| < \varepsilon;$$

that is, $|f(x) - f(x')| < \varepsilon$ for all $x' \in B_{x, \delta/2} \cap \partial R$. This shows that f is continuous at x, but, since x was any point of ∂R, f must be continuous on ∂R if the classical Dirichlet problem is solvable.

For the benefit of the reader who has cranked out the "solutions" of many Dirichlet problems involving discontinuous boundary data let us hasten to remark that our requirements of a solution are rather stringent. We shall later restore the good name of applied mathematics for such readers.

THEOREM **2.2** The solution of the classical Dirichlet problem for a nonempty open connected set R with compact closure and continuous boundary function f is unique if it exists.

Proof. Let h_1 and h_2 be two solutions. Suppose there is a point $z \in R$ such that $h_1(z) \neq h_2(z)$. We can assume that $h_1(z) > h_2(z)$. Then $\lim_{y \to x} [h_1(y) - h_2(y)] = 0$ for all $x \in \partial R$. Let

$$w = \begin{cases} h_1 - h_2 & \text{on} \quad R, \\ 0 & \text{on} \quad \partial R. \end{cases}$$

Since h_1 and h_2 are harmonic on R, w is harmonic on R, continuous on $\bar{R}$, and zero on ∂R. Since $w(z) > 0$ and w is continuous on the compact set $\bar{R}$, w attains its supremum at a point of R. By Corollary 1.13 w must be constant on R. Since $w = 0$ on ∂R and w is continuous on $\bar{R}$, this constant must be zero, but this contradicts the fact that $w(z) > 0$ and therefore $h_1 = h_2$.

2. Existence of a Solution for a Ball

In addition to constructing solutions of the classical Dirichlet problem for continuous boundary functions when R is a ball, we shall also construct functions harmonic on the ball which are not solutions of the classical Dirichlet problem. In doing so we shall make repeated use of the following equation:

$$\frac{1}{\sigma_n \rho} \int_{\partial B_{y,\rho}} \frac{\rho^2 - \|x - y\|^2}{\|z - x\|^n} \, d\sigma(z) = 1 \quad \text{for all } x \in B_{y,\rho}.$$

This follows from the fact that the constant function 1 is harmonic on a neighborhood of $\bar{B}_{y,\rho}$, that the left side of the equation is the Poisson integral of 1, and the right side is the value of 1 at x.

THEOREM **2.3** If μ is a signed measure of bounded variation defined on the Borel subsets of $\partial B_{y,\rho}$, then

$$h(x) = \frac{1}{\sigma_n \rho} \int_{\partial B_{y,\rho}} \frac{\rho^2 - \|x - y\|^2}{\|z - x\|^n} \, d\mu(z), \qquad x \in B_{y,\rho},$$

is harmonic on $B_{y,\rho}$.

Proof. Using Lemma 1.9, we can show that h has continuous second partials and that

$$\Delta h(x) = \frac{1}{\sigma_n \rho} \int_{\partial B_{y,\rho}} \Delta_{(x)} \frac{\rho^2 - \|y - x\|^2}{\|z - x\|^n} \, d\mu(z).$$

A tedious, but straightforward, differentiation shows that the integrand is zero for $x \in B_{y,\rho}$.

COROLLARY **2.4** If f is Borel measurable on $\partial B_{y,\rho}$ and integrable relative to surface area, then

$$h(x) = \frac{1}{\sigma_n \rho} \int_{\partial B_{y,\rho}} \frac{\rho^2 - \|y - x\|^2}{\|z - x\|^n} f(z) \, d\sigma(z)$$

is harmonic on $B_{y,\rho}$.

This corollary partially solves the classical Dirichlet problem for a ball since each continuous boundary function f determines a function h which is harmonic on the ball. If $\lim_{y \to x} h(y) = f(x)$ for each $x \in \partial B_{y,\rho}$, then we indeed have a solution. We now look at the boundary limits of the harmonic function h determined by the boundary function f.

In view of the repeated application of the Poisson integral formula, we introduce the following notation:

$$\mathbf{PI}(\mu, B)(x) = \frac{1}{\sigma_n \rho} \int_{\partial B} \frac{\rho^2 - \|y - x\|^2}{\|z - x\|^n} \, d\mu(z),$$

where $B = B_{y,\rho}$ and μ is a signed measure of bounded variation on ∂B. If μ is absolutely continuous relative to surface area on ∂B, then for each Borel set $M \subset \partial B$

$$\mu(M) = \int_M f(z) \, d\sigma(z)$$

for some integrable function f on ∂B and we shall let $\mathbf{PI}(\mu, B) = \mathbf{PI}(f, B)$. We shall also use this same notation even though f may be defined elsewhere. Note that $\mathbf{PI}(\mu, B)$ is a function which is defined only on B and that $\mathbf{PI}(1, B) = 1$. Moreover, $\mathbf{PI}(\mu, B)$ is linear in μ.

LEMMA **2.5** If f is Borel measurable on ∂B and integrable relative to surface area on ∂B, if $h = \mathbf{PI}(f, B)$ on B, and if there is a constant k such that $f \leq k$ a.e. (σ) in a neighborhood of $x_0 \in \partial B$, then $\limsup_{x \to x_0} h(x) \leq k$.

Proof. We first show that we can assume $k > 0$. If this is not the case, then consider $f - k + 1$. By hypothesis, $f - k + 1 \leq 1$ a.e. (σ) in a neighborhood of x_0 and $\limsup_{x\to x_0} \mathbf{PI}(f-k+1, B)(x) = \limsup_{x\to x_0} \mathbf{PI}(f, B)(x) - k + 1$. Thus, if we can prove the $k > 0$ case, then the general case can be reduced to it. Assuming $k > 0$, choose $\varepsilon > 0$ such that $f(z) \le k$ a.e. (σ) for $z \in B_{x_0,\varepsilon} \cap \partial B$. Then

$$h(x) = \frac{1}{\sigma_n \rho} \int_{\{\|z-x_0\|<\varepsilon\}\cap\partial B} \frac{\rho^2 - \|y-x\|^2}{\|z-x\|^n} f(z)\, d\sigma(z)$$

$$+ \frac{1}{\sigma_n \rho} \int_{\{\|z-x_0\|\ge\varepsilon\}\cap\partial B} \frac{\rho^2 - \|y-x\|^2}{\|z-x\|^n} f(z)\, d\sigma(z).$$

If $I_1(x)$ and $I_2(x)$ denote the first and second terms, respectively, then

$$I_1(x) \le \frac{k}{\sigma_n \rho} \int_{\{\|z-x_0\|<\varepsilon\}\cap\partial B} \frac{\rho^2 - \|y-x\|^2}{\|z-x\|^n}\, d\sigma\,(z)$$

$$\le \frac{k}{\sigma_n \rho} \int_{\partial B} \frac{\rho^2 - \|y-x\|^2}{\|z-x\|^n}\, d\sigma(z) = k\mathbf{PI}(1, B) = k.$$

Consider now $|I_2(x)|$. Suppose $x \in B_{x_0,\varepsilon/2}$. Then $\|z-x\| \ge \varepsilon/2$ when $\|z-x_0\| > \varepsilon$ for if not we would have $\|z-x_0\| \le \|z-x\| + \|x-x_0\| \le \varepsilon$. Then for $x \in B_{x_0,\varepsilon/2}$

$$|I_2(x)| \le \frac{1}{\sigma_n \rho} \int_{\{\|z-x_0\|\ge\varepsilon\}\cap\partial B} \frac{\rho^2 - \|y-x\|^2}{(\varepsilon/2)^n} |f(z)|\, d\sigma(z)$$

$$\le \frac{\rho^2 - \|y-x\|^2}{\sigma_n \rho(\varepsilon/2)^n} \int_{\partial B} |f(z)|\, d\sigma(z).$$

Since $\|y-x\| \to \rho$ as $x \to x_0$, $|I_2(x)| \to 0$ as $x \to x_0$. Therefore

$$\limsup_{x\to x_0} h(x) \le \limsup_{x\to x_0} I_1(x) + \limsup_{x\to x_0} I_2(x) \le k.$$

LEMMA **2.6** If f is Borel measurable on ∂B and integrable relative to surface area on ∂B and if $h = \mathbf{PI}(f, B)$ on B, then for $x_0 \in \partial B$

$$\limsup_{\substack{x\to x_0\\ x\in B}} h(x) \le \limsup_{\substack{x\to x_0\\ x\in\partial B}} f(x).$$

Proof. We can assume that the right side is not $+\infty$, for otherwise there is nothing to prove. If k is any number greater than the right member of the last inequality, then $f(x) < k$ for all $x \in \partial B$ in a neighborhood of $x_0 \in \partial B$. By Lemma 2.5

$$\limsup_{\substack{x\to x_0\\ x\in B}} h(x) \le k,$$

but, since k is any number greater than $\limsup_{\substack{x \to x_0 \\ x \in \partial B}} f(x)$, the theorem is proved.

LEMMA **2.7** If f is Borel measurable on ∂B and integrable relative to surface area on ∂B, if f is continuous at $x_0 \in \partial B$, and if $h = \mathbf{PI}(f, B)$ on B, then $\lim_{x \to x_0} h(x) = f(x_0)$.

Proof. By Lemma 2.6

$$\limsup_{x \to x_0} h(x) \leq f(x_0).$$

Since $\mathbf{PI}(-f, B) = -\mathbf{PI}(f, B)$, we also have

$$\limsup_{x \to x_0} -h(x) \leq -f(x_0)$$

or

$$\liminf_{x \to x_0} h(x) \geq f(x_0).$$

This shows that $\lim_{x \to x_0} h(x)$ exists and is equal to $f(x_0)$.

THEOREM **2.8** (*Schwarz*) The classical Dirichlet problem is solvable for a ball B and a continuous boundary function f. The solution is given by $\mathbf{PI}(f, B)$.

According to Theorem 1.5, harmonic functions are mean-valued. A partial converse is available.

COROLLARY **2.9** Let u be continuous on the open set R. If for each $x \in R$, $u(x) = L(u : x, \delta)$ for all sufficiently small δ, then u is harmonic on R; in particular, $u(x) = L(u : x, \delta) = A(u : x, \delta)$ whenever $\bar{B}_{x,\delta} \subset R$.

Proof. If $B = B_{x,\rho} \subset \bar{B}_{x,\rho} \subset R$, then the restriction of u to ∂B is continuous and the function v defined on $\bar{B}$ by

$$v(y) = \begin{cases} \mathbf{PI}(u, B)(y), & y \in B, \\ u(y), & y \in \partial B, \end{cases}$$

is continuous on $\bar{B}$ by Lemma 2.7. By definition $u - v = 0$ on ∂B. Since v is harmonic on B, for each $y \in B$, $v(y) = L(v : y, \delta)$ for all sufficiently small δ. Since the same is true of u by hypothesis, for each $y \in B$, $u(y) - v(y) = L(u - v : y, \delta)$ for all sufficiently small δ. Suppose there is a point $z \in B$ such that $u(z) - v(z) > 0$. Since $u - v$ is continuous on $\bar{B}$ and equal to zero on ∂B, $u - v$ attains its supremum on B. By Theorem 1.12 $u - v$ must be constant on B. Since $u - v = 0$ on ∂B, $u - v = 0$ on B, in contradiction to the assumption that $u(z) > v(z)$. Therefore $u \leq v$ on B. Applying the same argument to $v - u$, we obtain $u = v$ on B. Since v is harmonic on B, u is harmonic on B. Since B was an arbitrary ball with $\bar{B} \subset R$, u is harmonic on R.

3. Functions Harmonic on a Ball

It would appear from Theorem 2.3, that the class of functions harmonic on a ball is much more extensive than the class obtained by solving the classical Dirichlet problem; for example, if, in Theorem 2.3, μ is a unit measure concentrated on a point $z_0 \in \partial B_{y,\rho}$, we obtain a function that is harmonic on $B_{y,\rho}$ but does not arise from solving the classical Dirichlet problem, since its values approach $+\infty$ along a radial approach to z_0. We shall look at the converse of Theorem 2.3 to see what harmonic functions arise as the Poisson integral of signed measures of bounded variation. Before doing so, we prove two lemmas that may not be known to the beginner.

LEMMA **2.10** Let $\{\mu_i\}$ be a sequence of signed measures defined on the Borel subsets of the compact metric space S with $\|\mu_i\| \leq k$ for all i. Then there is a signed measure μ defined on the Borel subsets of S with $\|\mu\| \leq k$ and a subsequence $\{\mu_{i_j}\}$ which converges to μ in the w^*-topology.

Proof. Since S is compact metric, $C(S)$ is separable. Let $C_0 = \{f_1, f_2, \ldots\}$ be a countable dense subset of $C(S)$. The sequence of real numbers

$$\int f_1 \, d\mu_i$$

is bounded by $\|f_1\| \, k$. There is a subsequence $\{\mu_i^{(1)}\}$ of $\{\mu_i\}$ such that

$$\lim_{i \to \infty} \int f_1 \, d\mu_i^{(1)}$$

exists. Now consider the sequence of real numbers

$$\int f_2 \, d\mu_i^{(1)}$$

which is bounded by $\|f_2\| \, k$. There is a subsequence $\{\mu_i^{(2)}\}$ of $\{\mu_i^{(1)}\}$ such that

$$\lim_{i \to \infty} \int f_2 \, d\mu_i^{(2)}$$

exists. We successively select subsequences in this manner so that $\{\mu_i^{(j)}\}$ is a subsequence of $\{\mu_i^{(j-1)}\}$ and for each $j \geq 1$

$$\lim_{i \to \infty} \int f_j \, d\mu_i^{(j)}$$

exists. Since the diagonal sequence $\{\mu_i^{(i)}\}$ is a subsequence of every sequence in sight,

$$\lim_{i\to\infty} \int f_j \, d\mu_i^{(i)}$$

exists for each $j \geq 1$. Now consider any $f \in C(S)$ and any $\varepsilon > 0$. Let $f_{m_0} \in C_0$ be such that $\|f - f_{m_0}\| < \varepsilon/2k$. Then

$$\begin{aligned}
&\left|\int f \, d\mu_i^{(i)} - \int f \, d\mu_j^{(j)}\right| \\
&\quad \leq \left|\int (f - f_{m_0}) \, d\mu_i^{(i)}\right| + \left|\int f_{m_0} \, d\mu_i^{(i)} - \int f_{m_0} \, d\mu_j^{(j)}\right| + \left|\int (f_{m_0} - f) \, d\mu_j^{(j)}\right| \\
&\quad < \varepsilon + \left|\int f_{m_0} \, d\mu_i^{(i)} - \int f_{m_0} \, d\mu_j^{(j)}\right|.
\end{aligned}$$

Since $f_{m_0} \in C_0$, the sequence of absolute values on the right approaches zero as $i, j \to \infty$ and

$$\limsup_{i,\, j\to\infty} \left|\int f \, d\mu_i^{(i)} - \int f \, d\mu_j^{(j)}\right| \leq \varepsilon.$$

This shows that the sequence $\int f \, d\mu_i^{(i)}$, $i \geq 1$, is a Cauchy sequence and that $L(f) = \lim_{i\to\infty} \int f d\mu_i^{(i)}$ exists for each $f \in C(S)$. $L(f)$ is clearly a bounded linear functional on $C(S)$, since $|L(f)| = \lim_{i\to\infty} |\int f d\mu_i^{(i)}| \leq \|f\|\, k$. The Riesz representation theorem may be applied to assert the existence of the desired signed measure μ such that

$$\int f \, d\mu = \lim_{i\to\infty} \int f \, d\mu_i^{(i)}, \qquad f \in C(S)$$

and this is just the definition of w^*-convergence.

LEMMA **2.11** If $\{f_j\}$ is a sequence in $C(S)$ which converges uniformly to $f \in C(S)$ and $\{\mu_j\}$ is a sequence of signed measures on S with $\sup_{j\geq 1} \|\mu_j\| = k < +\infty$ which converges to μ in the w^*-topology, then

$$\lim_{j\to\infty} \int f_j \, d\mu_j = \int f \, d\mu.$$

Proof. If $\varepsilon > 0$, there is a j_ε such that

$$\left|\int f \, d\mu_j - \int f \, d\mu\right| < \frac{\varepsilon}{2}$$

and $\|f_j - f\| < \varepsilon/2k$ for all $j > j_\varepsilon$. For such j

$$\left|\int f_j\,d\mu_j - \int f\,d\mu\right| \le \left|\int (f_j - f)\,d\mu_j\right| + \left|\int f\,d\mu_j - \int f\,d\mu\right|$$

$$\le \|f_j - f\|\,\|\mu_j\| + \left|\int f\,d\mu_j - \int f\,d\mu\right|$$

$$< \varepsilon.$$

We now return to the class of functions harmonic on a ball; in particular, to those that arise from signed measures on the boundary of the ball.

LEMMA **2.12** If h is harmonic on $B = B_{y,\rho}$ and $L(|h| : y, \delta) \le k < +\infty$ for all $\delta < \rho$, there is a signed measure μ of bounded variation on ∂B such that $h = \mathbf{PI}(\mu, B)$ on B.

Proof. If $\delta < \rho$ and $x \in B_{y,\delta}$, then

$$h(x) = \frac{1}{\sigma_n \delta} \int_{\partial B_{y,\delta}} \frac{\delta^2 - \|x - y\|^2}{\|z - x\|^n} h(z)\,d\sigma(z).$$

If M is a Borel subset of $\bar{B}_{y,\rho}$ and $\delta < \rho$, define

$$\mu_\delta(M) = \int_{M \cap \partial B_{y,\delta}} h(z)\,d\sigma(z).$$

Then μ_δ is concentrated on $\partial B_{y,\delta}$ and

$$\|\mu_\delta\| = \int_{\partial B_{y,\delta}} |h(z)|\,d\sigma(z) = L(|h| : y, \delta)\sigma_n \delta^{n-1} \le k\sigma_n \rho^{n-1}$$

for all $\delta < \rho$. Let $\{\delta_j\}$ be a sequence of positive numbers such that $\delta_j \uparrow \rho$. By Lemma 2.10 a subsequence of the sequence $\{\mu_{\delta_j}\}$ converges in the w^*-topology to a signed measure μ with $\|\mu\| \le k\sigma_n \rho^{n-1}$. Since we only require a sequence $\delta_j \uparrow \rho$, we can assume that the μ_{δ_j}'s converge in the w^*-topology to μ. Since the signed measure μ_{δ_j} is concentrated on $\partial B_{y,\delta_j}$ and $\delta_j \uparrow \rho$, μ is concentrated on $\partial B_{y,\rho}$. Consider any $x \in B$. By dropping a finite number of terms of the sequence $\{\mu_{\delta_j}\}$, if necessary, we can assume that $\|y - x\| < \delta_j < \rho$ for all j. Then

(a)
$$h(x) = \frac{1}{\sigma_n \delta_j} \int_{\partial B_{y,\delta_j}} \frac{\delta_j^2 - \|y - x\|^2}{\|z - x\|^n}\,d\mu_{\delta_j}(z).$$

Since the signed measures μ_{δ_j} are concentrated in the spherical shell $\{z : \delta_1 \le \|z - y\| \le \rho\}$ we can consider them as signed measures on this shell. On this shell the sequence of integrands in (a) converges uniformly to $(\rho^2 - \|y - x\|^2)/\|z - x\|^n$. Applying Lemma 2.11 to (a), we obtain

$$h(x) = \frac{1}{\sigma_n \rho} \int_{\partial B_{y,\rho}} \frac{\rho^2 - \|x - y\|^2}{\|z - x\|^n} \, d\mu(z).$$

THEOREM **2.13** (*Herglotz*) The harmonic function h on $B = B_{y,\rho}$ is a difference of two positive harmonic functions if and only if there is a signed measure μ of bounded variation on ∂B such that $h = \mathbf{PI}(\mu, B)$.

Proof. Let $h = h_1 - h_2$, where h_1 and h_2 are positive functions harmonic on B. Suppose $0 < \delta < \rho$. Then

$$\begin{aligned} L(|h| : y, \delta) &\le L(h_1 : y, \delta) + L(h_2 : y, \delta) \\ &= h_1(y) + h_2(y) \end{aligned}$$

and the necessity follows from Lemma 2.12. If $h = \mathbf{PI}(\mu, B)$, where μ is a signed measure of bounded variation on ∂B, then $\mu = \mu^+ - \mu^-$, where μ^+, μ^- are finite measures on ∂B and $h = \mathbf{PI}(\mu^+, B) - \mathbf{PI}(\mu^-, B)$. The sufficiency follows from Theorem 2.3.

The natural question arises as to whether or not this corollary is valid for regions more general than balls. We return to this question in Chapter 12, where we take up the Martin boundary. There is also the question of the relationship between the limiting behavior of $h(x)$ as x approaches a boundary point and the measure μ on the boundary. Chapter 3 is devoted to the latter question.

4. Harnack's Inequality

The inequality proved in this paragraph is one of the fundamental tools of potential theory. We also consider an interesting variant of the inequality that will arise in connection with the study of boundary limits. The inequality of the following theorem is known as **Harnack's inequality**.

THEOREM **2.14** If $B_{y,\rho} \subset E^n$ and $0 < \delta < \rho$, then

$$\frac{h(x)}{h(x')} \le \frac{\rho^2}{\rho^2 - \delta^2} \left(\frac{\rho + \delta}{\rho - \delta} \right)^n$$

for all positive functions h harmonic on $B_{y,\rho}$ and all $x, x' \in \bar{B}_{y,\delta}$.

Proof. If $z \in \partial B_{y,\rho}$ and $x \in \bar{B}_{y,\delta}$, then

$$\rho - \delta \le \|z - x\| \le \rho + \delta$$

and

$$\frac{\rho^2 - \|x - y\|^2}{(\rho + \delta)^n} \le \frac{\rho^2 - \|x - y\|^2}{\|z - x\|^n} \le \frac{\rho^2 - \|x - y\|^2}{(\rho - \delta)^n}.$$

If we replace x with $x' \in \bar{B}_{y,\delta}$ and take the ratio of middle members, we obtain

$$\text{(b)} \qquad \frac{\rho^2 - \|x - y\|^2}{\|z - x\|^n} \le \left[\frac{\rho^2}{\rho^2 - \delta^2}\left(\frac{\rho + \delta}{\rho - \delta}\right)^n\right]\frac{\rho^2 - \|x' - y\|^2}{\|z - x'\|^n}.$$

If h is a positive function harmonic on $B_{y,\rho}$, then by Theorem 2.13 there is a measure μ such that $h = \mathbf{PI}(\mu, B_{y,\rho})$. If we integrate both sides of (b) by the measure μ, keeping x and x' fixed, and subsequently multiply both sides by $(\sigma_n \rho)^{-1}$, we then obtain the assertion of the theorem.

In the preceding theorem the two points x and x' were restricted to a compact subset of the ball $B_{y,\rho}$. This condition can be relaxed. Let $0 < \alpha < \pi/2$, $z \in \partial B_{y,\rho}$, and

$$C_{z,\alpha} = \left\{x : 1 \ge \frac{(x - z, y - z)}{\|x - z\|\,\|y - z\|} > \cos \alpha\right\},$$

$$C_{y,(\pi/2)-\alpha} = \left\{x : 1 \ge \frac{(x - y, z - y)}{\|x - y\|\,\|z - y\|} > \cos\left(\frac{\pi}{2} - \alpha\right)\right\},$$

$$D^s_{z,\alpha} = (C_{z,\alpha} \cap C_{y,(\pi/2)-\alpha}) \cup B_{y,\rho \sin \alpha}.$$

If U_z is a neighborhood of z, $D^s_{z,\alpha} \cap U_z$ is sometimes called a Stoltz domain but we shall refer to $D^s_{z,\alpha}$ as a **Stoltz domain.** The Stoltz domain $D^s_{z,\alpha}$ has the property that if $x \in D^s_{z,\alpha}$, then both x and the projection x' of x onto the line segment joining y to z lie in a ball $B' \subset D^s_{z,\alpha}$.

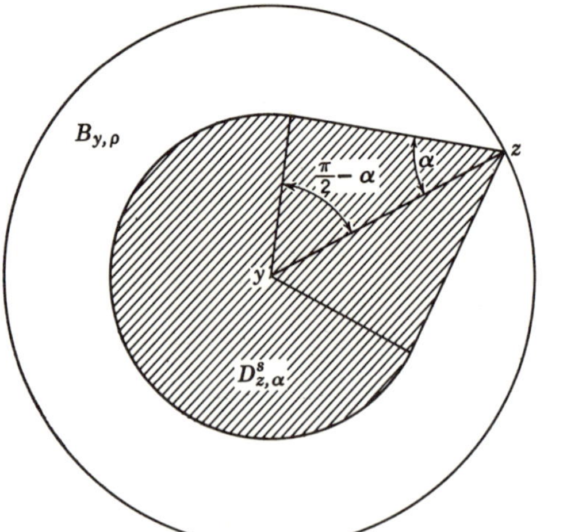

Figure 2

COROLLARY **2.15** Let $0 < \alpha < \pi/2$, $z \in \partial B_{y,\rho}$, and $D^s_{z,\alpha}$ the Stoltz domain corresponding to z and α. If B is a ball in $D^s_{z,\alpha}$, then there is a constant m depending only upon α such that $h(x)/h(x') \leq m$ for any positive harmonic function h on $B_{y,\rho}$ and any points $x, x' \in B$; the constant m is given by

$$m = \frac{1}{\cos^2 \alpha}\left(\frac{1 + \sin \alpha}{1 - \sin \alpha}\right)^n.$$

Proof. The Stoltz domain $D^s_{z,\alpha}$ has the property that there is a ball in $D^s_{z,\alpha}$ with center on the line segment joining z to y which contains B and which is internally tangent to the cone $C_{z,\alpha}$. We can therefore assume that B has the latter property. Since $0 < \alpha < \pi/2$, $\alpha + \varepsilon < \pi/2$ for sufficiently small positive ε. Consider any such ε and consider the ball B^ε which has the same center as B and is internally tangent to the cone $C_{z,\alpha+\varepsilon}$. If B has radius r, then B^ε has radius $(r/\sin \alpha)\sin(\alpha + \varepsilon)$. Applying Theorem 2.14 to $B \subset B^\varepsilon$,

$$\frac{h(x)}{h(x')} \leq \left[\frac{(r/\sin \alpha)^2 \sin^2(\alpha + \varepsilon)}{(r/\sin \alpha)^2 \sin^2(\alpha + \varepsilon) - r^2}\right]\left[\frac{(r/\sin \alpha)\sin(\alpha + \varepsilon) + r}{(r/\sin \alpha)\sin(\alpha + \varepsilon) - r}\right]^n$$

for all positive harmonic functions h on $B_{y,\rho}$ and all $x, x' \in B$. Letting $\alpha + \varepsilon \to \pi/2$, we obtain the constant m given above.

Harnack's inequality can be put into a neater form for applications.

THEOREM **2.16** (*Harnack's Inequality*) If h is a strictly positive function harmonic on an open connected set R and K is a compact subset of R, then there is a constant k, depending only upon R and K, such that $h(x)/h(y) \leq k$ for all $x, y \in K$.

Proof. Suppose there is no such constant. Then for every positive integer j there is a strictly positive function h_j harmonic on R and points x_j, $y_j \in K$ such that

$$\frac{h_j(x_j)}{h_j(y_j)} \geq j.$$

Since K is compact, the sequences $\{x_j\}$ and $\{y_j\}$ have subsequences with limits x and y, respectively, in K. We can assume that the sequences have this property. Since R is connected, we can join x and y by a polygonal path lying entirely in R. We can also find a chain of balls $B_1, \ldots, B_p$ with closures in R which cover the path, $x \in B_1$, $y \in B_p$, $B_i \cap B_{i+1} \neq \varnothing$. If for each i we apply Theorem 2.14 to $\bar{B}_i$ and a slightly larger concentric ball, there is a constant k_i such that $h(x')/h(y') \leq k_i$ whenever h is a positive function harmonic on R and $x', y' \in \bar{B}_i$. For each i let $z_i \in B_i \cap B_{i+1}$. Then for any $x' \in B_1$ and $y' \in B_p$

(c) $$\frac{h(x')}{h(z_1)}\frac{h(z_1)}{h(z_2)}\cdot\dots\cdot\frac{h(z_{p-1})}{h(y')} \le k_1 \cdot k_2 \cdot \dots \cdot k_p = k^*$$

for any positive function h harmonic on R. Now choose j_0 such that $x_{j_0} \in B_1$, $y_{j_0} \in B_p$ and

$$\frac{h_{j_0}(x_{j_0})}{h_{j_0}(y_{j_0})} \ge j_0 > k^*.$$

But this contradicts (c) with $x' = x_{j_0}$ and $y' = y_{j_0}$. Therefore the assumption that there is no such constant is incorrect.

5. Directed Families of Harmonic Functions

We have seen in Section 2 of this chapter that the Dirichlet problem is always solvable for a continuous boundary function on the boundary of a ball. In order to solve the Dirichlet problem for other types of regions, a method of approximation is used. This necessitates looking at properties of the limiting function of a convergent sequence of harmonic functions and generalized sequences of such functions.

LEMMA **2.17** (*Harnack*) If $\{h_\alpha : \alpha \in \mathfrak{A}\}$ is a net of functions harmonic on an open set R which converges uniformly on R to h, then h is harmonic on R.

Proof. Since the convergence is uniform on R, h is continuous on R. Suppose $x \in R$ and $\bar{B}_{x,\delta} \subset R$. Using the uniform convergence again,

$$h(x) = \lim_{\mathfrak{A}} \frac{1}{\sigma_n \delta^{n-1}} \int_{\partial B_{x,\delta}} h_\alpha \, d\sigma = \frac{1}{\sigma_n \delta^{n-1}} \int_{\partial B_{x,\delta}} h \, d\sigma = L(h : x, \delta).$$

By Corollary 2.9 h is harmonic on R.

THEOREM **2.18** If $\mathscr{F}$ is a family of harmonic functions on an open set R which is uniformly bounded on a neighborhood of each point of R and K is a compact subset of R, then the family $\mathscr{F}$ is equiuniformly continuous on K; moreover, each net $\{h_\alpha : \alpha \in \mathfrak{A}\}$ in $\mathscr{F}$ has a subnet which converges uniformly on K. If $\{h_\alpha : \alpha \in \mathfrak{A}\}$ is a convergent net of uniformly bounded harmonic functions on R, then $\lim_{\mathfrak{A}} h_\alpha$ is harmonic on R.

Proof. Since K is compact, the family $\mathscr{F}$ is uniformly bounded on some neighborhood W of K. Suppose $\sup\{|h(x)| : x \in W, h \in \mathscr{F}\} \le m$ and $\rho > 0$ is such that $\bar{B}_{x,\rho} \subset W$ for all $x \in K$. For $x, y \in K$,

$$|h(x) - h(y)| = \left| \frac{1}{v_n \rho^n} \int_{B_{x,\rho}} h(z)\, dz - \frac{1}{v_n \rho^n} \int_{B_{y,\rho}} h(z)\, dz \right|$$

$$\leq \frac{m}{v_n \rho^n} \int_{B_{x,\rho} \Delta B_{y,\rho}} dz.$$

Since the volume of $B_{x,\rho} \Delta B_{y,\rho}$ can be made arbitrarily small by making $\|x - y\|$ small, $|h(x) - h(y)|$ can be made arbitrarily small simultaneously for all $h \in \mathscr{F}$ and all $x, y \in K$ such that $\|x - y\|$ is small. This proves that the family $\mathscr{F}$ is equiuniformly continuous on K and, in particular, equicontinuous on R. The second assertion follows from the fact that the net $\{h_\alpha : \alpha \in \mathfrak{A}\}$, which is uniformly bounded and equicontinuous on K, has a subnet which converges uniformly on K by the Arzelà-Ascoli theorem. Now let $\{h_\alpha : \alpha \in \mathfrak{A}\}$ be a convergent net of uniformly bounded harmonic functions on R and let $h = \lim_{\mathfrak{A}} h_\alpha$. Let W be any open subset of R with compact closure $\overline{W} \subset R$. Then a subnet $\{h_\beta : \beta \in \mathfrak{B}\}$ converges uniformly on $\overline{W}$ to h. By Lemma 2.17 h is harmonic on W. Since W is arbitrary, h is harmonic on R.

THEOREM **2.19** (*Harnack*) If $\{h_j\}$ is a monotone increasing (decreasing) sequence of harmonic functions on an open connected set R, then $h(x) = \lim_{j\to\infty} h_j(x)$ is either identically $+\infty$ $(-\infty)$ or harmonic on R.

Proof. Suppose there is a point $x \in R$ such that $h(x) = \lim_{j\to\infty} h_j(x) < +\infty$. Let U be an open set containing x with compact closure $\overline{U} \subset R$. For each j, the non-negative harmonic function $h_j - h_1$ is either strictly positive or identically zero on R by the minimum principle. By Harnack's inequality there is a constant k such that

(d) $$h_j(y) - h_1(y) \leq k[h_j(x) - h_1(x)] \leq k[h(x) - h_1(x)]$$

for all $y \in \overline{U}$ provided $h_j - h_1$ is strictly positive; since (d) is trivially true if $h_j - h_1$ is identically zero on R, (d) holds for all j and all $y \in \overline{U}$. This implies that the sequence $\{h_j\}$ is uniformly bounded on U. By Theorem 2.18 h is harmonic on U and therefore on R.

The remainder of this section is devoted to families of functions that are not necessarily sequences. We shall henceforth use "function" to mean "extended real-valued function." Harmonic functions, of course, are always finite-valued.

Definition. A family $\mathscr{F}$ of functions defined on R is **right-directed** if for each pair $u, v \in \mathscr{F}$ there is a $w \in \mathscr{F}$ such that $u \leq w$ and $v \leq w$. There is a similar definition of **left-directed** obtained by reversing the inequalities.

LEMMA **2.20** (*Choquet*) Let $\{f_i : i \in I\}$ be a family of functions on an open set R. If $J \subset I$, let $f_J(x) = \inf_{i \in J} f_i(x)$, $x \in R$. Then there is a countable set $I_0 \subset I$ such that g l.s.c. on R and $g \leq f_{I_0}$ implies $g \leq f_I$.

Proof. Let $\{U_j\}$ be a countable base for the topology of R. We can assume that the f_i's take on values in $[-\pi/2, \pi/2]$, since we are concerned only with order properties of functions. This can be accomplished by replacing each f_i with arc tan f_i and noting that the arc tan function is an order-preserving map of the extended real number system. Consider $\alpha_j = \inf\{f_I(y) : y \in U_j\}$. Since there is a $y' \in U_j$ such that $f_I(y')$ is close to α_j and an $i \in I$ such that $f_i(y')$ is close to $f_I(y')$, for each j there is a $k_j \in I$ such that

$$\text{(e)} \qquad \inf_{y \in U_j} f_{k_j}(y) < \inf_{y \in U_j} f_I(y) + \frac{1}{j}.$$

Let $I_0 = \{k_j : j \geq 1\}$. Suppose g is l.s.c. on R and $g \leq f_{I_0}$, $\varepsilon > 0$, and $x \in R$. In some neighborhood of x, $g(y) > g(x) - \varepsilon/2$. Therefore there is a U_m such that (i) $x \in U_m$, (ii) $g(y) > g(x) - \varepsilon/2$ for all $y \in U_m$, and (iii) $1/m < \varepsilon/2$. Then $\inf_{y \in U_m} g(y) \geq g(x) - \varepsilon/2$ or

$$g(x) - \inf_{y \in U_m} g(y) \leq \frac{\varepsilon}{2}.$$

Since $g \leq f_{I_0} \leq f_{k_m}$,

$$\inf_{y \in U_m} g(y) - \inf_{y \in U_m} f_{k_m}(y) \leq 0.$$

Rewriting (e) with $j = m$,

$$\inf_{y \in U_m} f_{k_m}(y) - \inf_{y \in U_m} f_I(y) < \frac{1}{m} < \frac{\varepsilon}{2}.$$

Adding the last three inequalities,

$$g(x) - \inf_{y \in U_m} f_I(y) < \varepsilon.$$

Since $x \in U_m$,

$$g(x) < \inf_{y \in U_m} f_I(y) + \varepsilon \leq f_I(x) + \varepsilon.$$

There is, of course, a corresponding result for $f_I = \sup_{i \in I} f_i$ and u.s.c. functions. The following theorem generalizes Theorem 2.19.

LEMMA **2.21** If $\{h_i : i \in I\}$ is a right-directed (left-directed) family of functions harmonic on an open connected set R, then $h = \sup_{i \in I} h_i$ ($h = \inf_{i \in I} h_i$) is either identically $+\infty$ ($-\infty$) or harmonic on R.

Proof. If $i_0 \in I$, then $\sup\{h_i : i \in I\} = \sup\{h_i : i \in I, h_i \geq h_{i_0}\}$. We can therefore assume that the family $\{h_i : i \in I\}$ has a smallest element h_{i_0}. Suppose there is a point $x \in R$ such that $h(x) = \sup_{i \in I} h_i(x) < +\infty$. Let U be an open set containing x having compact closure $\bar{U} \subset R$. As in the proof of Theorem 2.19, the family $\{h_i : i \in I\}$ is uniformly bounded on U. Since this family is a net of harmonic functions having limit h, it follows from Theorem 2.18 that h is harmonic on U and therefore on R.

The next theorem is a generalization of the classical theorem of Dini which states that if a monotone sequence of continuous functions on a compact space converges to a continuous function, then the convergence is uniform.

LEMMA **2.22** (*Dini-Cartan*) Let $\{f_i : i \in I\}$ be a left-directed family of u.s.c. non-negative functions defined on a compact space X. If $\inf_{i \in I} f_i = 0$, then for each $\varepsilon > 0$ there is an $i_0 \in I$ such that $f_{i_0} < \varepsilon$.

Proof. For each $x \in X$, there is an i_x such that $f_{i_x}(x) < \varepsilon$. Since f_{i_x} is u.s.c., there is a neighborhood U_x of x such that $f_{i_x}(y) < \varepsilon$ for all $y \in U_x$. Since $\{U_x : x \in X\}$ is an open covering of X, there are a finite number of points $x_1, \ldots, x_p$ such that $X \subset \bigcup_{j=1}^{p} U_{x_j}$. Since the family $\{f_i : i \in I\}$ is left-directed, there is a function $f_{i_0} \leq f_{i_{x_j}}$, $j = 1, \ldots, p$. Consider any $x \in X$. Then $x \in U_{x_j}$ for some j and

$$f_{i_0}(x) \leq f_{i_{x_j}}(x) < \varepsilon.$$

The following lemma is a generalization of the Lebesgue monotone convergence theorem which will be needed later.

LEMMA **2.23** Let μ be a measure on the Borel subsets of a compact metric space X and let $\{f_i : i \in I\}$ be a left-directed family of finite-valued u.s.c. functions on X. Then

$$\inf_{i \in I} \int f_i \, d\mu = \int \inf_{i \in I} f_i \, d\mu.$$

Proof. Since the infimum of any family of u.s.c. functions is again u.s.c., the integral on the right is defined. If $\inf_{i \in I} f_i = 0$, then the integral on the right is zero. By the preceding lemma, given $\varepsilon > 0$, there is an $i_0 \in I$ such that $f_{i_0} < \varepsilon$. Thus,

$$0 \leq \inf_{i \in I} \int f_i \, d\mu \leq \int f_{i_0} \, d\mu \leq \varepsilon\mu(X).$$

Since ε is arbitrary, $\inf_{i \in I} \int f_i \, d\mu = 0$ and the lemma is true whenever $\inf_{i \in I} f_i = 0$. Now suppose $\inf_{i \in I} f_i \neq 0$. Since $g = \inf_{i \in I} f_i$ is u.s.c. and bounded above, there is a sequence $\{\phi_j\}$ of continuous functions such that $\phi_j \downarrow g$ and $\int \phi_n \, d\mu \downarrow \int g \, d\mu$. If $\lambda > \int g \, d\mu$, there is a continuous function $\phi \geq g$ such that $\int \phi \, d\mu \leq \lambda$. Since ϕ is continuous, $(f_i - \phi)$ is u.s.c. for each i and $(f_i - \phi)^+ = \max(f_i - \phi, 0)$ is u.s.c. Clearly, $\inf_{i \in I}(f_i - \phi)^+ = 0$. From the first part of the proof

$$\inf_{i \in I} \int (f_i - \phi) \, d\mu \leq \inf_{i \in I} \int (f_i - \phi)^+ \, d\mu = \int \inf_{i \in I} (f_i - \phi)^+ d\mu = 0.$$

Therefore

$$\inf_{i \in I} \int f_i \, d\mu \leq \int \phi \, d\mu \leq \lambda;$$

but since λ was any number greater than $\int g \, d\mu$,

$$\inf_{i \in I} \int f_i \, d\mu \leq \int \inf_{i \in I} f_i \, d\mu.$$

The assertion follows since the opposite inequality is trivially true.

6. The Kelvin Transformation

The mapping $x \rightsquigarrow x^\star$, where $x^\star$ is the inverse of x relative to the sphere $\partial B_{y,\rho}$, defined by

$$x^\star = y + \frac{\rho^2}{\|y - x\|^2}(x - y), \qquad x \neq y,$$

is particularly useful for solving the Dirichlet problem for unbounded regions and is known as an **inversion** relative to $\partial B_{y,\rho}$. Let R be an open subset of $E^n \sim \{y\}$ and let $R^\star$ be the image of R under this mapping. If $f^\star$ is a function defined on $R^\star$, the equation

$$f(x) = \frac{\rho^{n-2}}{r^{n-2}} f^\star\left(y + \frac{\rho^2}{r^2}(x - y)\right),$$

where $r = \|y - x\|$ and $r^{n-2} = 1$ if $n = 2$, defines a function f on R. The mapping $f^\star \rightsquigarrow f$ defined in this way is called the Kelvin transformation. It is easily seen that an inversion maps spheres or planes into spheres or planes, but not necessarily respectively.

THEOREM **2.24** The Kelvin transformation preserves positivity and harmonicity.

Proof. That positivity is preserved is obvious from the definition. The proof that f is harmonic on R whenever $f^\star$ is harmonic on $R^\star$ is accomplished by computing the Laplacian $\Delta_{(x)}$ of

$$\frac{\rho^{n-2}}{r^{n-2}} f^\star\left(y + \frac{\rho^2}{r^2}(x - y)\right),$$

using the chain rule for differentiation; the computation is straightforward but tedious.

As an application of this theorem we shall obtain a representation for a function which is harmonic and non-negative in a half-space.

THEOREM **2.25** If h is a positive harmonic function on the open set $R = \{(x_1, \ldots, x_n) : x_n > 0\}$, then there is a non-negative constant c and a measure μ on ∂R such that

$$h(x) = cx_n + \frac{2x_n}{\sigma_n} \int_{\partial R} \frac{1}{\|z - x\|^n} \, d\mu(z)$$

for $x \in R$.

Proof. Let $y = (0, 0, \ldots, 0, -1)$ and consider an inversion relative to $\partial B_{y,1}$. It is easily seen that the image of R under this mapping is the open set

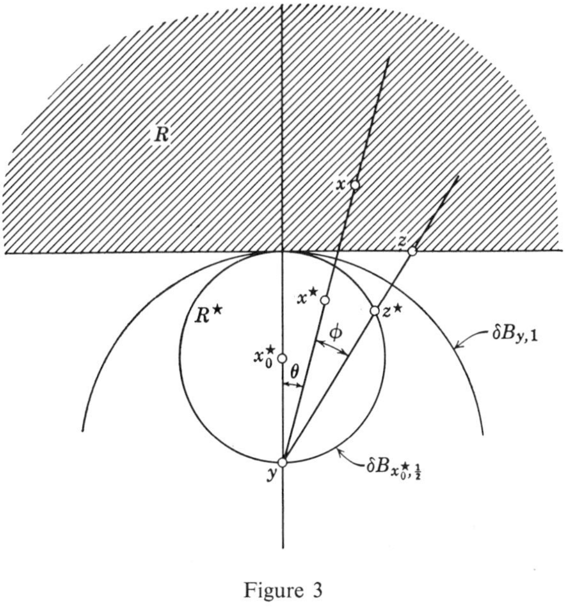

Figure 3

$R^\star = \{z^\star : \|z^\star - x_0^\star\| < 1/2\}$ where $x_0{}^\star = (0, 0, \ldots, 0, -1/2)$. Let $h^\star$ be the image of h under the Kelvin transformation relative to $\partial B_{y,1}$. Then $h^\star$ is a positive harmonic function on the ball $R^\star = \{z^\star : \| z^\star - x_0^\star\| < 1/2\}$ and we know that there is a measure $\mu^\star$ on $\partial R^\star$ such that

$$h^\star(x^\star) = \frac{2}{\sigma_n} \int_{\partial R^\star} \frac{\frac{1}{4} - \|x^\star - x_0^\star\|^2}{\|z^\star - x^\star\|^n} \, d\mu^\star(z^\star)$$

for $x^\star \in R^\star$. Let $c = (2/\sigma_n)\mu^\star(\{y\}) \geq 0$ and let $\mu_1^\star = \mu^\star|_{\partial R^\star \sim \{y\}}$. Then

$$\text{(f)} \qquad h^\star(x^\star) = c \frac{\frac{1}{4} - \|x^\star - x_0^\star\|^2}{\|y - x^\star\|^n} + \frac{2}{\sigma_n} \int_{\partial R^\star} \frac{\frac{1}{4} - \|x^\star - x_0^\star\|^2}{\|z^\star - x^\star\|^n} \, d\mu_1^\star(z^\star).$$

Letting θ denote the angle between the line segment joining $x^\star$ to y and the x_n-axis,

$$\tfrac{1}{4} - \|x^\star - x_0^\star\|^2 = -\|x^\star - y\|^2 + \|x^\star - y\| \cos \theta$$

by the law of cosines. Since $\|x^\star - y\| \, \|x - y\| = 1$ by definition of $x^\star$,

$$\tfrac{1}{4} - \|x^\star - x_0^\star\|^2 = \frac{1}{\|x - y\|^2} [-1 + \|x - y\| \cos \theta].$$

Since $\|x - y\| \cos \theta = x_n + 1$ where x_n is the nth component of x,

$$\text{(g)} \qquad \tfrac{1}{4} - \|x^\star - x_0^\star\|^2 = \frac{x_n}{\|x - y\|^2}.$$

Letting ϕ be the angle between the line segment joining $z^\star$ to y and the line segment joining $x^\star$ to y,

$$\begin{aligned}
\text{(h)} \quad \|z^\star - x^\star\|^2 &= \|z^\star - y\|^2 + \|x^\star - y\|^2 - 2\|z^\star - y\| \, \|x^\star - y\| \cos \phi \\
&= \frac{1}{\|z - y\|^2} + \frac{1}{\|x - y\|^2} - \frac{2 \cos \phi}{\|z - y\| \, \|x - y\|} \\
&= \frac{1}{\|z - y\|^2 \, \|x - y\|^2} \\
&\qquad \times [\|x - y\|^2 + \|z - y\|^2 - 2\|x - y\| \, \|z - y\| \cos \phi] \\
&= \frac{\|z - x\|^2}{\|z - y\|^2 \, \|x - y\|^2}.
\end{aligned}$$

By (f), (g), and (h)

$$h^\star(x^\star) = c x_n \|y - x\|^{n-2} + \frac{2x_n}{\sigma_n} \|y - x\|^{n-2} \int_{\partial R^\star} \frac{1}{\|z - x\|^n} \|z - y\|^n \, d\mu_1^\star(z^\star)$$

or

$$\frac{1}{\|y-x\|^{n-2}} h^{\star}\left(y+\frac{1}{r^2}(x-y)\right) = cx_n + \frac{2x_n}{\sigma_n}\int_{\partial R^{\star}} \frac{1}{\|z-x\|^n} \|z-y\|^n \, d\mu_1^{\star}(z^{\star}).$$

The left side of the last equation is just $h(x)$. Denoting the mapping $z^{\star} \rightsquigarrow z$ by $\mathbf{T}$, the measure $\mu_1^{\star}$ on $\partial R^{\star}$ induces a measure μ_1 on ∂R by $\mu_1 = \mu_1^{\star}\mathbf{T}^{-1}$. Defining $\mu(E) = \int_E \|z-y\|^n \, d\mu_1(z)$ for any Borel set E, the last equation can be written

$$\text{(j)} \qquad h(x) = cx_n + \frac{2x_n}{\sigma_n}\int_{\partial R} \frac{1}{\|z-x\|^n} \, d\mu(z),$$

the desired result.

We shall denote the right side of (j) by $\mathbf{PI}(c, \mu, R)(x)$ where R is the half-space $\{x = (x_1, \ldots, x_n) : x_n > 0\}$ and call $\mathbf{PI}(c, \mu, R)$ the Poisson integral of the pair (c, μ) for the half-space R. We will allow c to be any real number and μ to be a signed measure. As before, if μ is the indefinite integral of a measurable function f defined on the plane ∂R relative to Lebesgue measure, then we shall let $\mathbf{PI}(c, f, R) = \mathbf{PI}(c, \mu, R)$; that is,

$$\mathbf{PI}(c, f, R)(x) = cx_n + \frac{2x_n}{\sigma_n}\int_{\partial R} \frac{f(z)}{\|z-x\|^n} \, dz,$$

where $x = (x_1, \ldots, x_n)$ and $x_n > 0$. Although we shall not pursue the converse of the preceding theorem, we shall give a partial converse. In the following theorem R denotes the half-space $\{(x_1, \ldots, x_n) : x_n > 0\}$. Although this theorem is easily proved by using the Kelvin Transform, a direct proof will be given.

THEOREM **2.26** If f is a bounded measurable function on ∂R and c is any real number, then $h = \mathbf{PI}(c, f, R)$ is harmonic on R. If f is continuous at $z_0 \in \partial R$, then $\lim_{x \to z_0} h(x) = f(z_0)$. Moreover, $h(x)/\|x\|$ has the limit c as $\|x\| \to +\infty$ along the positive x_n-axis.

Proof. We first show that the integral on the right side of the equation

$$\text{(k)} \qquad h(x) = cx_n + \frac{2x_n}{\sigma_n}\int_{\partial R} \frac{f(z)}{\|z-x\|^n} \, dz$$

is finite for all $x \in R$. Since f is bounded, it suffices to consider the integral

$$\int_{\partial R} \frac{1}{\|z-x\|^n} \, dz = \int_{\partial R} \frac{1}{(\|z-x\|^2)^{n/2}} \, dz.$$

Let x^* be the projection of x onto ∂R. Then $\|z - x\|^2 = x_n^2 + \|z - x^*\|^2$. Now

$$\int_{\partial R} \frac{1}{(\|z - x\|^2)^{n/2}}\, dz = \int_{\partial R \cap \{\|z - x^*\| < 1\}} \frac{1}{(\|z - x\|^2)^{n/2}}\, dz + \int_{\partial R \cap \{\|z - x^*\| \geq 1\}} \frac{1}{(\|z - x\|^2)^{n/2}}\, dz$$

$$\leq \int_{\partial R \cap \{\|z - x^*\| < 1\}} \frac{1}{x_n^n}\, dz + \int_{\partial R \cap \{\|z - x^*\| \geq 1\}} \frac{1}{\|z - x^*\|^n}\, dz.$$

Recalling that ∂R is an $(n-1)$-dimensional Euclidean space and introducing spherical coordinates (r, θ) relative to x^* of the point $z \in \partial R$,

$$\int_{\partial R \cap \{\|z - x^*\| \geq 1\}} \frac{1}{\|z - x^*\|^n}\, dz = \int_1^\infty \int_{\|\theta\| = 1} r^{n-2} \frac{1}{r^n}\, d\sigma(\theta)\, dr = \sigma_{n-1}.$$

Therefore the integral in question is always finite. This same type of argument may be used to show that the Laplacian $\Delta_{(x)}$ may be applied to both sides of (k) to show that h is harmonic on R. The argument is straightforward but tedious and involves a justification of differentiation under the integral. If, in particular, we take $c = 0$ and $f = 1$ on ∂R, then

$$\frac{2x_n}{\sigma_n} \int_{\partial R} \frac{1}{\|z - x\|^n}\, dz$$

is harmonic on R. Note that the integral is unaffected by adding to x a vector in ∂R. Such an operation is a translation of x in a direction perpendicular to the x_n-axis. In other words, the integral $\int_{\partial R}(\|z - x\|^n)^{-1}\, dz$ is a function of x_n alone. Let

$$g(x_n) = \frac{2x_n}{\sigma_n} \int_{\partial R} \frac{1}{\|z - x\|^n}\, dz.$$

Since g is a harmonic function of x, $d^2g/dx_n^2 = 0$ and $g(x_n) = ax_n + b$. We now show that $a = 0$. Consider any point $x = (0, 0, \ldots, 0, x_n)$ with $x_n > 0$. Then

$$g(x_n) = \frac{2x_n}{\sigma_n} \int_{\partial R} \frac{1}{(\|z\|^2 + x_n^2)^{n/2}}\, dz_1, \ldots, dz_{n-1}$$

$$= \frac{2}{\sigma_n} \int_{\partial R} \frac{1}{(\|z/x_n\|^2 + 1)^{n/2}}\, d\left(\frac{z_1}{x_n}\right), \ldots, d\left(\frac{z_{n-1}}{x_n}\right)$$

$$= \frac{2}{\sigma_n} \int_{\partial R} \frac{1}{(\|z\|^2 + 1)^{n/2}}\, dz.$$

It follows that $ax_n + b = g(x_n) = g(1)$ for all $x_n > 0$ and therefore that $a = 0$ and $b = g(1)$. Letting (r, θ) denote the spherical coordinates relative to the origin in ∂R of the point $z \in \partial R$,

$$g(1) = \frac{2}{\sigma_n} \int_{\partial R} \frac{1}{(\|z\|^2 + 1)^{n/2}}\, dz = \frac{2}{\sigma_n} \int_0^\infty \int_{\|\theta\| = 1} \frac{r^{n-2}}{(r^2 + 1)^{n/2}}\, d\sigma(\theta)\, dr$$

$$= \frac{2\sigma_{n-1}}{\sigma_n} \int_0^\infty \frac{r^{n-2}}{(r^2 + 1)^{n/2}}\, dr.$$

By resorting to any table of integrals, the right side can be shown to be equal to 1. Therefore $g(x_n) = 1$ for all $x_n > 0$; that is,

$$\frac{2x_n}{\sigma_n} \int_{\partial R} \frac{1}{\|z - x\|^n}\, dz = 1$$

for all $x \in R$. This shows that $\mathbf{PI}(0, 1, R) = 1$ on R. It follows that

$$\lim_{\|x\| \to +\infty} \frac{2}{\sigma_n} \int_{\partial R} \frac{1}{\|z - x\|^n}\, dz = 0,$$

provided $x = (0, \ldots, 0, x_n)$. Then

$$\lim_{\|x\| \to +\infty} \frac{h(x)}{\|x\|} = c + \lim_{\|x\| \to +\infty} \frac{2}{\sigma_n} \int_{\partial R} \frac{f(z)}{\|z - x\|^n}\, dz = c,$$

provided $x = (0, \ldots, 0, x_n)$, since f is bounded. Suppose now that f is continuous at $z_0 \in \partial R$. To show that $\lim_{x \to z_0} h(x) = f(z_0)$ it suffices to prove as in the proof of Lemma 2.5 that $f(z) \le k$ in a neighborhood of z_0 implies that $\limsup_{x \to z_0} h(x) \le k$ where $k > 0$. Suppose $f(z) \le k$ for all $z \in \partial R \cap B_{z_0, \varepsilon}$. Since the term cx_n obviously approaches zero as $x \to z_0$, it suffices to consider

$$\frac{2x_n}{\sigma_n} \int_{\partial R \cap \{\|z - z_0\| < \varepsilon\}} \frac{f(z)}{\|z - x\|^n}\, dz + \frac{2x_n}{\sigma_n} \int_{\partial R \cap \{\|z - z_0\| \ge \varepsilon\}} \frac{f(z)}{\|z - x\|^n}\, dz.$$

Now

$$\frac{2x_n}{\sigma_n} \int_{\partial R \cap \{\|z - z_0\| < \varepsilon\}} \frac{f(z)}{\|z - x\|^n}\, dz \le k\mathbf{PI}(0, 1, R)(x) = k$$

for all $x \in R$. Suppose $|f(z)| \le m$ for all $z \in \partial R$. If $x \in B_{z_0, \varepsilon/2}$ and x^* is the projection of x onto ∂R, then

$$\frac{2x_n}{\sigma_n} \int_{\partial R \cap \{\|z - z_0\| \ge \varepsilon\}} \frac{f(z)}{\|z - x\|^n}\, dz \le m \frac{2x_n}{\sigma_n} \int_{\partial R \cap \{\|z - z_0\| \ge \varepsilon\}} \frac{1}{\|z - x^*\|^n}\, dz.$$

Introducing spherical coordinates (r, θ) relative to x^* of the point $z \in \partial R \sim B_{z_0, \varepsilon}$ and using the fact that $\|z - x^*\| \geq \varepsilon/2$ for $z \in \partial R \sim B_{z_0, \varepsilon}$ and $x \in B_{z_0, \varepsilon/2}$,

$$\int_{\partial R \cap \{\|z - z_0\| \geq \varepsilon\}} \frac{1}{\|z - x^*\|^n} \, dz \leq \int_{\partial R \cap \{\|z - x^*\| \geq \varepsilon/2\}} \frac{1}{\|z - x^*\|^n} \, dz$$

$$= \sigma_{n-1} \int_{\varepsilon/2}^{\infty} r^{n-2} \frac{1}{r^n} \, dr = \frac{2\sigma_{n-1}}{\varepsilon}.$$

Therefore

$$\frac{2x_n}{\sigma_n} \int_{\partial R \cap \{\|z - z_0\| \geq \varepsilon\}} \frac{f(z)}{\|z - x\|^n} \, dz \leq \frac{4m\sigma_{n-1}x_n}{\sigma_n \varepsilon}.$$

Since $x_n \to 0$ as $x \to z_0$,

$$\limsup_{x \to z_0} \frac{2x_n}{\sigma_n} \int_{\partial R \cap \{\|z - z_0\| \geq \varepsilon\}} \frac{f(z)}{\|z - x\|^n} \, dz \leq 0.$$

Therefore

$$\limsup_{x \to z_0} h(x) \leq \limsup_{x \to z_0} \frac{2x_n}{\sigma_n} \int_{\partial R \cap \{\|z - z_0\| < \varepsilon\}} \frac{f(z)}{\|z - x\|^n} \, dz \leq k,$$

as was to be proved.

This theorem illustrates the difficulties associated with the Dirichlet problem for unbounded regions. The Dirichlet problem for the half-space R has no unique solution, since the choice of c in **PI**(c, f, R) is arbitrary.

CHAPTER 3

Boundary Limit Theorems

If $h = \mathbf{PI}(\mu, B)$ on the ball B and μ is the indefinite integral of a continuous boundary function f with respect to surface area σ, we know that $\lim_{z \to x} h(z) = f(x)$ for all $x \in \partial B$. Since the derivative of μ with respect to σ is equal to f a.e. (σ), the boundary behavior of h is related to the derivative $d\mu/d\sigma$. Since there are various ways of defining the derivative $d\mu/d\sigma$, it is possible that we might get different boundary limit theorems, depending upon the derivative used. Generally speaking, a definition of derivative which imposes stringent conditions on μ in order that the derivative may exist will yield the strongest results about the boundary behavior. We use first a weak type derivative that will allow us to describe the behavior of h along radial approaches to the boundary. To allow a more general approach to the boundary it will be necessary to look into differentiation. Having done this, we shall take up the "nontangential" approach to the boundary.

1. The Radial Limit Theorem

As a matter of notational convenience we consider a ball $B = B_{0,\rho} \subset E^n$. It is not essential that 0 is the center of the ball. Consider any $z \in \partial B$ and $0 \leq \alpha \leq \pi/2$. We let $C_{z,\alpha}$ denote a cone with vertex at 0, axis coincident with the vector z, and half-angle α; that is,

$$C_{z,\alpha} = \left\{ y : \cos \alpha \leq \frac{(y, z)}{\|y\| \, \|z\|} \leq 1 \right\}.$$

The cone $C_{z,\alpha}$ determines a **closed polar cap** $K_{z,\alpha} = C_{z,\alpha} \cap \partial B$ having center z and spherical angle α. We denote the class of closed polar caps centered at x by $\mathfrak{K}_x$ and let $\mathfrak{K} = \bigcup_{x \in \partial B} \mathfrak{K}_x$. Let μ be a signed measure of bounded variation on the Borel subsets of ∂B and let σ denote surface area of the same Borel sets.

Definition. 1. If $\lim_{\sigma(K)\to 0} \mu(K)/\sigma(K)$ exists, where K is an arbitrary closed polar cap containing $x \in \partial B$, then this limit is defined to be the **derivative** of μ with respect to σ at the point x and is denoted by $D\mu(x)$.

2. If this same limit exists when K is restricted to being a closed polar cap centered at x, then the limit is defined to be the **symmetric derivative** of μ with respect to σ at the point x and is denoted by $D_s\mu(x)$. If $D\mu$ is defined at $x \in \partial B$, then clearly $D_s\mu$ is also defined at x and the two are equal.

The radial limit theorem which we now state uses the weaker of the two derivatives defined above and results in a weak boundary limit theorem.

THEOREM **3.1** Let μ be a signed measure of bounded variation defined on the Borel subsets of ∂B, let $h = \mathbf{PI}(\mu, B)$ on B, and let the symmetric derivative $D_s\mu$ be defined at $x \in \partial B$ and equal to a. Then $\lim_{\lambda\to 1^-} h(\lambda x)$ exists and is equal to a.

Proof. We first show that we can assume $a = 0$. Consider the signed measure $\mu - a\sigma$ which is also of bounded variation. Now $D_s(\mu - a\sigma)(x) = D_s\mu(x) - aD_s\sigma(x) = 0$ and $\mathbf{PI}(\mu - a\sigma, B) = \mathbf{PI}(\mu, B) - a\mathbf{PI}(\sigma, B) = h - a$. If we can show that $\lim_{\lambda\to 1^-}[h(\lambda x) - a] = 0$, we shall have $\lim_{\lambda\to 1^-} h(\lambda x) = a$. Therefore by replacing μ by $\mu - a\sigma$ we can assume that $a = 0$. Since $D_s\mu(x) = \lim_{\sigma(K)\to 0, K\in\mathfrak{K}_x} \mu(K)/\sigma(K) = 0$, given $\varepsilon > 0$ there is an angle ϕ such that $1 > \cos\phi > 1/2$ and $|\mu(K)| < \epsilon\sigma(K)$ whenever K is a closed polar cap centered at x and $K \subset K_{x,\phi}$. It is easily seen from the Lebesgue dominated convergence theorem that

(a) $$\lim_{\lambda\to 1^-} \frac{1}{\sigma_n\rho}\int_{\partial B \sim K_{x,\phi}} \frac{\rho^2 - \|\lambda x\|^2}{\|y - \lambda x\|^n}\, d\mu(y) = 0.$$

It remains to prove

(b) $$\lim_{\lambda\to 1^-} \frac{1}{\sigma_n\rho}\int_{K_{x,\phi}} \frac{\rho^2 - \|\lambda x\|^2}{\|y - \lambda x\|^n}\, d\mu(y) = 0.$$

Let γ denote the angle between λx and $y \in K_{x,\phi}$. Letting $J(\lambda)$ denote the integral in (b), by the law of cosines

$$J(\lambda) = \int_{K_{x,\phi}} \frac{\rho^2 - \|\lambda x\|^2}{(\rho^2 + \|\lambda x\|^2 - 2\rho\,\|\lambda x\|\cos\gamma)^{n/2}}\, d\mu(y).$$

The angle γ is, of course, a function of y. With λx fixed, the integrand is a function only of the angle γ. We use a "zonal region" method of evaluating the integral $J(\lambda)$. Let

$$F(\gamma) = \int_{K_{x,\gamma}} d\mu(y).$$

Then

$$J(\lambda) = \int_0^\phi \frac{\rho^2 - \|\lambda x\|^2}{(\rho^2 + \|\lambda x\|^2 - 2\rho\,\|\lambda x\| \cos\gamma)^{n/2}}\, dF(\gamma).$$

Integrating by parts,

$$\text{(c)} \qquad J(\lambda) = \frac{(\rho^2 - \|\lambda x\|^2)F(\gamma)}{(\rho^2 + \|\lambda x\|^2 - 2\rho\,\|\lambda x\| \cos\gamma)^{n/2}}\Bigg|_0^\phi + \frac{n}{2}\int_0^\phi \frac{(\rho^2 - \|\lambda x\|^2)(2\rho\,\|\lambda x\| \sin\gamma)F(\gamma)}{(\rho^2 + \|\lambda x\|^2 - 2\rho\,\|\lambda x\| \cos\gamma)^{(n+2)/2}}\, d\gamma.$$

We shall let $J(\lambda) = J_\phi(\lambda) - J_0(\lambda) + J_i(\lambda)$, where $J_\phi(\lambda)$ is the first term on the right evaluated at $\gamma = \phi$, $J_0(\lambda)$ is the first term on the right evaluated at $\gamma = 0$, and $J_i(\lambda)$ is the second term on the right. Then

$$J_\phi(\lambda) = \frac{(\rho^2 - \|\lambda x\|^2)F(\phi)}{(\rho^2 + \|\lambda x\|^2 - 2\rho\,\|\lambda x\| \cos\phi)^{n/2}}.$$

As $\lambda \to 1^-$, the denominator approaches $[2\rho^2(1 - \cos\phi)]^{n/2} > 0$, whereas the numerator approaches zero. Thus $\lim_{\lambda\to 1^-} J_\phi(\lambda) = 0$. Now consider

$$J_0(\lambda) = \frac{(\rho^2 - \|\lambda x\|^2)F(0)}{(\rho^2 + \|\lambda x\|^2 - 2\rho\,\|\lambda x\|)^{n/2}}.$$

Note that $F(0) = \mu(\{x\})$; but since $|\mu(K)| < \varepsilon\sigma(K)$ whenever K is a small closed polar cap centered at x, $\mu(\{x\}) = 0$ and $J_0(\lambda) = 0$. Trivially, $\lim_{\lambda\to 1^-} J_0(\lambda) = 0$. It remains only to consider $J_i(\lambda)$. Since

$$|F(\gamma)| = |\mu(K_{x,\gamma})| \le \varepsilon\sigma(K_{x,\gamma}) \qquad \text{for} \qquad \gamma \le \phi,$$

$$|J_i(\lambda)| \le \frac{n\varepsilon}{2}\int_0^\phi \frac{(\rho^2 - \|\lambda x\|^2)(2\rho\,\|\lambda x\| \sin\gamma)\sigma(K_{x,\gamma})}{(\rho^2 + \|\lambda x\|^2 - 2\rho\,\|\lambda x\| \cos\gamma)^{(n+2)/2}}\, d\gamma.$$

Integrating again by parts,

$$\text{(d)} \qquad |J_i(\lambda)| \le \varepsilon\left[-\frac{(\rho^2 - \|\lambda x\|^2)\sigma(K_{x,\gamma})}{(\rho^2 + \|\lambda x\|^2 - 2\rho\,\|\lambda x\| \cos\gamma)^{n/2}}\Bigg|_0^\phi + \int_0^\phi \frac{\rho^2 - \|\lambda x\|^2}{(\rho^2 + \|\lambda x\|^2 - 2\rho\,\|\lambda x\| \cos\gamma)^{n/2}}\, d_\gamma(\sigma(K_{x,\gamma}))\right].$$

The first term within the brackets can be shown to approach zero as $\lambda \to 1^-$ in just the same way that it was shown that $\lim_{\lambda\to 1^-}[J_\phi(\lambda) - J_0(\lambda)] = 0$. The integral within the brackets is just

$$\int_{K_{x,\phi}} \frac{\rho^2 - \|\lambda x\|^2}{\|y - \lambda x\|^n} \, d\sigma(y) \le \sigma_n \rho \mathbf{PI}(1, B) = \sigma_n \rho.$$

Going back to (d), $\limsup_{\lambda \to 1^-} |J_i(\lambda)| \le \sigma_n \rho \varepsilon$ and since ε is arbitrary

$$\lim_{\lambda \to 1^-} J(\lambda) = \lim_{\lambda \to 1^-} J_i(\lambda) = 0.$$

This proves (b). The equations (a) and (b) imply $\lim_{\lambda \to 1^-} h(\lambda x) = 0$.

As an example, suppose $B = B_{0,1} \subset E^2$ and define a function f on ∂B as follows:

$$f(x, y) = \begin{cases} 1 & (x, y) \in \partial B, \quad y \ge 0, \\ 0 & (x, y) \in \partial B, \quad y < 0. \end{cases}$$

Let $\mu(M) = \int_M f \, d\sigma$ for each Borel set $M \subset \partial B$. Then $D_s \mu$ exists and is equal to f except at the two points $(-1, 0)$, $(1, 0)$, where $D_s \mu$ is equal to $\frac{1}{2}$. Consider the function $h = \mathbf{PI}(\mu, B)$ on B which we know to be harmonic on B. We also know that $\lim_{(x', y') \to (x, y)} h(x', y') = f(x, y)$ for any $(x, y) \in \partial B$ different from $(-1, 0)$ and $(1, 0)$. From the preceding theorem we know that $\lim_{\lambda \to 1^-} h(\lambda x, \lambda y) = \frac{1}{2}$ if (x, y) is one of the two exceptional points.

2. Differentiation

In order to obtain significant results for a more general approach to the boundary, we shall have to replace the symmetric derivative by the derivative $D\mu$. We digress to develop some differentiation theory especially applicable to measures defined on the Borel subsets of a sphere. The main theorem is essentially the classical theorem of Lebesgue which states that a measure ν of bounded variation defined on the Borel subsets of E^n has a "generalized derivative" with respect to Lebesgue measure μ_n on E^n; that is, if C denotes a closed cube in E^n, then

$$\lim_{\substack{\mu_n(C) \to 0 \\ x \in C}} \frac{\nu(C)}{\mu_n(C)}$$

exists for almost all x with respect to Lebesgue measure.

A good deal of what follows is valid for spaces other than Euclidean spaces and we go to more general spaces. In what follows X is a metric space with metric d. If $F \subset X$, the diameter of F is denoted by $\delta(F) = \sup\{d(x, y) : x, y \in F\}$. If $F \subset X$ and $x \in X$, the distance from F to x is denoted by $d(F, x) = \inf\{d(y, x) : y \in F\}$. An ε-neighborhood of $F \subset X$ is denoted by $N(F, \varepsilon) = \{x : d(F, x) < \varepsilon\}$.

LEMMA **3.2** Let X be a compact metric space and let $A \subset X$. Suppose there is a family $\mathscr{F}$ of closed subsets of X with the property that for each

$x \in A$ there is a set $F \in \mathscr{F}$ of arbitrarily small positive diameter containing x. Then there is a countable family $\{F_j\}$ of disjoint elements of $\mathscr{F}$ such that (i) $A \subset \bigcup_j F_j$ if the family $\{F_j\}$ is finite and (ii)

$$\text{(e)} \qquad A \subset F_1 \cup \cdots \cup F_j \cup \bigcup_{k=j+1}^{\infty} N(F_k, 3\delta(F_k)) \quad \text{for every } j$$

if the family $\{F_j\}$ is infinite.

Proof. To define the family $\{F_j\}$ let F_1 be chosen arbitrarily from $\mathscr{F}$. Suppose $F_1, \ldots, F_k$ have been chosen. If $A \subset F_1 \cup \cdots \cup F_k$, then we are through. Otherwise let

$$\varepsilon_k = \sup\{\delta(F) : N(F, \delta(F)) \cap F_i = \varnothing, \qquad i = 1, \ldots, k, F \in \mathscr{F}\}.$$

Since there is a point $y \in A$ that is not in the closed set $\bigcup_{i=1}^{k} F_i$ and there are sets $F \in \mathscr{F}$ of arbitrarily small diameter containing y, the class defining ε_k is nonempty and therefore $\varepsilon_k > 0$. Let F_{k+1} be any element of $\mathscr{F}$ such that

$$N(F_{k+1}, \delta(F_{k+1})) \cap F_i = \varnothing, \qquad i = 1, 2, \ldots, k,$$

$$\delta(F_{k+1}) > \tfrac{2}{3}\varepsilon_k.$$

The family $\{F_j\}$ is so defined inductively. Suppose (e) is false for some integer $j > 1$. If

$$\text{(f)} \qquad x \in A \sim \left[F_1 \cup \cdots \cup F_j \cup \bigcup_{k=j+1}^{\infty} N(F_k, 3\delta(F_k))\right],$$

then $x \notin \bigcup_{i=1}^{j} F_i$ and there is an $F \in \mathscr{F}$ such that $x \in F$ with $\delta(F) > 0$ and $N(F, \delta(F)) \cap F_i = \varnothing$ $i = 1, 2, \ldots, j$. We now show that $N(F, \delta(F)) \cap F_k \neq \varnothing$ for some $k > j$. Suppose on the contrary that $N(F, \delta(F)) \cap F_k = \varnothing$ for all k. Then

$$\tfrac{2}{3}\,\delta(F) \leq \tfrac{2}{3}\varepsilon_{k-1} < \delta(F_k) \qquad \text{for all} \qquad k > 1.$$

For each $k > 1$, let $y_k \in F_k$. If $p < q$, then $N(F_q, \delta(F_q)) \cap F_p = \varnothing$ by definition of F_q. Since $N(y_q, 2\delta(F)/3) \subset N(F_q, \delta(F_q))$ and $y_p \notin N(F_q, \delta(F_q))$, $d(y_p, y_q) \geq 2\delta(F)/3$. The sequence $\{y_k\}$ therefore cannot have a convergent subsequence and we have a contradiction to the compactness of X. Therefore $N(F, \delta(F)) \cap F_k \neq \varnothing$ for some $k > j$. Let k_0 be the smallest such k. Then $N(F, \delta(F)) \cap F_{k_0-1} = \varnothing$ and $\delta(F) \leq \varepsilon_{k_0-1}$ by definition of ε_{k_0-1}. Therefore,

$$\delta(F_{k_0}) > \tfrac{2}{3}\varepsilon_{k_0-1} \geq \tfrac{2}{3}\delta(F).$$

From (f) and the fact that $k_0 > j$, $x \notin N(F_{k_0}, 3\delta(F_{k_0}))$. On the other hand, if y belongs to the nonempty set $N(F, \delta(F)) \cap F_{k_0}$, then

$$d(x, y) \leq 2\delta(F) < 3\delta(F_{k_0})$$

since $x \in F$. Since y also belongs to F_{k_0}, $x \in N(y, 3\delta(F_{k_0})) \subset N(F_{k_0}, 3\delta(F_{k_0}))$, and we have a contradiction. Therefore (e) is true for every integer $j > 1$. If (e) is true for $j = 2$, then obviously it is true for $j = 1$.

Definition. Let μ be a measure defined on the Borel subsets of a compact metric space X. A set $A \subset X$ is said to be **covered in the sense of Vitali by a family $\mathscr{F}$ of closed sets** if each $F \in \mathscr{F}$ has positive μ measure and there is a constant α such that each point of A is contained in elements $F \in \mathscr{F}$ of arbitrarily small positive diameter for which $\mu(N(F, 3\delta(F))/\mu(F) \leq \alpha$.

THEOREM **3.3** (*Vitali Covering Theorem*) Let μ be a complete measure defined on a σ-algebra containing the Borel subsets of a compact metric space X. If $A \subset X$ is covered in the sense of Vitali by a family $\mathscr{F}$ of closed sets, then there is sequence $\{F_j\}$ of disjoint elements of $\mathscr{F}$ such that $\mu(A \sim \bigcup_j F_j) = 0$.

Proof. If we apply Lemma 3.2 to the set A, we obtain a sequence $\{F_j\}$ of disjoint elements of $\mathscr{F}$ such that the theorem is trivially true if the sequence is finite and

$$A \sim \bigcup_{k=1}^{j} F_k \subset \bigcup_{k=j+1}^{\infty} N(F_k, 3\delta(F_k))$$

for each j if the sequence is infinite. Letting α be the constant in the definition of a Vitali covering,

$$\mu\left(\bigcup_{k=j+1}^{\infty} N(F_k, 3\delta(F_k))\right) \leq \sum_{k=j+1}^{\infty} \mu(N(F_k, 3\delta(F_k)))$$
$$\leq \alpha \sum_{k=j+1}^{\infty} \mu(F_k).$$

The last sum approaches zero as $j \to \infty$ since $\sum_k \mu(F_k) \leq \mu(X) < +\infty$. If $\varepsilon > 0$, then there is a j_ε such that

$$\mu\left(\bigcup_{k=j_\varepsilon+1}^{\infty} N(F_k, 3\delta(F_k))\right) < \varepsilon.$$

Since $A \sim \bigcup_{k=1}^{\infty} F_k \subset A \sim \bigcup_{k=1}^{j_\varepsilon} F_k \subset \bigcup_{k=j_\varepsilon+1}^{\infty} N(F_k, 3\delta(F_k))$ and

$$\mu\left(\bigcup_{k=j_\varepsilon+1}^{\infty} N(F_k, 3\delta(F_k))\right) < \epsilon,$$

$A \sim \bigcup_{k=1}^{\infty} F_k$ is a subset of a set of measure zero and therefore measurable with μ-measure zero.

The Vitali covering theorem is applied within the following context. For X we shall take ∂B where $B = B_{0,\rho} \subset E^n$. We shall use closed polar caps K for the family $\mathscr{F}$ of the Vitali covering theorem. To do so we shall have to show that the closed polar caps cover ∂B in the sense of Vitali. This requires some estimates of the surface area of closed polar caps.

Let $K_{z,\alpha}$, $0 \le \alpha < \pi/2$, be a closed polar cap on ∂B. We shall make free use of the fact that surface area is invariant under rotations of the sphere by assuming that $z = (0, 0, \ldots, 0, \rho)$. Then

$$\sigma(K_{z,\alpha}) = \int\cdots\int_{\xi_1^2+\cdots+\xi_{n-1}^2 \le \rho^2 \sin^2\alpha} \frac{\rho}{(\rho^2 - \xi_1^2 \cdots - \xi_{n-1}^2)^{1/2}}\, d\xi_1 \ldots d\xi_{n-1}$$

We can put this integral into a more convenient form by introducing spherical coordinates (θ, r) of $(\xi_1, \ldots, \xi_{n-1})$. Then

$$\sigma(K_{z,\alpha}) = \int_0^{\rho \sin\alpha} \rho\left[\int\cdots\int_{\|\theta\|=1} \frac{r^{n-2}}{(\rho^2 - r^2)^{1/2}} \frac{1}{(1-\theta_1^2 \cdots - \theta_{n-2}^2)^{1/2}}\, d\theta_1 \ldots d\theta_{n-2}\right] dr$$

$$= \rho\sigma_{n-1} \int_0^{\rho\sin\alpha} \frac{r^{n-2}}{(\rho^2 - r^2)^{1/2}}\, dr.$$

It is easily seen that $\delta(K_{z,\alpha}) = 2\rho \sin\alpha$. To simplify the notation let $K = K_{z,\alpha}$. Then there is an α_0 with $0 < \alpha_0 < \pi/2$ such that for $\alpha \le \alpha_0$

$$\sigma(N(K, 3\delta(K)) \le \rho\sigma_{n-1} \int_0^{7\rho\sin\alpha} \frac{r^{n-2}}{(\rho^2 - r^2)^{1/2}}\, dr$$

$$= \rho\sigma_{n-1} 7^{n-1} \int_0^{\rho\sin\alpha} \frac{r^{n-2}}{(\rho^2 - 49r^2)^{1/2}}\, dr.$$

Therefore

$$\frac{\sigma(N(K, 3\delta(K))}{\sigma(K)} \le \frac{7^{n-1} \int_0^{\rho\sin\alpha} [r^{n-2}/(\rho^2 - 49r^2)^{1/2}]dr}{\int_0^{\rho\sin\alpha} [r^{n-2}/(\rho^2 - r^2)^{1/2}]\, dr}.$$

Considering the integrals as functions of the upper limits and applying the second mean value theorem of the differential calculus, there is a ξ such that $0 < \xi < \rho \sin\alpha$ and the quotient on the right side of the last inequality is equal to

$$\frac{7^{n-1}[\xi^{n-2}/(\rho^2 - 49\xi^2)^{1/2}]}{\xi^{n-2}/(\rho^2 - \xi^2)^{1/2}}.$$

Therefore

$$\frac{\sigma(N(K, 3\delta(K)))}{\sigma(K)} \le 7^{n-1} \frac{(\rho^2 - \xi^2)^{1/2}}{(\rho^2 - 49\xi^2)^{1/2}}.$$

As $\alpha \to 0$, $\xi \to 0$ and the right member of this inequality approaches 7^{n-1}. This shows that there is an α_0' with $0 < \alpha_0' < \pi/2$ such that

$$\frac{\sigma(N(K, 3\delta(K)))}{\sigma(K)} \leq \text{constant}$$

whenever K is a closed polar cap on ∂B with spherical angle $\alpha \leq \alpha_0'$. The family $\mathscr{F}$ of such closed polar caps covers ∂B in the sense of Vitali.

LEMMA **3.4** Let μ be a measure defined on the Borel subsets of ∂B, where $B = B_{0,\rho}$, and let $0 < r < +\infty$.

(i) If for each z in a set $A \subset \partial B$

$$\liminf_{\substack{\sigma(K)\to 0 \\ z \in K \in \mathfrak{K}}} \frac{\mu(K)}{\sigma(K)} < r,$$

then each neighborhood of A contains an open set Q such that $\sigma(A \sim Q) = 0$ and $\mu(Q) < r\sigma(Q)$.

(ii) If for each z in a set $A \subset \partial B$

$$\limsup_{\substack{\sigma(K)\to 0 \\ z \in K \in \mathfrak{K}}} \frac{\mu(K)}{\sigma(K)} > r,$$

Then each neighborhood of A contains a Borel set A_0 such that $\sigma(A \sim A_0) = 0$ and $\mu(A_0) > r\sigma(A_0)$.

Proof. To prove (i) let U be an open set such that $A \subset U \subset \partial B$ and let $\mathscr{F}$ be the family of all closed polar caps $K \subset U$ satisfying $\mu(K) < r\sigma(K)$. Then $\mathscr{F}$ covers A in the sense of Vitali. By the Vitali covering theorem (applied to σ), there is a sequence $\{K_j\}$ of disjoint polar caps in $\mathscr{F}$ such that $\sigma(A \sim \bigcap_{j=1}^{\infty} K_j) = 0$. Let K_j^0 be the interior (relative to ∂B) of K_j. Since $\sigma(K_j \sim K_j^0) = 0$, $\sigma(A \sim \bigcup_{j=1}^{\infty} K_j^0) = 0$ and

$$\mu(K_j^0) \leq \mu(K_j) < r\sigma(K_j) = r\sigma(K_j^0).$$

If $Q = \bigcup_{j=1}^{\infty} K_j^0$, then Q is open, $Q \subset U$, and $\sigma(A \sim Q) = 0$. Moreover,

$$\mu(Q) = \sum_{j=1}^{\infty} \mu(K_j^0) < r \sum_{j=1}^{\infty} \sigma(K_j^0) = r\sigma(Q).$$

To prove (ii) let U be an open set such that $A \subset U \subset \partial B$. By hypothesis, the family $\mathscr{F}$ of closed polar caps $K \subset U$ such that $\mu(K) > r\sigma(K)$ covers A in the sense of Vitali. Again by the Vitali covering theorem, there is a sequence $\{K_j\}$ of disjoint closed polar caps in $\mathscr{F}$ such that $\sigma(A \sim \bigcup_{j=1}^{\infty} K_j) = 0$. Let $A_0 = \bigcup_{j=1}^{\infty} K_j$. Since

$$\mu(A_0) = \sum_{j=1}^{\infty} \mu(K_j) > r \sum_{j=1}^{\infty} \sigma(K_j) = r\sigma(A_0),$$

A_0 satisfies the requirements of (ii).

It is essential in the following lemma that the measure σ on ∂B is a complete measure.

LEMMA **3.5** Let μ be a measure on the Borel subsets of ∂B such that $\lim_{\sigma(K)\to 0,\, x\in K\in\mathfrak{K}} \mu(K)/\sigma(K)$ exists for almost all (σ) points $x \in \partial B$. Define $D\mu(x)$ to be this limit where it exists and define $D\mu(x)$ arbitrarily at points x where the limit does not exist. Then $D\mu$ is a Lebesgue measurable on ∂B.

Proof. For each positive integer m, let $\{K_{m1}, \ldots, K_{mp_m}\}$ be a finite collection of closed polar caps which covers ∂B and $\sigma(K_{mj}) < 1/m$. Define a sequence $\{f_m\}$ of functions on ∂B as follows. Let $K_{m0} = \varnothing$. If $x \in K_{mj} \sim \bigcup_{i=0}^{j-1} K_{mi}$, $j = 1, \ldots, p_m$, let $f_m(x) = \mu(K_{mj})/\sigma(K_{mj})$. Then each f_m is measurable and the sequence $\{f_m\}$ converges a.e. to $D\mu$.

THEOREM **3.6** (*Lebesgue*) Let μ be a signed measure of bounded variation defined on the Borel subsets of ∂B. Then $D\mu(x)$ exists for almost all (σ) points $x \in \partial B$ and is Lebesgue measurable.

Proof. By decomposing μ into its positive and negative variations, we can assume that μ is a measure. Let A be the set of points $x \in \partial B$ for which

$$\limsup_{\substack{\sigma(K)\to 0\\ x\in K\in\mathfrak{K}}} \frac{\mu(K)}{\sigma(K)} > \liminf_{\substack{\sigma(K)\to 0\\ x\in K\in\mathfrak{K}}} \frac{\mu(K)}{\sigma(K)}.$$

Let $A_{p,q}$ be the set of $x \in \partial B$ where

$$\limsup_{\substack{\sigma(K)\to 0\\ x\in K\in\mathfrak{K}}} \frac{\mu(K)}{\sigma(K)} > \frac{p+1}{q} > \frac{p}{q} > \liminf_{\substack{\sigma(K)\to 0\\ x\in K\in\mathfrak{K}}} \frac{\mu(K)}{\sigma(K)},$$

with p a non-negative integer and q a positive integer. Suppose $\sigma^*(A) > 0$ where σ^* is the outer measure induced by σ. Since $A = \cup A_{p,q}$, $\sigma^*(A_{p,q}) > 0$ for some p and q. If $\beta = \inf\{\sigma(M) : M \supset A_{p,q},\ M \text{ a Borel set}\}$, then $\beta > 0$. Given $\varepsilon > 0$, there is an open set U such that $A_{p,q} \subset U$ and $\sigma(U) < \beta + \varepsilon$. Apply part (i) of Lemma 3.4 with r replaced by p/q. Then there is an open set Q such that $Q \subset U$, $\sigma(A_{p,q} \sim Q) = 0$, and

$$\mu(Q) < \frac{p}{q}\sigma(Q) < \frac{p}{q}(\beta + \varepsilon).$$

By applying part (ii) of Lemma 3.4 to the set $A_{p,q}Q$ there is a Borel set $M \subset Q$ such that $\sigma(A_{p,q}Q \sim M) = 0$ and $\mu(M) > [(p+1)/q]\sigma(M)$. Since

$$A_{p,q} \subset M \cup (A_{p,q} \sim Q) \cup (A_{p,q}Q \sim M)$$

and

$$\sigma(A_{p,q} \sim Q) = 0 = \sigma(A_{p,q} Q \sim M),$$

$\sigma(M) \geq \beta$. Thus

$$\frac{p}{q}(\beta + \varepsilon) > \mu(Q) \geq \mu(M) > \frac{p+1}{q}\sigma(M) \geq \frac{p+1}{q}\beta.$$

By letting $\varepsilon \to 0$, we obtain $(p/q)\beta \geq [(p+1)/q]\beta$, a contradiction. Therefore $\sigma(A) = 0$ and $D\mu(x) = \lim_{\sigma(K)\to 0} \mu(K)/\sigma(K)$ exists as a finite or infinite limit for all $x \notin A$. Let $E = \{x \in \sim A : D\mu(x) = +\infty\}$. By (ii) of Lemma 3.4 for each positive integer j there is a Borel set $A_j \subset \partial B$ such that $\sigma(E \sim A_j) = 0$ and $\mu(A_j) > j\sigma(A_j)$. Since $\sigma^*(E) \leq \sigma^*(E \cap A_j) + \sigma^*(E \sim A_j)$, $\sigma^*(E) \leq \sigma(A_j)$ and $+\infty > \mu(\partial B) \geq \mu(A_j) > j\sigma(A_j) \geq j\sigma^*(E)$ for all positive integers j. Therefore $\sigma^*(E) = 0$ and $D\mu < +\infty$ a.e. (σ) on $\sim A$. The measurability of $D\mu$ follows from Lemma 3.5.

Although the previous theorem gives us the existence of $D\mu$ at almost all (σ) points of ∂B, the proof of the principal boundary limit theorem requires that we also exclude from consideration a set of points of surface area zero even though $D\mu$ may be defined at such points. This happens because of the necessity of making several reductions. The next two theorems pave the way for these reductions.

THEOREM **3.7** (*Lebesgue*) Let μ be a signed measure of bounded variation defined on the Borel subsets of ∂B.

(i) If f is an integrable function on ∂B and $\mu(M) = \int_M f\, d\sigma$ for each Borel set $M \subset \partial B$, then $D\mu = f$ a.e. (σ).

(ii) If μ is singular relative to σ, then $D\mu = 0$ a.e. (σ).

Proof. Suppose $\mu(M) = \int_M f\, d\sigma$ with f integrable. We can assume $f \geq 0$ by decomposing f into its positive and negative parts if necessary. Let

$$A = \{x : f(x) > D\mu(x)\}$$

and assume $\sigma(A) > 0$. There are then positive numbers α, β such that $A_{\alpha,\beta} = \{x : f(x) > \alpha > \beta > D\mu(x)\}$ and $\sigma(A_{\alpha,\beta}) > 0$. Since σ is a regular Borel measure, there is a compact set $C \subset A_{\alpha,\beta}$ with $\sigma(C) > 0$. Thus

$$\mu(C) = \int_C f\, d\sigma \geq \alpha\sigma(C).$$

Since $D\mu(x) = \lim_{\sigma(K)\to 0,\, x\in K\in\mathfrak{K}} \mu(K)/\sigma(K) < \beta$ for all $x \in C$, each neighborhood of C contains an open set U such that $\mu(U) < \beta\sigma(U)$ and $\sigma(C \sim U) = 0$ by part (i) of Lemma 3.4. Since $\sigma(C \sim U) = 0$ and μ is the indefinite integral of an integrable function, we can adjoin a small open set to U if necessary so that $C \subset U$ and $\mu(U) < \beta\sigma(U)$. Since C is the intersection of a sequence of such neighborhoods U, $\mu(C) \leq \beta\sigma(C) < \alpha\sigma(C) \leq \mu(C)$, a contradiction. Therefore

$\sigma(A) = 0$. In a similar manner (using part (ii) of Lemma 3.4) we can show that $\sigma(\{x : f(x) < D\mu(x)\}) = 0$. Therefore, $f = D\mu$ a.e. (σ). We now prove (ii). Suppose μ is singular relative to σ. Since $\mu = \mu^+ - \mu^-$ and the positive and negative variations of μ are also singular relative to σ, we can assume that μ is a measure and $D\mu \geq 0$. Then there is a measurable set $N \subset \partial B$ such that $\sigma(N) = 0$ and $\mu(\partial B \sim N) = 0$. Let $A = \{x : D\mu(x) > 0\}$ and assume $\sigma(A) > 0$. Since $D\mu$ is measurable and $\sigma(N) = 0$, for each $\varepsilon > 0$, there is a measurable set $E \subset A \sim N$ such that $\sigma(E) > 0$ and $D\mu(x) > \varepsilon$ for $x \in E$. Since μ is a regular Borel measure, we can assume that E is compact. By (ii) of Lemma 3.4 each neighborhood of E has μ-measure greater than $\varepsilon\sigma(E)$. Since E is the intersection of a sequence of such neighborhoods, $\mu(E) \geq \varepsilon\sigma(E) > 0$; but this contradicts the fact that $0 \leq \mu(E) \leq \mu(A \sim N) \leq \mu(\partial B \sim N) = 0$. Therefore $\sigma(A) = 0$ and $D\mu = 0$ a.e. (σ).

THEOREM **3.8** (*Lebesgue*) Let f be a measurable function on ∂B which is integrable relative to σ and let $\mu(M) = \int_M f\, d\sigma$ for each Borel set $M \subset \partial B$. Then

$$\lim_{\substack{\sigma(K)\to 0\\ x\in K\in\mathfrak{K}}} \frac{1}{\sigma(K)} \int_K |f(y) - f(x)|\, d\sigma(y) = 0.$$

for almost all (σ) points $x \in \partial B$.

Proof. Let $\{r_i\}$ be an enumeration of the rationals. By Theorem 3.7, there is a set $N_i \subset \partial B$ with $\sigma(N_i) = 0$ such that for $x \notin N_i$

$$\lim_{\substack{\sigma(K)\to 0\\ x\in K\in\mathfrak{K}}} \frac{1}{\sigma(K)} \int_K |f(y) - r_i|\, d\sigma(y) = |f(x) - r_i|.$$

Then $N = \bigcup_{i=1}^{\infty} N_i$ has surface area zero. Let $x \in \partial B \sim N$, let $\varepsilon > 0$, and choose r_j such that $|f(x) - r_j| < \varepsilon/2$. Then

$$\begin{aligned}
&\limsup_{\substack{\sigma(K)\to 0\\ x\in K\in\mathfrak{K}}} \frac{1}{\sigma(K)} \int_K |f(y) - f(x)|\, d\sigma(y)\\
&\qquad \leq \limsup_{\substack{\sigma(K)\to 0\\ x\in K\in\mathfrak{K}}} \frac{1}{\sigma(K)} \int_K \{|f(y) - r_j| + |r_j - f(x)|\}\, d\sigma(y)\\
&\qquad \leq \limsup_{\substack{\sigma(K)\to 0\\ x\in K\in\mathfrak{K}}} \frac{1}{\sigma(K)} \int_K |f(y) - r_j|\, d\sigma(y) + \frac{\varepsilon}{2}\\
&\qquad = |f(x) - r_j| + \frac{\varepsilon}{2} < \varepsilon.
\end{aligned}$$

3. Nontangential Limits

As in the preceding paragraph we shall consider a ball $B = B_{0,\rho}$ with center at the origin and remark again that it is not essential that the center be so chosen. It is not difficult to construct examples of functions h harmonic on $B \subset E^2$ which behave badly in a neighborhood of a point $z \in \partial B$. In fact, suppose $h = \mathbf{PI}(\delta_z, B)$ on B, where δ_z is a unit measure with support $\{z\}$. Then for $x \in B$

$$h(x) = \frac{1}{\sigma_n \rho} \frac{\rho^2 - \|x\|^2}{\|z - x\|^2}.$$

Suppose $x = \lambda z$ and we let $\lambda \uparrow 1$. Then

$$\begin{aligned} \lim_{\lambda \uparrow 1} h(\lambda z) &= \frac{1}{\sigma_n \rho} \lim_{\lambda \uparrow 1} \frac{\rho^2 - \lambda^2 \rho^2}{(1 - \lambda)^2 \rho^2} \\ &= \frac{1}{\sigma_n \rho} \lim_{\lambda \uparrow 1} \frac{1 + \lambda}{1 - \lambda} = +\infty. \end{aligned}$$

Thus, if $x \in B$ approaches z along the radial line to z, then $h(x) \to +\infty$. Suppose now we let $x \to z$ in the following manner. Since $B \subset E^2$, we can introduce polar coordinates (θ, r), where θ is the angle between the vectors x and z and $r = \|x\|$. We now let $x \to z$ subject to the condition $r = \rho \cos \theta$. In this case,

$$\lim_{\substack{x \to z \\ \|x\| = \rho \cos \theta}} h(x) = \lim_{\theta \to 0} \frac{1}{\sigma_n \rho} \frac{\rho^2 (1 - \cos^2 \theta)}{\rho^2 (1 - \cos^2 \theta)} = \frac{1}{\sigma_n \rho}.$$

Now $+\infty$ and $1/\sigma_n \rho$ are not the only possible limiting values; for example, if we let $x \to z$, subject to the conditions $r = \rho \cos^2 \theta$ and $\theta > 0$, the limiting value will be $2/\sigma_n \rho$ (in fact, any value between 0 and $+\infty$ is a possible limiting value). A little calculus will show that in both of the latter cases the approach to z is along a curve which is internally tangent to the circle ∂B at the point z. If anything positive is to be said about the limiting behavior of h at z, then tangential approaches to z must be excluded.

Suppose u is a function defined on the ball $B = B_{0,\rho}$ and $0 \le \theta < \pi/2$. Consider a fixed $z \in \partial B$ and a cone C with vertex z, with axis the line segment from 0 to z, and with half-angle θ; that is,

$$C = \left\{ x : 1 \ge \frac{(-z, x - z)}{\|z\| \, \|x - z\|} \ge \cos \theta \right\}.$$

The **nontangential limit** $\theta - \lim_{x \to z} u(x)$ is defined as follows

$$\theta - \lim_{x \to z} u(x) = \lim_{\substack{x \to z \\ x \in C \cap B}} u(x),$$

provided the limit on the right exists. Throughout the following discussion θ denotes a fixed angle such that $0 \leq \theta < \pi/2$.

THEOREM 3.9 (*Fatou*)   Let μ be a signed measure of bounded variation on the Borel subsets of the boundary of $B = B_{0,\rho}$ and let $h = \mathbf{PI}(\mu, B)$ on B. Then $\theta - \lim_{x \to z} h(x)$ exists and is equal to $D_\mu(z)$ a.e. (σ).

Proof. We first show that it suffices to prove

(g) ν is a measure on ∂B, $D\nu(z) = 0 \Rightarrow \theta - \lim_{x \to z} \mathbf{PI}(\nu, B)(x) = 0$.

Assume that (g) is true. Consider the decomposition $\mu = \mu_a + \mu_s$ of μ relative to σ, where μ_a and μ_s are the absolutely continuous and singular parts of μ, respectively. Now $\mu_s = \mu_s^+ - \mu_s^-$, where μ_s^+ and μ_s^- are the positive and negative variations of μ_s, respectively. By (ii) of Theorem 3.7 $D\mu_s^+ = 0$ a.e. (σ) and $D\mu_s^- = 0$ a.e. (σ). Since μ_s^+ and μ_s^- are both measures, (g) implies $\theta - \lim_{x \to z} \mathbf{PI}(\mu_s^\pm, B)(x) = 0$ a.e. (σ). Now consider μ_a which can be written

$$\mu_a(M) = \int_M f \, d\sigma$$

for each Borel set M and some integrable function f defined on ∂B. By (i) of Theorem 3.7 $D\mu_a = f$ a.e. (σ). Now for each Borel set M

$$(\mu_a - f(z)\sigma)(M) = \int_M [f(y) - f(z)] \, d\sigma(y)$$

and

$$|\mu_a - f(z)\sigma|(M) = \int_M |f(y) - f(z)| \, d\sigma(y).$$

By Theorem 3.8 $D|\mu_a - f(z)\sigma|(z) = 0$ for almost all (σ) points $z \in \partial B$. Let N_a be the set of $z \in \partial B$ such that $D|\mu_a - f(z)\sigma|(z) \neq 0$. For $z \notin N_a$, $D|\mu_a - f(z)\sigma|(z) = 0$ and by (g)

$$\theta - \lim_{x \to z} \mathbf{PI}(|\mu_a - f(z)\sigma|, B)(x) = 0.$$

Since

$$|\mathbf{PI}(\mu_a - f(z)\sigma, B)(z)| \leq \mathbf{PI}(|\mu_a - f(z)\sigma|, B)(z),$$

$$\theta - \lim_{x \to z}[\mathbf{PI}(\mu_a, B)(x) - f(z)] = 0$$

for $z \notin N_a$. The set N_a has σ-measure zero. We include in the set N_a those points z where $D\mu_a(z) \neq f(z)$. This does not affect the equation $\sigma(N_a) = 0$. Then for $z \notin N_a$

$$\theta - \lim_{x \to z} \mathbf{PI}(\mu_a, B)(x) = f(z) = D\mu_a(z).$$

In other words,

$$\theta - \lim_{x \to z} \mathbf{PI}(\mu_a, B)(x) = D\mu_a(z)$$

a.e. (σ), from which follows

$$\begin{aligned} \theta - \lim_{x \to z} h(x) &= \theta - \lim_{x \to z} \mathbf{PI}(\mu, B)(x) \\ &= \theta - \lim_{x \to z} [\mathbf{PI}(\mu_a, B)(x) + \mathbf{PI}(\mu_s^+, B)(x) - \mathbf{PI}(\mu_s^-, B)(x)] \\ &= D\mu(z) \end{aligned}$$

for almost all z. It therefore suffices to prove (g). Suppose then that v is a measure on ∂B and $Dv(z) = 0$. Letting $h = \mathbf{PI}(v, B)$ on B, we know by Theorem 3.1 that $\lim_{\lambda \to 1^-} h(\lambda z) = 0$. If C is a cone with vertex at z, the vector z as axis, and half-angle θ, then by Corollary 2.15 there is a constant m depending only upon θ such that $h(x^*)/h(x) \leq m$ whenever x and x^* simultaneously lie in a ball which is contained in $C \cap B$. It is assumed, of course, that the cone C is properly truncated. Consider any $x \in C \cap B$ and close to z. Let x' be the projection of x onto the axis of C. Then $h(x)/h(x') \leq m$ and

$$|h(x') - h(x)| = h(x') \left| 1 - \frac{h(x)}{h(x')} \right| \leq h(x')(1 + m).$$

Since $h(x') \to 0$ as $x \to z$, $h(x) \to 0$ as $x \to z$.

CHAPTER 4

Superharmonic Functions

Functions u having continuous second partials on the open set R for which $\Delta u \leq 0$ on R are of interest in their own right; in particular, solutions of Poisson's equation $\Delta u = -f$, where f is a non-negative function on R, are of this type. The formal definition of superharmonic function will allow discontinuous superharmonic functions.

1. Definitions of Superharmonic Functions

We shall give a preliminary definition of superharmonic functions and establish a property of such functions which is used later for the formal definition.

Preliminary Definition. The function u is superharmonic on the open set R if u has continuous second partials on R and $\Delta u \leq 0$ on R.

LEMMA **4.1** If u is superharmonic on the open set $R \subset E^n$, then $u(x) \geq L(u : x, \delta)$ whenever $\bar{B}_{x,\delta} \subset R$.

Proof. Suppose first that $\Delta u < 0$ on $\bar{B} = \bar{B}_{x,\delta} \subset R$. Let

$$h = \begin{cases} u & \text{on} \quad \partial B, \\ \mathbf{PI}(u, B) & \text{on} \quad B. \end{cases}$$

The function h is continuous on $\bar{B}$ and harmonic on B. If we can show that $u \geq h$ on B, the result will follow for the case under consideration, since then $u(x) \geq h(x) = \mathbf{PI}(u, B)(x) = L(u : x, \delta)$. Let $w = u - h$. Then $\Delta w = \Delta u - \Delta h < 0$ on B, $w = 0$ on ∂B, and w is continuous on $\bar{B}$. Suppose w attains its minimum on $\bar{B}$ at x_0. If $x_0 \in B$, then

$$\left.\frac{\partial^2 w}{\partial x_i^2}\right|_{x_0} \geq 0, \qquad i = 1, \ldots, n,$$

and $\Delta w(x_0) \geq 0$, a contradiction. Therefore, $w = u - h \geq 0$ on $\bar{B}$ and $u \geq h$ on $\bar{B}$. Now suppose only that $\Delta u \leq 0$ on B. Letting $q(x) = \|x\|^2$, $\Delta q = 2n$. For each $\varepsilon > 0$, $\Delta(u - \varepsilon q) = \Delta u - \varepsilon \Delta q < 0$ on R. By the first part of the proof,

$$u(x) - \varepsilon q(x) \geq L(u - \varepsilon q : x, \delta) = L(u : x, \delta) - \varepsilon L(q : x, \delta).$$

Letting $\varepsilon \to 0$, we obtain $u(x) \geq L(u : x, \delta)$.

Before defining superharmonic functions in terms of averages, we shall make the following convention concerning averages. Let f be a function on $\partial B_{x,\delta}$. If $f \geq 0$ and is integrable relative to surface area on $\partial B_{x,\delta}$, then $L(f : x, \delta)$ is defined and finite; if $f \geq 0$ and not integrable, then we let $L(f : x, \delta) = +\infty$. More generally, if $f = f^+ - f^-$ is Borel measurable on $\partial B_{x,\delta}$, then we let $L(f : x, \delta) = L(f^+ : x, \delta) - L(f^- : x, \delta)$ provided one of the latter averages is finite. If f is l.s.c. on $\partial B_{x,\delta}$, then f is bounded below on $\partial B_{x,\delta}$ and $L(f : x, \delta)$ is always defined, possibly having the value $+\infty$ but not the value $-\infty$.

Definition. The extended real-valued Borel measurable function u defined on an open set R is (i) **super-mean-valued** at $x \in R$ if $L(u : x, \delta)$ is defined and $u(x) \geq L(u : x, \delta)$ whenever $\bar{B}_{x,\delta} \subset R$; (ii) **super-mean-valued** on R if super-mean-valued at each point of R; (iii) **locally super-mean-valued** at $x \in R$ if there is a $\delta_x > 0$ such that $B_{x,\delta_x} \subset R$, $L(u : x, \delta)$ is defined for all $\delta < \delta_x$, and $u(x) \geq L(u : x, \delta)$ for all $\delta < \delta_x$; and (iv) **locally super-mean-valued** on R if locally super-mean-valued at each point of R.

There are analogous "sub-mean-valued" definitions obtained by reversing the inequalities. A function which is both sub-mean-valued and super-mean-valued in any of the above senses is said to be **mean-valued**. The phrases "satisfies the superaveraging principle at $x \in R$," "satisfies the superaveraging principle on R," "satisfies the superaveraging principle locally at $x \in R$," and "satisfies the superaveraging principle locally" are also used in place of the corresponding phrases in (i), (ii), (iii) and (iv), respectively, of the above definition.

Definition. If R is an open subset of E^n, $\mathscr{L}_R$ will denote the class of extended real-valued functions on R satisfying

(i) u is not identically $+\infty$ on any component of R,
(ii) $u > -\infty$ on R,
(iii) u is l.s.c. on R.

Note that for any $u \in \mathscr{L}_R$, $L(u : x, \delta)$ is defined whenever $\bar{B}_{x,\delta} \subset R$ since u is bounded below on $\partial B_{x,\delta}$ by the l.s.c. of u.

Definition. An extended real-valued function u on an open set R is **superharmonic** on R if $u \in \mathscr{L}_R$ and u is super-mean-valued on R; the function u is **subharmonic** on R if $-u$ is superharmonic on R.

The set of functions superharmonic on R is denoted by $\mathscr{S}_R$, whereas the set of non-negative superharmonic functions on R is denoted by $\mathscr{S}_R^+$. It is easily seen that $u + h \in \mathscr{S}_R$ if $u \in \mathscr{S}_R$ and h is harmonic on R; but it is not obvious that $\mathscr{S}_R$ is closed under addition. The latter problem is dealt with later. Before giving examples of superharmonic functions, we shall look at alternative definitions that make it easier to verify superharmonicity.

THEOREM **4.2** If R is an open connected set, $u \in \mathscr{L}_R$, and for each $x \in R$ there is a δ_x such that $B_{x,\delta_x} \subset R$ and $u(x) \geq L(u : x, \delta)$ (or $u(x) \geq A(u : x, \delta)$) whenever $\delta < \delta_x$, then u satisfies the minimum principle on R.

Proof. Suppose there is a point $x_0 \in R$ such that $u(x_0) = \inf_R u$. Since $u \in \mathscr{L}_R$, $-\infty < u(x_0) = \inf_R u < +\infty$. Let $M = \{x : u(x) = \inf_R u\}$, which is a relatively closed subset of R by the l.s.c. of u. We shall show that M is also open. Consider any $y \in M$. Then there is a δ_y such that $B_{y,\delta_y} \subset R$ and $u(y) \geq L(u : y, \delta)$ whenever $\delta < \delta_y$. Suppose there is a point $z \in B_{y,\delta_y} \sim M$. Let $\rho = \|y - z\|$. Since $y \in M$, $u(y) \leq L(u : y, \rho)$. Therefore $u(y) = L(u : y, \rho)$ or $L(u - u(y) : y, \rho) = 0$. Since $u - u(y) \geq 0$ on $\partial B_{y,\rho}$, $u - u(y) = 0$ a.e. (σ) on $\partial B_{y,\rho}$. But since $u(z) > u(y)$, there is an α such that $u(z) > \alpha > u(y)$. By the l.s.c. of u there is a neighborhood U_z of z such that $u > \alpha > u(y)$ on $U_z \cap \partial B_{y,\rho}$. Since the latter set has positive surface area, $u - u(y) > 0$ on a set of positive surface area, a contradiction. This shows that $B_{y,\delta_y} \subset M$; that is, M is open. Therefore $M = \varnothing$ or $M = R$ by the connectedness of R with the first possibility being excluded from the onset. It follows that $M = R$ and that u is constant if it attains its infimum. The proof requires only minor modifications if volume averages are used in place of spherical averages.

COROLLARY **4.3** If u is superharmonic on the open connected set R, then u satisfies the minimum principle on R. If R is a bounded open set, u is superharmonic on R, and $\liminf_{z \to x} u(z) \geq 0$ for all $x \in \partial R$, then $u \geq 0$ on R.

Proof. That u satisfies the minimum principle follows directly from the definition of superharmonic function and Theorem 4.2. To prove the second assertion, it suffices to prove that $u \geq 0$ on each component of R; that is, we can assume that R is connected. Suppose there is a point $y \in R$ such that $u(y) < 0$. Then u is not a constant function. Define a function v on $\bar{R}$ by

$v(x) = \liminf_{z \to x} u(z)$, $x \in \bar{R}$. Then v is l.s.c. on $\bar{R}$, $v \geq 0$ on ∂R, $v(y) < 0$. Since v is l.s.c. on $\bar{R}$, it must attain a negative minimum on $\bar{R}$, in fact, on R; but this contradicts the minimum principle. Therefore $u \geq 0$ on R.

THEOREM **4.4** Let u be superharmonic on the open set R and let W be an open subset of R with compact closure $\bar{W} \subset R$. If h is continuous on $\bar{W}$, harmonic on W, and $u \geq h$ on ∂W, then $u \geq h$ on W.

Proof. If h has the above properties, then it has the same properties on each component of W. By considering the components of W, we can assume that W is connected. Consider the function $u - h$ on $\bar{W}$. On ∂W, $u - h \geq 0$. Since h is harmonic on W, $u - h$ is superharmonic on W and cannot attain its infimum on W by the preceding corollary. As a l.s.c. function on the compact set $\bar{W}$, $u - h$ attains its infimum on $\bar{W}$ and, in fact, on ∂W. Since $u - h \geq 0$ on ∂W, $u - h \geq 0$ on $\bar{W}$.

Theorem 4.4 is used to formulate an alternative definition.

Second Definition. An extended real-valued function $u \in \mathscr{L}_R$ is superharmonic on the open set R if u has the following property:

(P) If W is an open subset of R with compact closure $\bar{W} \subset R$, h is continuous on $\bar{W}$, h is harmonic on W, and $u \geq h$ on ∂W, then $u \geq h$ on W.

Recall that $\mathscr{S}_R$ is the class of functions superharmonic on R, according to the original definition. Now let $\mathscr{S}'_R$ be the class of functions $u \in \mathscr{L}_R$ satisfying (P). By Theorem 4.4 $\mathscr{S}_R \subset \mathscr{S}'_R$. We now show that $\mathscr{S}'_R \subset \mathscr{S}_R$. Consider any $u \in \mathscr{S}'_R$ and any $\bar{B}_{x,\delta} \subset R$. Since u is l.s.c. on $\partial B_{x,\delta}$, there is a sequence $\{\phi_j\}$ of continuous functions on $\partial B_{x,\delta}$ such that $\phi_j \uparrow u$ on $\partial B_{x,\delta}$. Let

$$h_j = \begin{cases} \phi_j & \text{on} \quad \partial B_{x,\delta}, \\ \mathbf{PI}(\phi_j, B_{x,\delta}) & \text{on} \quad B_{x,\delta}. \end{cases}$$

Then h_j is continuous on $\bar{B}_{x,\delta}$, harmonic on $B_{x,\delta}$, and $u \geq \phi_j = h_j$ on $\partial B_{x,\delta}$. Since u satisfies (P), $u \geq h_j$ on $B_{x,\delta}$. Therefore

$$u(x) \geq h_j(x) = \mathbf{PI}(\phi_j, B_{x,\delta})(x) = L(\phi_j : x, \delta)$$

with the latter equality holding since x is the center of the ball $B_{x,\delta}$. Since $\phi_j \uparrow u$ on $\partial B_{x,\delta}$, $u(x) \geq L(u : x, \delta)$ by the Lebesgue monotone convergence theorem. This proves that $\mathscr{S}'_R \subset \mathscr{S}_R$ and that the two definitions of superharmonic function are equivalent. A disadvantage of the original definition is that one must verify the inequality $u(x) \geq L(u : x, \delta)$ for every ball $B_{x,\delta}$ with $\bar{B}_{x,\delta} \subset R$.

THEOREM **4.5** If R is an open set, $u \in \mathscr{L}_R$, and u is locally super-mean-valued on R, then u is superharmonic on R.

Proof. We show that u is superharmonic according to the second definition. We need only show that (P) is satisfied. Let W be an open subset of R with compact closure $\overline{W} \subset R$ and let h be continuous on $\overline{W}$, harmonic on W, and $u \geq h$ on ∂W. By considering each of the components of W, we can assume that W is connected. Consider $u - h$ which is l.s.c. on $\overline{W}$ and attains its infimum on $\overline{W}$. Since h is mean-valued on R, $u - h$ is locally super-mean-valued on R. By Theorem 4.2 and the fact that $u \geq h$ on ∂W, $u - h$ cannot attain its infimum at a point of W. Therefore, $u - h \geq 0$ on W. This shows that $u \in \mathscr{S}'_R = \mathscr{S}_R$.

Theorem 4.5 can be used to give yet another definition of superharmonic function.

Third Definition. An extended real-valued function u on the open set R is superharmonic if $u \in \mathscr{L}_R$ and u is locally super-mean-valued on R.

Letting $\mathscr{S}''_R$ denote the class of functions $u \in \mathscr{L}_R$ which are locally super-mean-valued on R, we have just seen that $\mathscr{S}''_R \subset \mathscr{S}_R = \mathscr{S}'_R$. Trivially, $\mathscr{S}_R \subset \mathscr{S}''_R$ and therefore $\mathscr{S}_R = \mathscr{S}'_R = \mathscr{S}''_R$ and all three definitions are equivalent. The last definition is usually the easiest for determining if a specific function is superharmonic. We now give some examples.

EXAMPLE **1** If u is harmonic on R, then u is both subharmonic and superharmonic on R.

EXAMPLE **2** The fundamental harmonic function u_y with pole y is superharmonic on E^n. We know that u_y is harmonic on $E^n \sim \{y\}$ and therefore locally super-mean-valued on $E^n \sim \{y\}$; u_y is also super-mean-valued at y since $+\infty = u_y(y) \geq L(u_y : y, \delta)$ for any $\delta > 0$. Since $u_y \in \mathscr{L}_{E^n}$, u_y is superharmonic on E^n.

EXAMPLE **3** If u is harmonic on the open set R, then $|u|$ is subharmonic on R and $-|u|$ is superharmonic on R. This follows from the fact that $|u|$ is finite-valued, continuous on R, and $|u(x)| = |L(u : x, \delta)| \leq L(|u| : x, \delta)$ whenever $\overline{B}_{x,\delta} \subset R$.

EXAMPLE **4** If u is harmonic on the open set R, then $u^+ = u \vee 0$ is subharmonic on R and $u^- = (-u) \vee 0$ is subharmonic on R. Again finiteness and continuity of u^+ and u^- follow from that of u. Since $u(x) = L(u : x, \delta) \leq L(u^+ : x, \delta)$ wherever $\overline{B}_{x,\delta} \subset R$, $u^+(x) \leq L(u^+ : x, \delta)$ whenever $\overline{B}_{x,\delta} \subset R$. This shows that u^+ is subharmonic on R.

EXAMPLE 5 Let $f(z)$ be a function of the complex variable z which is analytic on an open subset R of the complex plane. Then $|f(z)|$ is subharmonic on R. This follows from the Cauchy integral formula

$$f(z) = \frac{1}{2\pi} \oint_c \frac{f(\zeta)}{\zeta - z}\, d\zeta,$$

where $c = \{\zeta : |z - \zeta| = \delta\} \subset R$ and $\{\zeta : |z - \zeta| \leq \delta\} \subset R$. Then

$$\begin{aligned} |f(z)| &\leq \frac{1}{2\pi} \int_c \frac{|f(\zeta)|}{|\zeta - z|}\, |d\zeta| = \frac{1}{2\pi\delta} \int_c |f(\zeta)|\, |d\zeta| \\ &= L(|f| : z, \delta). \end{aligned}$$

Examples 3 and 4 are just special cases of a more general statement that a convex function of a harmonic function is a subharmonic function. To be more specific, let ϕ be a real-valued function defined on a finite or infinite interval (a, b). The function ϕ is **convex** if $\phi(\lambda x + (1 - \lambda)y) \leq \lambda\phi(x) + (1 - \lambda)\phi(y)$ whenever $a < x < y < b$ and $0 < \lambda < 1$. The function ψ is **concave** on (a, b) if $-\psi$ is convex thereon. We shall assume that ϕ is convex in the following discussion. It is easy to show that such a function is continuous on (a, b). Moreover, the right and left derivates $D_x^+\phi$ and $D_x^-\phi$ are both defined for each $x \in (a, b)$ with $-\infty < D_x^-\phi \leq D_x^+\phi < +\infty$. One can show that for each $x \in (a, b)$, there is a linear function χ such that $\chi \leq \phi$ and $\chi(x) = \phi(x)$. Leaving ϕ for a moment, let $(\Omega, \mathscr{B}, \mu)$ be a probability measure space; that is, $\mathscr{B}$ is a σ-algebra of subsets of Ω and μ is a measure on $\mathscr{B}$ of total mass 1. Consider a real-valued function f taking on values in the domain (a, b) of ϕ, measurable relative to $\mathscr{B}$, and integrable relative to μ. Then $\alpha = \int f\, d\mu$ is defined and is a number in the interval (a, b). Let χ be a linear function on (a, b) such that $\chi(\alpha) = \phi(\alpha)$ and $\chi \leq \phi$. Suppose $\chi(x) = px + q$ on (a, b). Then $\phi(f) \geq \chi(f) = pf + q$ and therefore

$$\int \phi(f)\, d\mu \geq \int (pf + q)\, d\mu = p\alpha + q = \chi(\alpha) = \phi(\alpha) = \phi\left(\int f\, d\mu\right).$$

The inequality

$$\phi\left(\int f\, d\mu\right) \leq \int \phi(f)\, d\mu$$

is known as **Jensen's inequality**. If ϕ is convex on $(-\infty, +\infty)$, if f is integrable relative to μ, and if f can take on the values $+\infty$ and $-\infty$, then Jensen's inequality still holds if ϕ is defined arbitrarily at $-\infty$ and $+\infty$. This can be seen by applying Jensen's inequality to $\int_{\sim F} f\, d\mu$ where the set $F = \{w : |f(w)| = +\infty\}$ has μ measure zero.

If f is a function on $\partial B_{x,\delta}$ with $L(|f|: x, \delta) < +\infty$ and ϕ is convex on an open interval containing the range of f, then $\phi(L(f: x, \delta)) \leq L(\phi(f): x, \delta)$. Example 3 is a specific application of this inequality with $\phi(\zeta) = |\zeta|$ for $\zeta \in (-\infty, +\infty)$ and Example 4 is also with $\phi(\zeta) = \zeta \vee 0$ for $\zeta \in (-\infty, +\infty)$.

THEOREM **4.6** If h is harmonic on the open set R and ϕ is a convex function with domain an interval containing the range of h, then $\phi(h)$ is subharmonic on R.

Proof. $\phi(h)$ is finite-valued and continuous on R since h is. If $\bar{B}_{x,\delta} \subset R$, then $h(x) = L(h: x, \delta)$ implies $\phi(h(x)) = \phi(L(h: x, \delta)) \leq L(\phi(h): x, \delta)$ and this shows that $\phi(h)$ is subharmonic.

THEOREM **4.7** If u is subharmonic on the open set R and ϕ is an increasing convex function with domain an interval containing the range of u, then $\phi(u)$ is subharmonic on R.

Proof. Since ϕ is finite-valued, $\phi(u)$ is finite-valued. Since a convex function is continuous, $\phi(u)$ is also u.s.c. By Jensen's inequality and that fact that ϕ is increasing, $\phi(u(x)) \leq \phi(L(u: x, \delta)) \leq L(\phi(u): x, \delta)$ whenever $\bar{B}_{x,\delta} \subset R$.

EXAMPLE 7 If h is harmonic on the open set R and $p > 1$, then $|h|^p$ is subharmonic on R. We know that $|h|$ is subharmonic on R. Since

$$\phi(\zeta) = \begin{cases} 0 & \text{for} \quad \zeta < 0, \\ \zeta^p & \text{for} \quad \zeta \geq 0, \end{cases}$$

is an increasing convex function, $|h|^p$ is subharmonic on R.

2. Generalized Laplacian

Initially superharmonic functions were those smooth functions u for which $\Delta u \leq 0$. At the beginning of this chapter we proved that such functions are super-mean-valued and then used this property to define a wider class of superharmonic functions. According to the following theorem, we have not obtained any new smooth superharmonic functions.

THEOREM **4.8** Let u be a function defined on an open set $R \subset E^n$ having continuous second partials. Then u is superharmonic on R if and only if $\Delta u \leq 0$ on R.

Proof. The sufficiency part is just Lemma 4.1. Suppose u is superharmonic on R and has continuous second partials on R. Then Δu is continuous on R. Let $R_1 = \{x : \Delta u(x) > 0\}$. R_1 is an open subset of R. Suppose $R_1 \neq \varnothing$. Then $\Delta(-u) < 0$ on R_1. By the sufficiency part $-u$ is superharmonic on R_1 or, in other words, u is subharmonic on R_1. Thus we have both $u(x) \leq L(u : x, \delta)$ and $u(x) \geq L(u : x, \delta)$ whenever $\bar{B}_{x,\delta} \subset R_1$. By Corollary 2.9 u is harmonic on R_1 and this means that $\Delta u = 0$ on R_1, a contradiction. Therefore $R_1 = \varnothing$ and $\Delta u \leq 0$ on R.

On the surface it would appear that the differential calculus is being phased out in favor of treatment by means of averages. Recall, however, that an ordinary derivative is nothing but a limit of averages. The same is true of the Laplacian as we now show.

Let R be an open subset of E^n and let x be a fixed point of R. Consider the function v defined on R as follows

$$v(y) = \frac{\|y - x\|^2}{2n}.$$

Then $v(x) = 0$, $\Delta v = 1$, and

$$L(v : x, \delta) = \frac{1}{\sigma_n \delta^{n-1}} \int_{\partial B_{x,\delta}} \frac{\|y - x\|^2}{2n} \, d\sigma(y) = \frac{\delta^2}{2n}.$$

Consider any function u defined on R having continuous second partials. Consider only those balls $B = B_{x,\delta}$ for which $\bar{B}_{x,\delta} \subset R$. Since Δu and Δv are continuous on $\bar{B}$, $\Delta(u + tv) = \Delta u + t\,\Delta v > 0$ on $\bar{B}$ for sufficiently large t, in fact, for $t \geq \sup_{\bar{B}} (-\Delta u/\Delta v) = -\inf_{\bar{B}} (\Delta u/\Delta v)$. For such t

$$u(x) + tv(x) \leq L(u + tv : x, \delta) = L(u : x, \delta) + tL(v : x, \delta);$$

that is,

$$-t[L(v : x, \delta) - v(x)] \leq L(u : x, \delta) - u(x).$$

Since $t = -\inf_{\bar{B}} (\Delta u/\Delta v)$ is one of the values of t for which this is true, $\Delta v = 1$, $v(x) = 0$, and $L(v : x, \delta) = \delta^2/2n$,

$$\inf_{\bar{B}} \Delta u \leq 2n \frac{L(u : x, \delta) - u(x)}{\delta^2}.$$

By choosing t so that $\Delta(u + tv) < 0$ on $\bar{B}$ we can show that

$$\inf_{\bar{B}} \Delta u \leq 2n \frac{L(u : x, \delta) - u(x)}{\delta^2} \leq \sup_{\bar{B}} \Delta u.$$

Since Δu is continuous at x, the extreme members approach $\Delta u(x)$ as $\delta \downarrow 0$. Therefore

$$\Delta u(x) = \lim_{\delta \downarrow 0} 2n \frac{L(u : x, \delta) - u(x)}{\delta^2}.$$

Looking at the Laplacian from this point of view, our definitions in terms of averages are quite natural. Note that the limit on the right might exist even if u does not have continuous second partials. For an arbitrary function u defined on R the **generalized Laplacian** $\tilde{\Delta} u$ is given by

$$\tilde{\Delta} u(x) = \lim_{\delta \downarrow 0} 2n \frac{L(u : x, \delta) - u(x)}{\delta^2}$$

whenever the limit on the right exists. If u has continuous second partials on R, then $\Delta u = \tilde{\Delta} u$ on R.

3. Properties of Superharmonic Functions

If u and v are superharmonic on an open set R, then it is not evident that $u + v$ is superharmonic on R because of the requirement that $u + v$ not be identically $+\infty$ on any component of R. To show that $u + v$ is superharmonic, we must show that v cannot be identically $+\infty$ on the set where u is finite. The set of points on which a superharmonic function is infinite is a focal point in much of the remaining chapters. The following theorem not only limits the infinities of a superharmonic function but also shows that superharmonic functions could have been defined using volume averages rather than spherical averages.

THEOREM **4.9** If R is an open set, $u \in \mathscr{L}_R$, and for each $x \in R$ there is a $\delta_x > 0$ such that $B_{x,\delta_x} \subset R$ and $u(x) \geq A(u : x, \delta)$ whenever $\delta < \delta_x$, then u is superharmonic on R. Moreover, if u is superharmonic on R and $x \in R$, then $u(x) \geq A(u : x, \delta)$ whenever $B_{x,\delta} \subset R$.

Proof. Let $u \in \mathscr{L}_R$ satisfy the requirements of the first assertion. Also let W be an open set with compact closure $\overline{W} \subset R$ and let h be continuous on $\overline{W}$, harmonic on W, and $u \geq h$ on ∂W. As usual, we can assume that W is connected. By Theorem 4.2 $u - h$ satisfies the minimum principle on W. Since $u - h$ is l.s.c. on $\overline{W}$ and $u > h \geq 0$ on ∂W, $u - h \geq 0$ on W by the minimum principle. It follows that u is superharmonic on R by the second definition. Suppose now that u is superharmonic on R, $x \in R$, and $B_{x,\delta} \subset R$. We show that $u(x) \geq A(u : x, \delta)$. If $u(x) = +\infty$, the inequality is trivially true. Assume $u(x) < +\infty$. Since $u(x) \geq L(u : x, \rho)$ for $0 < \rho < \delta$,

$$\sigma_n \rho^{n-1} u(x) \geq \int_{\partial B_{x,\rho}} u(z)\, d\sigma(z), \qquad 0 < \rho < \delta.$$

Integrating over $(0, \delta)$,

$$\frac{\sigma_n \delta^n}{n} u(x) > \int_0^\delta \int_{\partial B_{x,\rho}} u(z)\, d\sigma(z)\, d\rho = \int_{B_{x,\delta}} u(z)\, dz.$$

Since $v_n = \sigma_n / n$, $u(x) \geq A(u : x, \delta)$ whenever $B_{x,\delta} \subset R$.

This theorem conveys some information about the infinities of a superharmonic function u; namely that u is finite a.e. on $B_{x,\delta}$ relative to Lebesgue measure whenever it is finite at the center of the ball. This is sufficient reason to believe that u is finite a.e. on R relative to Lebesgue measure.

THEOREM **4.10** If u is superharmonic on the open set R, then u is finite a.e. on R relative to Lebesgue measure and Lebesgue integrable on each compact set $K \subset R$.

Proof. It suffices to show that u is finite a.e. on each of the components of R. We might as well assume that R is connected. Since u is not identically $+\infty$ on R, there is at least one point of R where u is finite. Let

$$M = \{x \in R : u \text{ is finite a.e. on } B_{x,\delta} \subset R \text{ for some } \delta > 0\}.$$

M is nonempty since there is at least one point of R where u is finite and, according to Theorem 4.9, u is finite a.e. on each ball in R having this point as its center. We first show that M is open. Suppose $x \in M$. Then u is finite a.e. on $B_{x,\delta} \subset R$ for some $\delta > 0$. Consider any $y \in B_{x,\delta}$ and the ball $B_{y,\rho}$, where $2\rho = \min\{\|y - x\|, \delta - \|y - x\|\}$. Then $B_{y,\rho} \subset B_{x,\delta}$ and u is finite a.e. on $B_{y,\rho}$; that is, $y \in M$. This shows that $B_{x,\delta} \subset M$ and that M is open. We next show that M is relatively closed in R. Let $\{x_j\}$ be a sequence in M with $x_j \to x \in R$ as $j \to \infty$. Since R is open, there is an $\varepsilon > 0$ such that $B_{x,\varepsilon} \subset R$. Choose i so that $x_i \in B_{x,\varepsilon/2}$. Since $x_i \in M$, there is a $\delta > 0$ such that u is finite a.e. on $B_{x_i,\delta}$. In particular, u is finite a.e. on $B_{x_i,\delta} \cap B_{x,\varepsilon/2}$ which has positive Lebesgue measure. It follows that there is a point $z \in B_{x_i,\delta} \cap B_{x,\varepsilon/2}$ such that $u(z) < +\infty$. By Theorem 4.9 u is finite a.e. on $B_{z,\varepsilon/2} \subset R$. Since $x \in B_{z,\varepsilon/2}$, there is a ball $B_{x,\rho} \subset B_{z,\varepsilon/2}$ on which u is finite a.e. This shows that $x \in M$ and that M is relatively closed. Since $M \neq \varnothing$, $M = R$ by the connectedness of R. By definition of M to each $x \in M = R$ there corresponds a $B_{x,\delta_x} \subset R$ on which u is finite a.e. Since such open sets cover R, a countable number of them suffice to cover R. Since u is finite a.e. on each element of a countable covering of R, u is finite a.e. on R. To prove that u is Lebesgue integrable on each compact set $K \subset R$, cover K be a finite number of balls with centers x_i and radii δ_i such that $\bar{B}_{x_i,\delta_i} \subset R$, $i = 1, \ldots, p$. Since u is finite a.e. on arbitrarily small balls containing each x_i, we can assume that $u(x_i) < +\infty$. This may require replacing x_i by a nearby x_i^* and increasing δ_i slightly. Then $+\infty > u(x_i) \geq A(u : x_i, \delta_i)$ for each i. Since u is bounded below on $\bigcup_{i=1}^{p} \bar{B}_{x_i,\delta_i}$ we can assume that $u \geq 0$ on this set. Then

$$-\infty < \int_K u(z)\,dz \leq \sum_{i=1}^{p} \int_{B_{x_i,\delta_i}} u(z)\,dz \leq \sum_{i=1}^{p} \nu_n \delta_i^n u(x_i) < +\infty$$

and u is integrable on K.

THEOREM **4.11** If u is superharmonic on the open set R and B is a ball with $\bar{B} \subset R$, then u is integrable relative to surface area on ∂B, $\mathbf{PI}(u, B)$ is harmonic on B, and $u \geq \mathbf{PI}(u, B)$ on B.

Proof. We can assume that $u \geq 0$ on ∂B. Since u is l.s.c. on ∂B, there is a sequence $\{\phi_j\}$ of non-negative continuous functions on ∂B such that $\phi_j \uparrow u$ on ∂B. Let

$$v_j = \begin{cases} \mathbf{PI}(\phi_j, B) & \text{on} \quad B, \\ \phi_j & \text{on} \quad \partial B. \end{cases}$$

Since $u \geq v_j = \phi_j$ on ∂B, v_j is harmonic on B and v_j is continuous on $\bar{B}$, $u \geq v_j$ on B by the second definition of superharmonic functions. Now $\{v_j\}$ is an increasing sequence of functions harmonic on B and $v = \lim_{j \to \infty} v_j$ is either identically $+\infty$ or harmonic on B by Theorem 2.19. Since u is finite a.e. on R and $u \geq v$, v is harmonic on B. It also follows from the Lebesgue monotone convergence theorem that $v = \mathbf{PI}(u, B)$ on B. This proves the last two assertions. If $B = B_{x,\delta}$, then $+\infty > v(x) = \mathbf{PI}(u, B)(x) = L(u : x, \delta)$ and u is integrable relative to surface area on ∂B.

Now that we know there are some limitations on the infinities of a superharmonic function, we can discuss the sum of two superharmonic functions.

THEOREM **4.12** If u and v are superharmonic on an open set R and $c > 0$, then

(i) cu is superharmonic on R,
(ii) $u + v$ is superharmonic on R,
(iii) $u \wedge v = \min(u, v)$ is superharmonic on R.

Proof. (i) is obvious from the definition of superharmonic function. To prove (ii) note that $u + v$ cannot be identically $+\infty$ on any component U of R since each component has positive Lebesgue measure and both u and v are simultaneously finite a.e. on U. Moreover, $u + v > -\infty$ on R since both u and v have this property and $u + v$ is l.s.c. on R since the sum of any two l.s.c. functions is l.s.c. Since $u(x) \geq L(u : x, \delta)$ and $v(x) \geq L(v : x, \delta)$ whenever $\bar{B}_{x,\delta} \subset R$, $u(x) + v(x) \geq L(u + v : x, \delta)$ whenever $\bar{B}_{x,\delta} \subset R$. This concludes the proof of (ii). Rather than prove (iii) completely, we shall show only that $u \wedge v$ is super-mean-valued on R, the other properties being easy to verify. Suppose $\bar{B}_{x,\delta} \subset R$. Then

$$u(x) \geq L(u : x, \delta) \geq L(u \wedge v : x, \delta)$$

and

$$v(x) \geq L(v : x, \delta) \geq L(u \wedge v : x, \delta).$$

Therefore

$$(u \wedge v)(x) = \min(u(x), v(x)) \geq L(u \wedge v : x, \delta).$$

THEOREM **4.13** If u is both subharmonic and superharmonic on an open set R, then u is harmonic on R.

Proof. Clearly, $-\infty < u < +\infty$. Since u is both l.s.c. and u.s.c., u is continuous on R. Since $u(x) = L(u : x, \delta)$ whenever $\bar{B}_{x,\delta} \subset R$, u is harmonic on R by Corollary 2.9.

THEOREM **4.14** If u is superharmonic on the open set R, then $u(x) = \liminf_{y \to x, y \neq x} u(y)$ for $x \in R$.

Proof. If $x \in R$,

$$u(x) = \liminf_{y \to x} u(y) \leq \liminf_{y \to x,\, y \neq x} u(y).$$

If $u(x) = +\infty$, all three are equal and the theorem is trivially true. We can assume then that $u(x) < +\infty$. Since

$$u(x) \geq A(u : x, \delta) \geq \inf\{u(y) : y \in B_{x,\delta},\, y \neq x\}$$

whenever $\bar{B}_{x,\delta} \subset R$, $u(x) \geq \liminf_{y \to x,\ y \neq x} u(y).$

This together with the first inequality establishes the equality.

We conclude this section by considering monotone sequences of superharmonic functions; more generally, we shall consider directed families of superharmonic functions.

THEOREM **4.15** If $\{u_i : i \in I\}$ is a right-directed family of superharmonic functions on the open set R, then on each component of R the function $u = \sup_{i \in I} u_i$ is either superharmonic or identically $+\infty$.

Proof. Since each u_i is l.s.c. and the supremum of any collection of l.s.c. functions is again l.s.c., u is l.s.c. on R. If $i_0 \in I$, then $u \geq u_{i_0} > -\infty$ on R. If $\bar{B}_{x,\delta} \subset R$, then $u_i(x) \geq L(u_i : x, \delta)$ and by Lemma 2.23

$$u(x) = \sup_{i \in I} u_i(x) \geq \sup_{i \in I} L(u_i : x, \delta) = L\left(\sup_{i \in I} u_i : x, \delta\right)$$
$$= L(u : x, \delta).$$

Therefore u is super-mean-valued on R and is superharmonic on each component of R, provided it is not identically $+\infty$ thereon.

The preceding theorem naturally applies to monotone increasing sequences of superharmonic functions. The analogous theorem for monotone decreasing sequences is not true. Consider, for example, the superharmonic function

u_y on E^3 with pole y. Define $u_j = j^{-1}u_y$. Then $\{u_j\}$ is a monotone decreasing sequence of superharmonic functions. Letting $u = \inf_{j \geq 1} u_j$, $u = 0$ except at y where it is equal to $+\infty$. Clearly, u is not superharmonic because it fails to be l.s.c. at y. Not only is the semi-continuity one of the difficulties in the general case, but the requirement that the lower envelope not take on the value $-\infty$ is another difficulty. The lower envelope, however, comes close to being superharmonic provided we eliminate the latter difficulty.

THEOREM **4.16** If $\{u_i : i \in I\}$ is a left-directed family of superharmonic functions on the open set R which is locally bounded below, then the lower regularization $\hat{u}$ of $u = \inf_{i \in I} u_i$ is superharmonic on R.

Proof. It follows from the local boundedness that $\hat{u}$ cannot take on the value $-\infty$. Since $u_i \geq u \geq \hat{u}$, $i \in I$, $\hat{u}$ is not identically $+\infty$ on any component of R. By definition, $\hat{u}$ is l.s.c. on R. We shall show that $\hat{u}$ is superharmonic according to the second definition. Suppose W is an open subset of R with compact closure $\overline{W} \subset R$, h is continuous on $\overline{W}$, h is harmonic on W, and $\hat{u} \geq h$ on ∂W. Then $u_i \geq h$ on ∂W for each $i \in I$ and therefore $u_i \geq h$ on W for each $i \in I$ by the second definition of superharmonic function. It follows that $u \geq h$ on W and that $\hat{u} \geq \hat{h} = h$ on W. Therefore $\hat{u}$ is superharmonic on R by the second definition.

The set of points at which $\hat{u} \neq u$ in the preceding theorem is not too large. We shall return to this set of points later and show, among other things, that $\hat{u}$ and u are equal a.e. relative to Lebesgue measure.

The modification of a superharmonic function so as to be harmonic over part of its domain is a fundamental operation which is justified by the following lemma.

LEMMA **4.17** Let u be superharmonic on the open set R and let $\overline{B}_{x,\delta} \subset R$. If $B = B_{x,\delta}$, define

$$v = \begin{cases} \mathbf{PI}(u, B) & \text{on} \quad B, \\ u & \text{on} \quad R \sim B. \end{cases}$$

Then $u \geq v$ on R, v is harmonic on B, and v is superharmonic on R.

Proof. $\mathbf{PI}(u, B)$ is defined and $v \leq u$ on B by Theorem 4.11. Clearly, $u \geq v$ on R and v is harmonic on B. About all that remains is to show that v is l.s.c. and super-mean-valued at points of ∂B. Consider any $y \in \partial B$. By Lemma 2.6,

$$\liminf_{\substack{z \to y \\ z \in B}} v(z) \geq \liminf_{\substack{z \to y \\ z \in \partial B}} u(z) \geq \liminf_{z \to y} u(z) = u(y) = v(y).$$

Since $v = u$ on $R \sim B$,

$$\liminf_{\substack{z \to y \\ z \in R \sim B}} v(z) = \liminf_{\substack{z \to y \\ z \in R \sim B}} u(z) \geq \liminf_{z \to y} u(z) = u(y) = v(y).$$

Therefore $\liminf_{z \to y} v(z) \geq v(y)$ and v is l.s.c. at y. Finally, we show that v is super-mean-valued at $y \in \partial B$. Suppose $\bar{B}_{y, \rho} \subset R$. Then

$$v(y) = u(y) \geq L(u : y, \rho) \geq L(v : y, \rho),$$

since $u \geq v$ on R. Since v is obviously locally super-mean-valued and l.s.c. at points of $R \sim \partial B$, v is superharmonic on R by the third definition of superharmonic function.

4. Approximation of Superharmonic Functions

In this section we show that superharmonic functions can be approximated by smooth superharmonic functions. The following lemma is the basis for such approximations.

LEMMA **4.18** If u is superharmonic on the open set R, $x \in R$, and $\delta_x = d(x, \sim R)$, then $L(u : x, \delta)$ and $A(u : x, \delta)$ are monotone decreasing functions of δ on $[0, \delta_x)$ which are right continuous at $\delta = 0$ provided we define $L(u : x, 0) = A(u : x, 0) = u(x)$; moreover, $A(u : x, \delta)$ is continuous on $[0, \delta_x)$.

Proof. Suppose $0 \leq \delta < \delta_1 < \delta_x$. Define a function h on B_{x, δ_1} by $h = \mathbf{PI}(u, B_{x, \delta_1})$. Then $u \geq h$ on B_{x, δ_1} by Theorem 4.11. Since $\bar{B}_{x, \delta} \subset B_{x, \delta_1}$ and h is harmonic on the latter ball,

$$L(u : x, \delta) \geq L(h : x, \delta) = h(x) = L(u : x, \delta_1).$$

This shows that $L(u : x, \delta)$ is monotone decreasing on $[0, \delta_x)$. Since u is l.s.c. at x, $\{y : u(y) > u(x) - \varepsilon\}$ is a neighborhood of x for each $\varepsilon > 0$ and $L(u : x, \delta) \geq L(u(x) - \varepsilon : x, \delta) = u(x) - \varepsilon$ for all sufficiently small δ. This shows that $L(u : x, \delta) \uparrow u(x)$ as $\delta \downarrow 0$. For any $\delta \leq \delta_1 < \delta_x$, $A(u : x, \delta)$ is finite and

$$\begin{aligned} A(u : x, \delta) &= \frac{1}{v_n \delta^n} \int_{B_{x, \delta}} u(z)\, dz \\ &= \frac{1}{v_n \delta^n} \int_0^\delta \rho^{n-1} \left(\int_{\|\theta\| = 1} u(x + \rho\theta)\, d\sigma(\theta) \right) d\rho \\ &= \frac{\sigma_n}{v_n \delta^n} \int_0^\delta \rho^{n-1} L(u : x, \rho)\, d\rho. \end{aligned}$$

This shows, among other things, that $\rho^{n-1} L(u : x, \rho)$ is integrable on $[0, \delta_1]$. Since the indefinite integral of an integrable function is absolutely continuous, $A(u : x, \delta)$ is continuous on $(0, \delta_1)$. The above equation also gives a representation of $A(u : x, \delta)$ on any closed interval $[a, b] \subset (0, \delta_x)$ as a product of two bounded absolutely continuous functions on $[a, b]$ and it follows directly from the definition of such functions that $A(u : x, \delta)$ is absolutely continuous on $[a, b]$. This implies that $A(u : x, \delta)$ is differentiable a.e. on $(0, \delta_x)$. Recall also that the derivative of an indefinite integral exists and is equal to the integrand a.e. relative to Lebesgue measure. Then for almost all $\delta \in (0, \delta_1)$

$$\begin{aligned} D_\delta A(u : x, \delta) &= \frac{\sigma_n}{v_n \delta} L(u : x, \delta) - \frac{n\sigma_n}{v_n \delta^{n+1}} \int_0^\delta \rho^{n-1} L(u : x, \rho)\, d\rho \\ &= \frac{n}{\delta}[L(u : x, \delta) - A(u : x, \delta)], \end{aligned}$$

since $v_n = \sigma_n/n$. In view of the fact that $L(u : x, \rho)$ is monotone decreasing on $(0, \delta)$

$$\begin{aligned} A(u : x, \delta) &= \frac{\sigma_n}{v_n \delta^n} \int_0^\delta \rho^{n-1} L(u : x, \rho)\, d\rho \\ &\geq \frac{\sigma_n}{v_n \delta^n} \left(\int_0^\delta \rho^{n-1}\, d\rho \right) L(u : x, \delta) \\ &= L(u : x, \delta) \end{aligned}$$

and $D_\delta A(u : x, \delta) \leq 0$ for almost all $\delta \in (0, \delta_x)$. If $0 < a < \delta$,

$$A(u : x, \delta) - A(u : x, a) = \int_a^\delta D_\rho A(u : x, \rho)\, d\rho$$

and $A(u : x, \delta)$ is monotone decreasing function of δ on $(0, \delta_x)$. Since $A(u : x, 0) = u(x) \geq A(u : x, \delta) \geq L(u : x, \delta)$ for any $\delta \in (0, \delta_x)$, $A(u : x, \delta)$ is monotone decreasing on $[0, \delta_x]$ and right continuous at 0.

Recall that if R is an open set and $\alpha > 0$, then $R_\alpha = \{y : d(y, \sim R) > \alpha\}$. If $\alpha, \beta > 0$, then it is easily seen from the triangle inequality that $R_{\alpha+\beta} \subset (R_\alpha)_\beta$.

THEOREM **4.19** If u is superharmonic on the open set R, let $u_\delta(x) = A(u : x, \delta)$ for $x \in R_\delta$. Then u_δ is a continuous superharmonic function on R_δ for each δ and u_δ increases to u as $\delta \downarrow 0$; if, in addition, u is harmonic on the open set $U \subset R$, then u_δ is harmonic on U_δ for each δ.

Proof. The continuity of u_δ on R_δ follows from Theorem 1.14. Suppose $\delta, \rho > 0$ and $x \in R_{\delta+\rho}$. Since $x \in R_{\delta+\rho} \subset (R_\delta)_\rho$, $B_{x,\rho} \subset R_\delta$; similarly

$B_{x,\delta} \subset R_\rho$. It follows that both $A(u_\delta : x, \rho)$ and $A(u_\rho : x, \delta)$ are defined. By Fubini's theorem,

$$A(u_\delta : x, \rho) = \frac{1}{v_n \rho^n} \int_{B_{x,\rho}} u_\delta(y)\, dy$$

$$= \frac{1}{v_n \rho^n} \int \chi_{B_{x,\rho}}(y) \left(\frac{1}{v_n \delta^n} \int \chi_{B_{y,\delta}}(z) u(z)\, dz \right) dy$$

$$= \frac{1}{v_n^2 \rho^n \delta^n} \int u(z) \left(\int \chi_{B_{x,\rho}}(y) \chi_{B_{y,\delta}}(z)\, dy \right) dz.$$

Since $z \in B_{y,\delta}$ if and only if $y \in B_{z,\delta}$, the integral within the last parentheses is just the volume of $B_{z,\delta} \cap B_{x,\rho}$ which is a symmetric function of δ and ρ. Therefore, $A(u_\delta : x, \rho)$ is a symmetric function of δ and ρ for $x \in R_{\delta+\rho}$. Now consider any $x \in R_\delta$ and choose r such that $x \in R_{\delta+r}$. Then $x \in R_{\delta+r} \subset R_{\delta+\rho}$ for all $\rho < r$ and

$$A(u_\delta : x, \rho) = A(u_\rho : x, \delta) \leq A(u : x, \delta) = u_\delta(x)$$

for all $\rho < r$. It follows that u_δ is superharmonic on R_δ by Theorem 4.9. The assertion that the u_δ's increase to u follows from Lemma 4.18. If, in addition, u is harmonic on the open set U, then $u_\delta(x) = A(u : x, \delta) = u(x)$ for $x \in U_\delta$ and u_δ is harmonic on U_δ.

According to the preceding theorem, averaging a superharmonic function over balls of a fixed radius yields a smoother superharmonic function at the expense of going to a slightly smaller region. We would expect to obtain an even smoother function, say one with continuous first partials derivatives, by averaging and reducing the domain again.

THEOREM **4.20** Let u be superharmonic on the open set R and let V be an open set with compact closure $\overline{V} \subset R$. Then there is an increasing sequence $\{v_j\}$ of superharmonic functions on V having continuous second partials such that $u = \lim_{j\to\infty} v_j$ on V.

Proof. Let $3\delta = d(\overline{V}, \sim R)$ and let $\{\delta_j\}$ be a decreasing sequence in $(0, \delta)$ with $\lim_{j\to\infty} \delta_j = 0$. For any $\rho > 0$ and any superharmonic function w on R, define w_ρ on R_ρ as in the preceding theorem. Fix j for the time being. Now u_{δ_j} is superharmonic and $u_{\delta_j} \leq u$ on R_{δ_j}, $(u_{\delta_j})_{\delta_j}$ is superharmonic and $(u_{\delta_j})_{\delta_j} \leq u_{\delta_j} \leq u$ on $(R_{\delta_j})_{\delta_j}$, and $((u_{\delta_j})_{\delta_j})_{\delta_j}$ is superharmonic and $((u_{\delta_j})_{\delta_j})_{\delta_j} \leq (u_{\delta_j})_{\delta_j} \leq u_{\delta_j} \leq u$ on $((R_{\delta_j})_{\delta_j})_{\delta_j} \supset V$. Letting $v_j = ((u_{\delta_j})_{\delta_j})_{\delta_j}$ on $((R_{\delta_j})_{\delta_j})_{\delta_j}$, v_j is superharmonic and has continuous second partials on V by Theorem 1.14. Since $w_{\delta_j} \leq w_{\delta_{j+1}}$ on R_{δ_j} for any superharmonic function w on R,

$$\begin{aligned} v_j = ((u_{\delta_j})_{\delta_j})_{\delta_j} &\le ((u_{\delta_{j+1}})_{\delta_j})_{\delta_j} \\ &\le ((u_{\delta_{j+1}})_{\delta_{j+1}})_{\delta_j} \\ &\le ((u_{\delta_{j+1}})_{\delta_{j+1}})_{\delta_{j+1}} = v_{j+1} \end{aligned}$$

on $((R_{\delta_j})_{\delta_j})_{\delta_j} \supset V$ and the sequence $\{v_j\}$ is monotone increasing on V. Since $w_{\delta_j} \le w_{\delta_{j+k}}$ on R_{δ_j} for any superharmonic function w on R and positive integers j, k,

$$((u_{\delta_{j+k}})_{\delta_j})_{\delta_j} \le ((u_{\delta_{j+k}})_{\delta_{j+k}})_{\delta_{j+k}} = v_{j+k}$$

on $((R_{\delta_{j+k}})_{\delta_j})_{\delta_j}$. Since $u_{\delta_{j+k}} \uparrow u$ on R as $k \uparrow \infty$ by the preceding theorem, $((u_{\delta_{j+k}})_{\delta_j})_{\delta_j} \to (u_{\delta_j})_{\delta_j}$ on $(R_{\delta_j})_{\delta_j}$ as $k \to \infty$. Therefore $(u_{\delta_j})_{\delta_j} \le \lim_{k\to\infty} v_k$ on $(R_{\delta_j})_{\delta_j}$. Repeating this argument twice, we obtain $u \le \lim_{k\to\infty} v_k$ on R. Since $v_k = ((u_{\delta_k})_{\delta_k})_{\delta_k} \le u$ on $((R_{\delta_k})_{\delta_k})_{\delta_k}$, $\lim_{k\to\infty} v_k \le u$ on R; that is, $u = \lim_{k\to\infty} v_k$ on R. Restricting the functions v_j to V, we obtain the desired sequence.

THEOREM **4.21** Let u be superharmonic on the open set R and harmonic on $R \sim K$ where K is a compact subset of R. Then there is an increasing sequence $\{u_j\}$ of superharmonic functions on R such that (i) each u_j has continuous second partials on R, (ii) $\lim_{j\to\infty} u_j = u$, and (iii) if V is any neighborhood of K with compact closure $\overline{V} \subset R$, then $u_j = u$ on $R \sim \overline{V}$ for sufficiently large j.

Proof. Let $6\delta = d(K, \sim R)$ and let $\{\delta_j\}$ be a decreasing sequence in $(0, \delta)$ with $\lim_{j\to\infty} \delta_j = 0$. Then $K \subset R_{3\delta_j}$ and $\partial R_{3\delta_j} \subset R \sim K$ for each j. It follows that $((u_{\delta_j})_{\delta_j})_{\delta_j} = u$ on a neighborhood of $\partial R_{3\delta_j}$ for each j. Define

$$u_j = \begin{cases} ((u_{\delta_j})_{\delta_j})_{\delta_j} & \text{on} \quad R_{3\delta_j}, \\ u & \text{on} \quad R \sim R_{3\delta_j}. \end{cases}$$

Since $((u_{\delta_j})_{\delta_j})_{\delta_j}$ is superharmonic on $R_{3\delta_j}$ and equal to u on a neighborhood of $\partial R_{3\delta_j}$, each u_j is superharmonic on R, has continuous second partials on R, and $u_j = u$ on $R \cap (\sim K)_{3\delta_j}$. As in the preceding proof, the u_j's increase to u on R. Now let V be a neighborhood of K with compact closure $\overline{V} \subset R$. For sufficiently large j, $R \cap (\sim K)_{3\delta_j} \supset R \sim \overline{V}$ and $u_j = u$ on $R \sim \overline{V}$ for such j.

THEOREM **4.22** Let u be superharmonic on the open set R and suppose $\partial B_{x,\rho} \subset R$ for $r_1 \le \rho \le r_2$. Then $L(u : x, \rho)$ is a concave function of $-\log \rho$ if $n = 2$ and a concave function of ρ^{-n+2} if $n \ge 3$ on (r_1, r_2); in particular, $L(u : x, \rho)$ is a continuous function of ρ on (r_1, r_2). If, in addition, u is harmonic on R, then

$$L(u : x, \rho) = \left[\frac{1}{\sigma_2} \int_{\partial B_{x, r_1}} D_{\mathbf{n}} u \, d\sigma(z)\right] \log \rho + \text{const.}$$

on (r_1, r_2) if $n = 2$ and

$$L(u : x, \rho) = \left[-\frac{1}{\sigma_n(n-2)}\int_{\partial B_{x,r_1}} D_{\mathbf{n}}u \, d\sigma(z)\right]\rho^{-n+2} + \text{const.}$$

on (r_1, r_2) if $n \geq 3$.

Proof. We shall first prove the theorem under the additional assumption that u has continuous second partials on R. If A_ρ denotes the closed annulus $\bar{B}_{x,\rho} - B_{x,r_1} \subset R$, $r_1 < \rho \leq r_2$, then u has a bounded normal derivative on ∂A_ρ. Since u is superharmonic on R, $\Delta u \leq 0$ on R. Applying Green's theorem to the region A_ρ,

$$0 \geq \int_{A_\rho} \Delta u \, dz = \int_{\partial B_{x,\rho}} D_{\mathbf{n}} u \, d\sigma(z) - \int_{\partial B_{x,r_1}} D_{\mathbf{n}} u \, d\sigma(z)$$

for $r_1 < \rho \leq r_2$. Using the fact that u has a bounded normal derivative on ∂A_ρ,

$$\begin{aligned}
\int_{\partial B_{x,\rho}} D_{\mathbf{n}} u \, d\sigma(z) &= \int_{\|\theta\|=1} \rho^{n-1} D_r u(x + r\theta)|_{r=\rho} \, d\sigma(\theta) \\
&= \rho^{n-1} \int_{\|\theta\|=1} \lim_{\eta \to 0} \frac{u(x + (\rho + \eta)\theta) - u(x + \rho\theta)}{\eta} \, d\sigma(\theta) \\
&= \rho^{n-1} \lim_{\eta \to 0} \frac{1}{\eta}\left[\int_{\|\theta\|=1} u(x + (\rho + \eta)\theta) \, d\sigma(\theta)\right. \\
&\qquad \left. - \int_{\|\theta\|=1} u(x + \rho\theta) \, d\sigma(\theta)\right] \\
&= \sigma_n \rho^{n-1} D_\rho L(u : x, \rho).
\end{aligned}$$

Therefore

$$\sigma_n \rho^{n-1} D_\rho L(u : x, \rho) = \int_{A_\rho} \Delta u \, dz + \int_{\partial B_{x,r_1}} D_{\mathbf{n}} u \, d\sigma(z).$$

Note that if u is harmonic on R the first integral on the right is zero and the second integral is a constant; an integration with respect to ρ results in the final assertion of the theorem. Returning to the general case, suppose $n \geq 3$. Letting $\zeta = 1/\rho^{n-2}$, $\rho(\zeta)$ the inverse function, and k the last term on the right side of the preceding equation,

$$-\sigma_n(n-2) D_\zeta L(u : x, \rho(\zeta)) = \int_{A_{\rho(\zeta)}} \Delta u \, dz + k.$$

As ρ increases on the interval (r_1, r_2), ζ decreases and $\int_{A_\rho} \Delta u \, dz$ decreases. It follows that $D_\zeta L(u : x, \rho(\zeta))$ increases as ζ decreases. Therefore $L(u : x, \rho(\zeta))$

is a concave function of $\zeta = 1/\rho^{n-2}$; that is, $L(u : x, \rho)$ is a concave function of ρ^{-n+2}. The proof of the $n = 2$ case is accomplished in the same way by letting $\zeta = -\log \rho$. Consider now an arbitrary superharmonic function u on R and the $n \geq 3$ case. By Theorem 4.20 there is a sequence $\{u_j\}$ of superharmonic functions defined on a neighborhood of $\bar{B}_{x, r_2} \sim B_{x, r_1}$ and having continuous second partials with $u_j \uparrow u$. For each $j \geq 1$, there is a concave function ψ_j defined on an interval such that $L(u_j : x, \rho) = \psi_j(\rho^{-n+2})$ for $\rho \in (r_1, r_2)$. Note that the sequence $\{\psi_j\}$ is increasing. It is easily seen that $\psi = \lim_{j \to \infty} \psi_j$ is also a concave function and that $L(u : x, \rho) = \psi(\rho^{-n+2})$ for $\rho \in (r_1, r_2)$. The same argument applies to the $n = 2$ case. The continuity of $L(u : x, \rho)$ on (r_1, r_2) follows from the fact that a concave function on an open interval is necessarily continuous.

CHAPTER 5

Green Functions

The results obtained so far stem from Green's identity and judicious choice of one of the two functions appearing therein. As indicated earlier, Green's identity holds for regions other than balls. Theorem 1.4, in particular, is valid for open sets with smooth boundaries. In exactly the same way as we proved Theorem 1.4, we can show that if u is harmonic on a neighborhood of the closure of a bounded open set R having smooth boundary, then for $x \in R$ and $n \geq 3$

$$u(x) = \frac{1}{\sigma_n(n-2)} \int_{\partial R} \left[\frac{1}{\|x-z\|^{n-2}} D_{\mathbf{n}} u - u D_{\mathbf{n}}\left(\frac{1}{\|x-z\|^{n-2}}\right)\right] d\sigma(z).$$

In deriving the Poisson integral formula we essentially chose a function v_x, depending upon $x \in R$, harmonic on a neighborhood of $\bar{R}$ so that

$$0 = \frac{1}{\sigma_n(n-2)} \int_{\partial R} (v_x D_{\mathbf{n}} u - u D_{\mathbf{n}} v_x)\, d\sigma(z).$$

By adding these two equations we obtain

$$u(x) = \frac{1}{\sigma_n(n-2)} \int_{\partial R} \left[\left(\frac{1}{\|x-z\|^{n-2}} + v_x\right) D_{\mathbf{n}} u - u D_{\mathbf{n}}\left(\frac{1}{\|x-z\|^{n-2}} + v_x\right)\right] \times d\sigma(z).$$

In the case where R is a ball, it was possible to choose the harmonic function v_x so that

$$\frac{1}{\|x-z\|^{n-2}} + v_x(z) = 0$$

for $z \in \partial R$ and the above integral representation became just the Poisson integral representation. Whether or not this same procedure would work for an open set R with smooth boundary depends upon whether or not the Dirichlet problem corresponding to the boundary function

$$f_x(z) = \frac{1}{\|x - z\|^{n-2}}, \qquad z \in \partial R,$$

can be solved for each $x \in R$. This is one possible approach to Green functions but we shall not pursue this line further in favor of a direct study of functions of the type (for $n \geq 3$)

$$\frac{1}{\|x - z\|^{n-2}} + v_x(z),$$

where v_x is harmonic on R.

1. Green Function for the Ball

As indicated in the introduction, we have already constructed the Green function for the ball. We shall formalize the discussion by defining the Green function for the ball.

Definition. If $n = 2$, the Green function for the ball $B = B_{y,\rho}$ is given by

$$G_B(x, z) = \begin{cases} \log \dfrac{\|y - x\|}{\rho} \dfrac{\|z - x^\star\|}{\|z - x\|} & \text{for } z \in B \sim \{x\}, \quad x \neq y, \\ \log \dfrac{\rho}{\|z - y\|} & \text{for } z \in B \sim \{x\}, \quad x = y, \\ +\infty & \text{for } z = x, \end{cases}$$

where $x^\star$ is the inverse of x with respect to ∂B. If $n \geq 3$, the Green function is given by

$$G_B(x, z) = \begin{cases} \dfrac{1}{\|z - x\|^{n-2}} - \left(\dfrac{\rho}{\|x - y\|}\right)^{n-2} \dfrac{1}{\|z - x^\star\|^{n-2}} & \\ \qquad\qquad \text{for } z \in B \sim \{x\}, \quad x \neq y, \\ \dfrac{1}{\|z - y\|^{n-2}} - \dfrac{1}{\rho^{n-2}} \qquad \text{for } z \in B \sim \{x\}, \quad x = y, \\ +\infty \qquad\qquad \text{for } z = x. \end{cases}$$

With these definitions, the integral representation of Theorem 1.7 can be rewritten using G_B; that is, if h is harmonic on a neighborhood of the closure of $B = B_{y,\rho}$ and $x \in B$, then for $n = 2$

(a) $$h(x) = -\frac{1}{2\pi}\int_{\partial B} h(z) D_{\mathbf{n}} G_B(x, z)\, d\sigma(z).$$

For $x \neq y$ this equation is precisely (i) of Theorem 1.7. It remains only to show that (a) holds when $x = y$. In this case at points on ∂B

$$D_{\mathbf{n}} G_B(y, z) = D_{\mathbf{n}} \log \frac{\rho}{\|z - y\|} = D_r \log \frac{\rho}{r}\bigg|_{r=\rho} = \frac{-1}{\rho}$$

and (a) can be written

$$h(y) = \frac{1}{2\pi\rho}\int_{\partial B} h(z)\, d\sigma(z).$$

But this just states that $h(y)$ is the average of its values over ∂B. Similarly, for $n \geq 3$

$$h(x) = -\frac{1}{\sigma_n(n-2)}\int_{\partial B} h(z) D_{\mathbf{n}} G_B(x, z)\, d\sigma(z),$$

$x \in B$, whenever h is harmonic on a neighborhood of $\bar{B}$.

It is also worth noting that $\lim_{z \to z_0} G_B(x, z) = 0$ for all $z_0 \in \partial B$ and fixed $x \in B$. It should also be noted that $G_B(x, z)$ has been defined by continuity for $x = y$. This arises from the fact that $x^\star$ is not defined when $x = y$. Since for $x \neq y$

$$x^\star = y + \frac{\rho^2}{\|x - y\|^2}(x - y),$$

$$\frac{\|y - x\|}{\rho}\frac{\|z - x^\star\|}{\|z - x\|} = \frac{\|y - x\| \left\|(z - y) - \dfrac{\rho^2}{\|y - x\|^2}(x - y)\right\|}{\rho\|z - x\|}$$

$$\to \frac{\rho}{\|z - y\|}$$

as $x \to y$; that is, $G_B(y, z) = \lim_{x \to y} G_B(x, z)$, $z \in B \sim \{y\}$. Lastly, it should be noted that

$$\mathbf{PI}(\mu, B)(x) = \begin{cases} -\dfrac{1}{2\pi}\displaystyle\int_{\partial B} D_{\mathbf{n}} G_B(x, z)\, d\mu(z), & n = 2 \\ -\dfrac{1}{\sigma_n(n-2)}\displaystyle\int_{\partial B} D_{\mathbf{n}} G_B(x, z)\, d\mu(z), & n \geq 3. \end{cases}$$

These remarks along with Theorem 1.4 allow us to state the calculus version of the Riesz decomposition theorem for superharmonic functions.

THEOREM **5.1** If u has continuous second partials on a neighborhood of the closure of a ball $B = B_{y,\rho}$, then

(i) for each $x \in B$ and $n = 2$

$$u(x) = -\frac{1}{2\pi}\int_{\partial B} u(z)D_{\mathbf{n}}\, G_B(x, z)\, d\sigma(z) - \frac{1}{2\pi}\int_B G_B(x, z)\, \Delta u(z)\, dz;$$

(ii) for each $x \in B$ and $n \geq 3$

$$u(x) = -\frac{1}{\sigma_n(n-2)}\int_{\partial B} u(z)D_{\mathbf{n}}\, G_B(x, z)\, d\sigma(z)$$

$$- \frac{1}{\sigma_n(n-2)}\int_B G_B(x, z)\, \Delta u(z)\, dz.$$

Proof. We shall only prove (i). Suppose first that $x \neq y$. If we apply Green's identity to u and the function

$$v_x(z) = \log\left(\frac{\|y - x\|\, \|z - x^\star\|}{\rho}\right),$$

we obtain

$$-\int_B v_x\, \Delta u\, dz = \int_{\partial B} (uD_{\mathbf{n}}\, v_x - v_x\, D_{\mathbf{n}}\, u)\, d\sigma(z),$$

since $\Delta v_x = 0$ on B; that is,

$$0 = \frac{1}{2\pi}\int_{\partial B} (v_x\, D_{\mathbf{n}}\, u - uD_{\mathbf{n}}\, v_x)\, d\sigma(z) - \frac{1}{2\pi}\int_B v_x\, \Delta u\, dz.$$

Adding this equation to the equation of part (i) of Theorem 1.4,

$$u(x) = \frac{1}{2\pi}\int_{\partial B}\left[\left(\log\frac{\|y - x\|}{\rho}\frac{\|z - x^\star\|}{\|z - x\|}\right)D_{\mathbf{n}}\, u - uD_{\mathbf{n}}\left(\log\frac{\|y - x\|}{\rho}\frac{\|z - x^\star\|}{\|z - x\|}\right)\right]$$
$$\times\, d\sigma(z)$$
$$-\frac{1}{2\pi}\int_B \Delta u\left(\log\frac{\|y - x\|}{\rho}\frac{\|z - x^\star\|}{\|z - x\|}\right) dz.$$

Since

$$\frac{\|y - x\|}{\rho}\frac{\|z - x^\star\|}{\|z - x\|} = 1$$

for $z \in \partial B$, we have for $x \in B \sim \{y\}$

$$u(x) = -\frac{1}{2\pi}\int_{\partial B} u(z)D_{\mathbf{n}}\, G_B(x, z)\, d\sigma(z) - \frac{1}{2\pi}\int_B \Delta u(z)G_B(x, z)\, dz.$$

This equation can be shown to hold when $x = y$ by repeating the above argument with $v_x(z) = \log \rho$.

Definition. If μ is a signed measure on B, the **Green potential** of μ is the function

$$u(x) = \int G_B(x, z)\, d\mu(z), \qquad x \in B,$$

provided the integral is everywhere defined on B.

COROLLARY **5.2** If u has continuous second partials on a neighborhood of $\bar{B}$, then u is the sum of a function harmonic on B and the Green potential of a signed measure μ on B.

We return to this corollary in Chapter 6 in order to obtain a similar decomposition for superharmonic functions.

LEMMA **5.3** G_B is symmetric on $B \times B$; moreover, for each $x \in B$, $G_B(x, \cdot)$ is strictly positive on B, superharmonic on B, and $\lim_{z \to z_0} G_B(x, z) = 0$ for all $z_0 \in \partial B$.

Proof. We shall prove the $n = 2$ case only, the $n \geq 3$ case being essentially the same. Let $B = B_{y,\rho}$. If $x = z = y$, then $G_B(x, z) = G_B(z, x) = +\infty$. If $x = y$ and $z \in B \sim \{x\}$, then $G_B(x, z) = \log (\rho/\|z - x\|)$,

$$G_B(z, x) = \log \frac{\|y - z\|}{\rho} \frac{\|x - z^\star\|}{\|x - z\|} = \log \frac{\|x - z^\star\|}{\rho} = \log \frac{\rho}{\|x - z\|},$$

and $G_B(x, z) = G_B(z, x)$. Suppose now that $x \neq y$, $z \neq y$, and $z \in B \sim \{x\}$. Let ϕ be the angle between the line segment joining $x^\star$ to y (equivalently x to y) and the line segment joining z to y (equivalently $z^\star$ to y). Then

$$\begin{aligned} \|z - x^\star\|^2 &= \|z - y\|^2 + \|y - x^\star\|^2 - 2\|z - y\|\,\|y - x^\star\| \cos \phi \\ &= \|z - y\|^2 + \frac{\rho^4}{\|y - x\|^2} - 2\|z - y\| \frac{\rho^2}{\|y - x\|} \cos \phi \end{aligned}$$

and

$$\|y - x\|^2 \|z - x^\star\|^2 = \|z - y\|^2 \|y - x\|^2 + \rho^4 - 2\rho^2 \|z - y\|\,\|y - x\| \cos \phi.$$

Similarly,

$$\|y - z\|^2 \|x - z^\star\|^2 = \|x - y\|^2 \|y - z\|^2 + \rho^4 - 2\rho^2 \|x - y\|\,\|y - z\| \cos \phi$$

for $x \neq y$ and $z \in B \sim \{x\}$, and it follows that

$$G_B(x, z) = \log \frac{\|y - x\|}{\rho} \frac{\|z - x^\star\|}{\|z - x\|} = \log \frac{\|y - z\|}{\rho} \frac{\|x - z^\star\|}{\|x - z\|} = G_B(z, x).$$

This completes the proof that G_B is symmetric. That $G_B(x, \cdot)$ approaches zero at points of ∂B follows directly from the definition and equation (g) of Chapter 1. It is also apparent from the definition that $G_B(x, \cdot)$ is superharmonic on B. By Corollary 4.3 $G_B(x, \cdot) \geq 0$ on B. Since $G_B(x, \cdot)$ satisfies the minimum principle, it cannot be zero at any point of B, for otherwise it would have to be identically zero, which it is not.

THEOREM **5.4** (*Bôcher*) Let V be a neighborhood of $x_0 \in E^n$ and let h be defined, harmonic, and non-negative on $V \sim \{x_0\}$. Then either

(i) h can be defined at x_0 so as to be harmonic on V or

(ii) there is a constant $c > 0$ such that

$$h = cu_{x_0} + v$$

where v is harmonic on V.

Proof. Let $B = B_{x_0, \delta}$ be such that $\bar{B} \subset V$. Define

$$u(x) = \begin{cases} +\infty & \text{if} \quad x = x_0, \\ h(x) + u_{x_0}(x) & \text{if} \quad x \in V \sim \{x_0\}. \end{cases}$$

Since h and u_{x_0} are harmonic on $V \sim \{x_0\}$, the same is true of u. We shall show that u is superharmonic on V. By definition u is continuous at x_0 in the extended sense since $h(x) \geq 0$ and $u_{x_0}(x) \to +\infty$ as $x \to x_0$, $x \in V \sim \{x_0\}$. Note that u is locally super-mean-valued at x_0 since $u(x_0) = +\infty \geq L(u : x, \rho)$, $\rho < \delta$. Therefore u is superharmonic on V. By Theorem 4.21 there is a sequence $\{u_j\}$ of functions that are superharmonic on V and have continuous second partials such that $u_j \uparrow u$. Moreover, outside each small neighborhood of x_0, $u_j = u$ and is harmonic for sufficiently large j. The rest of the proof will be carried out for $n \geq 3$. By Theorem 5.1 for each $x \in B$

$$u_j(x) = -\frac{1}{\sigma_n(n-2)} \int_{\partial B} u_j(z) D_{\mathbf{n}} G_B(x, z)\, d\sigma(z) - \frac{1}{\sigma_n(n-2)} \int_B G_B(x, z)\, \Delta u_j(z)\, dz.$$

Since $u_j = u$ outside any small neighborhood of x_0, for all sufficiently large j

$$u_j(x) = -\frac{1}{\sigma_n(n-2)} \int_{\partial B} u(z) D_{\mathbf{n}} G_B(x, z)\, d\sigma(z) - \frac{1}{\sigma_n(n-2)} \int_B G_B(x, z)\, \Delta u_j(z)\, dz$$

for all large j Since $u_j \uparrow u$,

(b) $$u(x) = -\frac{1}{\sigma_n(n-2)}\int_{\partial B} u(z) D_{\mathbf{n}}\, G_B(x, z)\, d\sigma(z)$$
$$-\frac{1}{\sigma_n(n-2)} \lim_{j\to\infty} \int_B G_B(x, z)\, \Delta u_j(z)\, dz.$$

For each j and each Borel set $M \subset \bar{B}$, define

$$\mu_j(M) = -\frac{1}{\sigma_n(n-2)} \int_M \Delta u_j\, dz.$$

Since u_j is superharmonic, $\Delta u_j \leq 0$ and μ_j is a measure on $\bar{B}$. Moreover, if M is contained in the complement of a small neighborhood of x_0, then $\mu_j(M) = 0$ for large j, since u_j is harmonic on M for such j; that is, $\lim_{j\to\infty} \mu_j(M) = 0$ if $M \subset B \sim B_{x_0,\rho}$ for some ρ. Now consider any $y \in B$, $y \neq x_0$, and a neighborhood W of x_0 whose closure does not contain y and $\bar{W} \subset B$. Since $u(y) < +\infty$, it follows from (b) that

$$\lim_{j\to\infty} \int_B G_B(y, z)\, d\mu_j(z) < +\infty.$$

Since $\mu_j(\bar{B} \sim \bar{W}) = 0$ for large j,

(c) $$\lim_{j\to\infty} \int_W G_B(y, z)\, d\mu_j(z) < +\infty.$$

By Lemma 5.3 $G_B(y, \cdot) > 0$ on $\bar{W}$ and by the continuity of $G_B(y, \cdot)$ on $\bar{W}$ there is a constant $m > 0$ such that $G_B(y, x) \geq m$ for $x \in \bar{W}$. Therefore

$$\int_W G_B(y, z)\, d\mu_j(z) \geq m\mu_j(\bar{W})$$

and this shows that $\sup_j \mu_j(\bar{W}) < +\infty$. Now

$$\|\mu_j\| = \mu_j(\bar{B}) = \mu_j(\bar{W}) + \mu_j(\bar{B} \sim \bar{W}) = \mu_j(\bar{W})$$

for large j and there is a constant k such that $\|\mu_j\| \leq k$ for all j. By Lemma 2.10 there is a measure μ and a subsequence $\{\mu_{j_k}\}$ such that $\mu_{j_k} \xrightarrow{w^*} \mu$. We can assume that $\mu_j \xrightarrow{w^*} \mu$ by replacing the sequence with the subsequence if necessary. Since μ is zero on the complement of each small neighborhood of x_0, μ is actually a positive mass concentrated on the point x_0. If $x \neq x_0$, then $G_B(x, \cdot)$ is continuous at x_0 and

$$\lim_{j\to\infty} \int_B G_B(x, z)\, d\mu_j(z) = \mu(\{x_0\}) G_B(x, x_0)$$

and (b) becomes

$$u(x) = \mathbf{PI}(u, B)(x) + \mu(\{x_0\})G_B(x, x_0)$$

for $x \neq x_0$ It follows from the definition of u that

$$h(x) = \mathbf{PI}(u, B)(x) + \mu(\{x_0\})G_B(x, x_0) - u_{x_0}(x)$$

for $x \neq x_0$. Since $G_B(\cdot, x_0) = G_B(x_0, \cdot)$ by Lemma 5.3, $G_B(\cdot, x_0)$ can be expressed as a sum of u_{x_0} and a harmonic function. Therefore for $x \neq x_0$

$$h(x) = v(x) + [\mu(\{x_0\}) - 1]u_{x_0},$$

where v is harmonic on B. If $\mu(\{x_0\}) - 1 = 0$, then h can be defined at x_0 to be harmonic on B by putting $h(x_0) = v(x_0)$. We cannot have $\mu(\{x_0\}) - 1 < 0$ for in that case we would have $h(x) \to -\infty$ as $x \to x_0$ which contradicts the fact that $h \geq 0$ on $V \sim \{x_0\}$. Let $c = \mu(\{x_0\}) - 1$, if the latter is positive, to get (ii).

We shall digress in the remainder of this section in order to prove a theorem which will be needed later. The basic problem is that of extending a harmonic function. One case is particularly easy. Let R be the half-space defined by $R = \{(x_1, \ldots, x_n) : x_n > 0\}$; any other half-space would serve just as well. If y is any point in E^n, we will let y^r denote the reflection of y across ∂R; that is, $y^r = (y_1, \ldots, y_{n-1}, -y_n)$ if $y = (y_1, \ldots, y_n)$. If $E \subset E^n$, E^r will denote the reflection of E across ∂R; that is, $E^r = \{y^r : y \in E\}$. Now let K be a compact subset of R and let h be a function which is continuous on $\bar{R} \sim K$, harmonic on $R \sim K$, and $h = 0$ on ∂R. Define a function h^* on $E^n \sim (K \cup K^r)$ by

$$h^*(y) = \begin{cases} h(y), & y \in R \sim K, \\ 0, & y \in \partial R, \\ -h(y^r), & y \in (R \sim K)^r. \end{cases}$$

Using Corollary 2.9, we easily see that h^* is harmonic on $E^n \sim (K \cup K^r)$. We can therefore extend h harmonically across ∂R. We shall use this result to obtain a less trivial result. In the course of the proof of the following theorem, we shall use the fact that the Kelvin transformation is idempotent; that is, if f is a real-valued function on the open set R and $f^\star$ is its Kelvin transform on $R^\star$, then the Kelvin transform $f^{\star\star}$ of $f^\star$ is just f on R (it is assumed, of course, that R does not contain the reference point of the inversion). This property of the Kelvin transformation follows directly from the definition of the transformation.

THEOREM **5.5** Let $0 < \delta < \rho$. If h is continuous on $\bar{B}_{x,\rho} \sim B_{x,\delta}$, harmonic on the interior, and $h = 0$ on $\partial B_{x,\rho}$, then there is an $\varepsilon > 0$ such that h has a harmonic extension to $B_{x,\rho+\varepsilon} \sim \bar{B}_{x,\delta}$.

Proof. The proof will be given for the $n = 2$ case only, the $n \geq 3$ case being essentially the same. We can assume that $h \geq 0$ on $\bar{B}_{x,\rho} \sim B_{x,\delta}$ for the following reason. Since h is continuous on $\partial B_{x,\delta}$, $v_\alpha(z) = h(z) + \alpha \log(\rho/\|x - z\|)$, $z \in \bar{B}_{x,\rho} \sim B_{x,\delta}$, is strictly positive on $\partial B_{x,\delta}$ for sufficiently large α, zero on $\partial B_{x,\rho}$, and harmonic on $B_{x,\rho} \sim \bar{B}_{x,\delta}$. Therefore $v_\alpha > 0$ on $B_{x,\rho} \sim B_{x,\delta}$ for some large α by the minimum principle. If v_α can be extended harmonically across $\partial B_{x,\rho}$, the same is true of h since $\alpha \log(\rho/\|x - z\|)$ is harmonic on $\sim \bar{B}_{x,\delta}$. We therefore assume that $h \geq 0$ on $\bar{B}_{x,\rho} \sim B_{x,\delta}$. Now let y be any point of $\partial B_{x,\rho}$ and consider an inversion relative to $\partial B_{y,2\rho}$. Under this inversion the ball $B_{x,\rho}$ maps onto a half-space R with the plane ∂R externally tangent to $B_{x,\rho}$, $B = B_{x,\delta}$ maps onto a ball $B^\star$ with closure $\bar{B}^\star \subset R$, and $B_{x,\rho} \sim \bar{B}_{x,\delta}$ maps onto $R \sim \bar{B}^\star$. The Kelvin transform $h^\star$ of h is continuous on $R \sim B^\star$, harmonic on $R \sim \bar{B}^\star$, and zero on ∂R. Note that $h^\star \geq 0$. Now restrict $h^\star$ to $R \sim \bar{B}^\star$. By the remarks preceding this theorem $h^\star$ has a harmonic extension, which is also denoted by $h^\star$, to $E^n \sim (\bar{B}^\star \cup \bar{B}^{\star r})$, where $\bar{B}^{\star r}$ is the reflection of $\bar{B}^\star$ across ∂R. Note that $\sim \bar{B}^{\star r}$ maps onto a ball containing $\bar{B}_{x,\rho}$ in its interior under the inversion. Also note that $h^\star \leq 0$ on $(R \sim \bar{B}^\star)^r$. It follows that there is an $\varepsilon > 0$ such that $B_{x,\rho+\varepsilon}$ is a subset of the image of $\sim \bar{B}^{\star r}$ under the inversion. Let h^* be the restriction of the Kelvin transform of $h^\star$ to $B_{x,\rho+\varepsilon} \sim \bar{B}_{x,\delta} \sim \{y\}$ (recall that y is not the image of any point of $E^n \sim (\bar{B}^\star \cup \bar{B}^{\star r})$. Then h^* is a harmonic extension of h to $B_{x,\rho+\varepsilon} \sim \bar{B}_{x,\delta} \sim \{y\}$. Since $h^\star \leq 0$ on $(R \sim \bar{B}^\star)^r$, $h^* \leq 0$ on $B_{x,\rho+\varepsilon} \sim \bar{B}_{x,\rho}$. In view of the fact that $\lim_{\substack{z \to y \\ z \in B_{x,\rho} \sim \bar{B}_{x,\delta}}} h(z) = 0$, h^* is bounded above on $W \sim \{y\}$ for some neighborhood W of y. It follows from Theorem 5.4 that $h^* = w + cu_y$ $W \sim \{y\}$ for some $c \leq 0$ and some harmonic function w on W. Now $c = 0$ for otherwise we would have $\lim_{\substack{z \to y \\ z \in B_{x,\rho} \sim \bar{B}_{x,\delta}}} h^*(z) = -\infty$. It follows that h^* can be defined at y so as to be harmonic on W and therefore h can be extended harmonically to $B_{x,\rho+\varepsilon} \sim \bar{B}_{x,\delta}$.

2. Green Function for Open Regions

Although the Green function for a ball was defined directly, we shall use a different definition for dealing with arbitrary open sets. This will require showing that the definition in the case of a ball coincides with the definition given below. Generally speaking, a Green function does not always exist and it will be our task to determine when it does exist.

Definition. Let R be an open subset of E^n. A **Green function**, if it exists, for the open set R is an extended real-valued function G_R on $R \times R$ with the following properties:

(i) $G_R(x, \cdot) = u_x + h_x$ where for each $x \in R$, h_x is harmonic on R.
(ii) $G_R \geq 0$.
(iii) if for $x \in R$, v_x is a non-negative superharmonic function and is the sum of u_x and a superharmonic function, then $v_x \geq G_R(x, \cdot)$.

In other words, for each $x \in R$, $G_R(x, \cdot)$ is minimal in the class of non-negative superharmonic functions of the form $u_x + w_x$ where w_x is superharmonic on R.

We shall now show that the Green function for the ball satisfies (i), (ii), and (iii). It is apparent from the definition of G_B that (i) is satisfied. By Lemma 5.3, (ii) is satisfied. Suppose now that for each $x \in B$, $v_x = u_x + w_x$ where $v_x \geq 0$ and w_x is superharmonic on B. Moreover, let $G_B(x, \cdot) = u_x + h_x$ where h_x is harmonic on B for each $x \in B$. Since both u_x and h_x have limits at points z_0 of ∂B,

$$\begin{aligned}\liminf_{z \to z_0} &[u_x(z) + w_x(z) - (u_x(z) + h_x(z))] \\ &= \liminf_{z \to z_0} [u_x(z) + w_x(z)] - \lim_{z \to x} G_B(x, z) \\ &= \liminf_{z \to z_0} [u_x(z) + w_x(z)] \geq 0 \quad \text{for} \quad z_0 \in \partial B;\end{aligned}$$

that is,

$$\liminf_{z \to z_0} (w_x(z) - h_x(z)) \geq 0 \quad \text{for} \quad z_0 \in \partial B.$$

Now $w_x - h_x$ is superharmonic on B and by Corollary 4.3, $w_x - h_x \geq 0$ on R. Therefore $G_B(x, \cdot) \leq v_x$ on B and we have shown that G_B satisfies (iii). We were justified in calling G_B the Green function for the ball B.

THEOREM **5.6** The Green function G_R for the open set R is unique if it exists.

Proof. Let G_R^* be a second Green function for R. By (iii) of the definition, $G_R \geq G_R^*$ and $G_R^* \geq G_R$.

LEMMA **5.7** If the open set R has a Green function G_R, then $\inf_{y \in R} G_R(x, y) = 0$ for each $x \in R$.

Proof. Suppose $c = \inf_{y \in R} G_R(x, y) > 0$. Then $G_R(x, \cdot) - c \geq 0$. Moreover, $G_R(x, \cdot) - c$ is non-negative, superharmonic, and representable as the sum of u_x and a harmonic function. Since $G_R(x, \cdot) - c < G_R(x, \cdot)$, the minimality of G_R is contradicted.

LEMMA **5.8** If the open set R has a Green function G_R and B is a ball containing $x \in R$ with $\bar{B} \subset R$, then $G_R(x, \cdot)$ is bounded outside B.

Proof. Outside B, $G_R(x, \cdot)$ is continuous and non-negative. Note that $G_R(x, \cdot)$ is strictly positive on the component of R containing x for if $G_R(x, y) = 0$ for some point y in this component then $G_R(x, \cdot)$ would be identically zero thereon by the minimum principle, in contradiction to the fact that $G_R(x, x) = +\infty$. Since $\bar{B} \subset R$ and B is a subset of the component of R containing x, $G_R(x, \cdot)$ is strictly positive on $\bar{B}$. Let $m = \sup_{y \in \partial B} G_R(x, y)$. Then $m > 0$ by the minimum principle. Define

$$g(x, y) = \begin{cases} G_R(x, y), & y \in B, \\ \min[G_R(x, y), 2m], & y \in R \sim B. \end{cases}$$

Since $g(x, \cdot)$ agrees with $G_R(x, \cdot)$ on B, $g(x, \cdot)$ is superharmonic on B. Since $g(x, \cdot)$ is the minimum of two superharmonic functions on $R \sim \bar{B}$, $g(x, \cdot)$ is superharmonic on $R \sim \bar{B}$. Since $G_R(x, \cdot) \le m < 2m$ on ∂B and $G_R(x, \cdot)$ is continuous at points of ∂B, $g(x, \cdot) = G_R(x, \cdot)$ on a neighborhood of ∂B and therefore superharmonic thereon. This shows that $g(x, \cdot)$ is superharmonic on R. Since $g(x, \cdot) - u_x = G_R(x, \cdot) - u_x$ on $B \sim \{x\}$ and $G_R(x, \cdot) - u_x$ can be defined at x so as to be harmonic on B, $g(x, \cdot) - u_x$ can be defined at x so as to be superharmonic on R; $[g(x, \cdot) - u_x]$ will denote the extended function. Since $0 \le g(x, \cdot) = u_x + [g(x, \cdot) - u_x]$, where the bracket function is superharmonic on R, $G_R(x, \cdot) \le g(x, \cdot)$ by the definition of the Green function. But since $g(x, \cdot) \le 2m$ on $R \sim B$, $G_R(x, \cdot) \le 2m$ on $R \sim B$.

Definition. A family $\mathscr{F}$ of superharmonic functions on an open set R is **saturated** if

(i) $u, v \in \mathscr{F} \Rightarrow u \wedge v \in \mathscr{F}$
(ii) $u \in \mathscr{F} \Rightarrow u^* \in \mathscr{F}$, where for some ball B with $\bar{B} \subset R$

$$u^* = \begin{cases} u & \text{on} \quad R \sim B, \\ \mathbf{PI}(u, B) & \text{on} \quad B. \end{cases}$$

The following theorem will not only allow us to prove the existence of the Green function for certain regions but it will also allow us to solve the Dirichlet problem for certain regions.

THEOREM **5.9** If $\mathscr{F}$ is a saturated family of superharmonic functions on an open set R, then $\inf_{u \in \mathscr{F}} u$ is either identically $-\infty$ or harmonic on each component of R.

Proof. Let $v = \inf_{u \in \mathscr{F}} u$. Consider any ball B with $\bar{B} \subset R$. Since the assertion of the theorem concerns the components of R, we can assume that R is connected. If $u \in \mathscr{F}$, let

$$u^* = \begin{cases} u & \text{on} \quad R \sim B, \\ \mathbf{PI}(u, B) & \text{on} \quad B. \end{cases}$$

Then $u^* \leq u$, u^* is harmonic on B, and $v = \inf_{u \in \mathscr{F}} u^*$ on B. Since $u, w \in \mathscr{F}$ implies that $u \wedge w \in \mathscr{F}$, $\mathscr{F}$ is left-directed. Since $(u \wedge w)^* \leq u^* \wedge w^*$, the family $\{u^* : u \in \mathscr{F}\}$ is left-directed also. By Lemma 2.21 $v = \inf_{u \in \mathscr{F}} u^*$ is identically $-\infty$ on B or harmonic on B. It follows that v is either subharmonic on R or identically $-\infty$ on R; if not identically $-\infty$ on R, then v is subharmonic on R and harmonic on each ball $B \subset R$.

Definition. If R is an open subset of E^n and $x \in R$, define

$$\mathfrak{B}_x = \{v_x : v_x \geq 0,\ v_x = u_x + w_x,\ w_x \text{ superharmonic on } R\}.$$

The class $\mathfrak{B}_x$ will generally depend upon x and R and it may very well be that $\mathfrak{B}_x = \varnothing$ for some $x \in R$.

LEMMA **5.10** $\mathfrak{B}_x$ is a saturated family of superharmonic functions on $R \sim \{x\}$.

Proof. We shall assume that $\mathfrak{B}_x \neq \varnothing$ for otherwise there is nothing to prove. Suppose $v, v' \in \mathfrak{B}_x$. Then $v, v' \geq 0$, $v = u_x + u$, and $v' = u_x + u'$, where u, u' are superharmonic on R. Then $v \wedge v' \in \mathfrak{B}_x$, since $v \wedge v' = u_x + (u \wedge u') \geq 0$ and $u \wedge u'$ is superharmonic on R. Suppose $\bar{B} \subset R \sim \{x\}$. Since u_x is harmonic on B,

$$v^* = \begin{cases} u + u_x & \text{on } R \sim B \\ \mathbf{PI}(u + u_x, B) = \mathbf{PI}(u, B) + u_x & \text{on } B \end{cases}$$

belongs to $\mathfrak{B}_x$ since $v^* = u^* + u_x \geq 0$, where

$$u^* = \begin{cases} u & \text{on } R \sim B \\ \mathbf{PI}(u, B) & \text{on } B \end{cases}$$

is superharmonic on R.

LEMMA **5.11** If $\mathfrak{B}_x \neq \varnothing$ for each $x \in R$, then R has a Green function $G_R(x, \cdot) = \inf\{v_x : v_x \in \mathfrak{B}_x\}$.

Proof. Let $\underline{v}_x = \inf\{v_x : v_x \in \mathfrak{B}_x\}$ and consider a fixed $x \in R$. Since $\mathfrak{B}_x$ is a saturated family of superharmonic functions over $R \sim \{x\}$ and $\underline{v}_x \geq 0$, $\underline{v}_x$ is harmonic on $R \sim \{x\}$ by Theorem 5.9. According to Theorem 5.4, there is a constant $c \geq 0$ such that $\underline{v}_x = cu_x + h_x$ on $R \sim \{x\}$, where h_x is harmonic on R. We shall show that $c = 1$. Consider a ball $B = B_{x,\delta}$ with $\bar{B} \subset R$ and the Green function $G_B(x, \cdot)$. It follows from (iii) of the definition of the Green function G_B that $v_x \geq G_B(x, \cdot)$ on B for all $v_x \in \mathfrak{B}_x$. It follows that $\underline{v}_x \geq G_B(x, \cdot)$ on B and that $c \neq 0$. Since $cu_x + h_x = \underline{v}_x \geq G_B(x, \cdot) = u_x + h_x^*$ on B, where h_x^* is harmonic on B,

$$(h_x^* - h_x) \leq (c - 1)u_x \quad \text{on } B \sim \{x\}.$$

If $c < 1$, then the harmonic function on the left would have a limit $-\infty$ at x, a contradiction. Therefore $c \geq 1$ and

$$\frac{\underline{v}_x}{c} = u_x + H_x$$

on R where $H_x = h_x/c$, a harmonic function on R. Therefore, $\underline{v}_x/c \in \mathfrak{B}_x$ and

$$\underline{v}_x \leq \frac{\underline{v}_x}{c} \leq \underline{v}_x .$$

This shows that $c = 1$. Therefore, $G_R(x, y) = \underline{v}_x(y)$ is the Green function for R.

THEOREM **5.12** Let R be an open subset of E^n. If (i) $n > 2$, (ii) $n = 2$ and R is bounded, or (iii) $n = 2$ and R is a subset of a set with a Green function, then R has a Green function.

Proof.

(i) For $n > 2$ $u_x \in \mathfrak{B}_x$ for $x \in R$.
(ii) If $n = 2$ and R is bounded, then there is a constant c such that $v_x = u_x + c \in \mathfrak{B}_x$ for all $x \in R$.
(iii) If $n = 2$, $R \subset R^*$, and R^* has a Green function g^*, then

$$v_x = g^*(x, \cdot)\,|_R \in \mathfrak{B}_x$$

for all $x \in R$.

THEOREM **5.13** $G(x, y) = u_x(y)$ is the Green function for E^n, $n > 2$.

Proof. Let $G(x, y)$ be the Green function for E^n, $n > 2$ which exists by the preceding theorem. Then $G(x, \cdot) = u_x + h_x$ where h_x is harmonic on E^n. Since $u_x \in \mathfrak{B}_x$, $u_x + h_x \leq u_x$ and $h_x \leq 0$. By Theorem 1.11 $h_x = c$ with $c \leq 0$. Suppose $c < 0$. Since $u_x(y) \to 0$ as $\|y\| \to +\infty$, $G(x, y) < 0$ for $\|y\|$ large and this contradicts the fact that $\inf_{y \in E^n} G(x, y) = 0$ by Lemma 5.7. Therefore $c = 0$ and $h_x = 0$.

The question of existence of the Green function for an open set $R \subset E^n$ is completely answered in the $n > 2$ case by (i) of Theorem 5.12. The $n = 2$ case requires additional investigation. Before doing so we shall state two theorems needed in later chapters.

THEOREM **5.14** If $R_1 \subset R_2$ are open subsets of E^n having Green functions G_{R_1} and G_{R_2}, respectively, then $G_{R_1} \leq G_{R_2}$ on $R_1 \times R_1$.

Proof. Let the class $\mathfrak{B}_x$ be defined relative to R_1. Then $G_{R_2}(x, \cdot) \in \mathfrak{B}_x$ for each $x \in R_1$ and $G_{R_1}(x, \cdot) \leq G_{R_2}(x, \cdot)$ for each $x \in R_1$.

THEOREM **5.15** Let $\{R_j\}$ be a monotone increasing sequence of open sets, let $R = \bigcup_{j=1}^{\infty} R_j$, and let R_j have a Green function G_j. If R has a Green function G_R, then $G_j \uparrow G_R$ on $R \times R$.

Proof. Let $G_\infty = \lim_{j \to \infty} G_j$ which is defined since the G_j's form a monotone increasing sequence by the preceding theorem. Although G_j is only defined on $R_j \times R_j \subset R \times R$, G_∞ is defined on $R \times R$. Since $G_j \leq G_R$ on $R_j \times R_j$ by the same theorem, $G_\infty \leq G_R$ on $R \times R$. Now $G_\infty(x, \cdot)$ is harmonic on $R \sim \{x\}$, since the limit of a monotone increasing sequence of harmonic functions is harmonic or $+\infty$ on each component of $R \sim \{x\}$ and the latter possibility is excluded since $G_\infty(x, \cdot) \leq G_R(x, \cdot) < +\infty$ on $R \sim \{x\}$. Moreover, $G_\infty(x, \cdot) - u_x$ is harmonic on $R \sim \{x\}$ and can be defined at x so as to be harmonic on R since it is the limit of a monotone sequence $\{G_j(x, \cdot) - u_x\}$ of functions with these properties. Since $G_j(x, \cdot) \geq 0$ on R_j, $G_\infty(x, \cdot) \geq 0$ on R. Consider the representation

$$G_\infty(x, \cdot) = u_x + [G_\infty(x, \cdot) - u_x].$$

Since the term in brackets can be defined at x so as to be harmonic on R, $G_R(x, \cdot) \leq G_\infty(x, \cdot)$ by the minimality of G_R. Therefore $G_R = G_\infty$.

THEOREM **5.16** If the open set R is not everywhere dense in E^2, then it has a Green function.

Proof. Since R is not everywhere dense there is some point such that some ball about the point does not intersect R. Suppose, for example, that R does not intersect the closed unit ball $\bar{B}_{0,1}$. Consider any $x, y \in R$. Since x^{-1} and y^{-1} (considering x and y as complex numbers) are in $B_{0,1}$,

$$\left| \frac{1}{x} - \frac{1}{y} \right| < 2.$$

Then

$$\log \frac{2}{|1/x - 1/y|} = \log \frac{2|xy|}{|x - y|} > 0.$$

Consider

$$v_x(y) = \log \frac{2}{|1/x - 1/y|} = \log 2|x| + \log|y| - \log|x - y|.$$

We have just shown that $v_x \geq 0$ on R for each $x \in R$. For x fixed in R, the first term on the right is a constant harmonic function on R and the second term on the right is the fundamental harmonic function and is harmonic on R. Hence

$$v_x(y) = (\text{harmonic function on } R) + \log \frac{1}{|x - y|} \quad \text{on } R.$$

It follows that $\mathfrak{B}_x \neq \emptyset$ for each $x \in R$ and the Green function for R exists by Lemma 5.11.

THEOREM **5.17** E^2 does not have a Green function.

Proof. Suppose E^2 does have a Green function

$$G(x, y) = -\log\|x - y\| + h_x(y),$$

where h_x is harmonic on E^2 for each x. Consider a fixed x. Now $h_x(y) \to +\infty$ as $\|y\| \to +\infty$ since $-\log\|x - y\| \to -\infty$ as $\|y\| \to +\infty$. If N is an arbitrary positive integer, then there is a ball about 0 such that $h_x \geq N$ on the boundary of the ball. By the minimum principle, $h_x \geq N$ on the ball. Therefore $h_x(0) = +\infty$ but this contradicts the fact that a harmonic function is finite-valued.

Although we have not pinned down those open subsets of E^2 which have Green functions, we have sufficient conditions for the existence and for the nonexistence. We give one last condition for nonexistence.

THEOREM **5.18** E^2 less a closed set all of whose points are isolated has no Green function.

Proof. Let $R = E^2 \sim$ (closed set all of whose points are isolated). Suppose R has a Green function G. Consider any $x \in R$. Since R is open there is a neighborhood U of x outside of which $G(x, \cdot)$ is bounded by Lemma 5.8. It can be assumed that U does not contain any of the points not in R. Consider any $y \notin R$. Then there is a neighborhood V of y which does not contain x and does not contain any of the other points of $E^2 \sim R$ such that $G(x, \cdot)$ is bounded and harmonic on $V \sim \{y\}$. Therefore $G(x, \cdot)$ can be defined on $E^2 \sim R$ in such a way as to be harmonic on $E^2 \sim \{x\}$. Writing $G(x, \cdot) = u_x + h_x$, where h_x is harmonic on R, this means that h_x can be extended to E^2 as a harmonic function. The class $\mathfrak{B}_x$ relative to E^2 is then nonempty. The same can be seen to be true of $\mathfrak{B}_{x_0}$ relative to E^2 for arbitrary $x_0 \in E^2$ by means of a translation mapping x into x_0 since harmonic functions map into harmonic functions and u_x maps into u_{x_0} under such a transformation. It would then follow that E^2 has a Green function in contradiction to the preceding theorem.

3. Harmonic Minorants

The principal result of this section is that the Green function is a symmetric function of its arguments. To prove this we shall introduce the concept of a greatest harmonic minorant and a smoothing operation. The smoothing operation had its origin in the work of Poincaré and is known as "balayage" or "sweeping out." We return to balayage in a later chapter.

Let R be an open subset of E^n, W an open subset of R, and let u be a superharmonic function on R. Moreover, let $\{B_j\}$ be a sequence of balls such that

(i) $\bar{B}_j \subset W$ for all j,
(ii) $W = \bigcup_{j=1}^{\infty} B_j$,
(iii) for each j, B_j occurs infinitely often in the sequence $\{B_j\}$.

Define

$$u_1 = \begin{cases} u & \text{on} \quad R \sim B_1, \\ \mathbf{PI}(u, B_1) & \text{on} \quad B_1, \end{cases}$$

and, inductively,

$$u_j = \begin{cases} u_{j-1} & \text{on} \quad R \sim B_j, \\ \mathbf{PI}(u_{j-1}, B_j) & \text{on} \quad B_j. \end{cases}$$

By Lemma 4.17 $u_j \leq u_{j-1}$ on R, u_j is harmonic on B_j, and u_j is superharmonic on R. Since the sequence $\{u_j\}$ is monotone decreasing, we can define $u_\infty = \lim_{j \to \infty} u_j$ on R.

Definition. u_∞ is called the **reduction of u over W relative to R.**

We shall postpone showing that u_∞ is independent of the sequence $\{B_j\}$ used to define it.

THEOREM **5.19** If u is superharmonic on the open set R, W is an open subset of R, and u_∞ is the reduction of u over W, then u_∞ is either harmonic or identically $-\infty$ on each component of W and $u_\infty = u$ on $R \sim W$.

Proof. We can assume that W is connected by consideration of components. Let $\{B_j\}$ be the sequence of balls and $\{u_j\}$ the corresponding sequence of superharmonic functions defining u_∞. Consider a fixed j. Since B_j occurs infinitely often in the sequence of balls, there is a subsequence $\{B_{j_k}\}$ such that $B_j = B_{j_k}$ for all $k \geq 1$. On B_j we have $u_\infty = \lim_{k \to \infty} u_{j_k}$. Since u_{j_k} is harmonic on $B_{j_k} = B_j$, u_∞ is either harmonic or identically $-\infty$ on B_j by Theorem 2.19. It follows from the definition of subharmonic function that u_∞ is either subharmonic or identically $-\infty$ on W; if not identically $-\infty$ on W, then

u_∞ is subharmonic on W and therefore harmonic on W since it is harmonic on each ball B_j. Since $u_j = u$ on $R \sim W$ for each j, $u_\infty = u$ on $R \sim W$.

Of course, if $W = R$, the reduction of the superharmonic function u over R is either harmonic or identically $-\infty$ on each component of R. If $W \neq R$, we must be careful not to jump to the conclusion that u_∞ is superharmonic on R even if R is connected and u_∞ is not identically $-\infty$. The u_∞ is the limit of a decreasing sequence of superharmonic functions but u_∞ may not be a l.s.c. function.

LEMMA **5.20** If u is superharmonic on the open set R, W is an open subset of R, v is harmonic on W, and $v \leq u$ on W, then $v \leq u_\infty \leq u$ on W, where u_∞ is the reduction of u over W.

Proof. Let $\{B_j\}$ be a sequence of balls defining u_∞. On B_1, $v = \mathbf{PI}(v, B) \leq \mathbf{PI}(u, B_1) = u_1$. Moreover, $v \leq u = u_1$ on $W \sim B_1$. Thus, $v \leq u_1$ on W and by induction, $v \leq u_j$ on W for all j.

Definition. If u is superharmonic on the open set R, h is harmonic on R, and $h \leq u$ on R, then h is called a **harmonic minorant** of u. The function h is a **greatest harmonic minorant** of u if h is a harmonic minorant and $h \geq v$ whenever v is a harmonic minorant of u.

THEOREM **5.21** If u is superharmonic on the open set R and has a harmonic minorant on R, then u has a unique greatest harmonic minorant, namely u_∞ the reduction of u over R.

Proof. Let h be a harmonic minorant of u. By the preceding lemma $h \leq u_\infty \leq u$ on R. Since h is harmonic on R, u_∞ cannot take on the value $-\infty$ on R and is therefore harmonic on R by Theorem 5.19. Since h can be any harmonic minorant of u, we have also shown that u_∞ is a greatest harmonic minorant of u. Uniqueness is clear from the definition of greatest harmonic minorant.

In defining the reduction of a superharmonic function u over an open set $W \subset R$, we did not show that the reduction u_∞ is independent of the sequence of balls used. If u has a harmonic minorant on W, then this is the case since $u_\infty|_W$ is the greatest harmonic minorant of u over W. Since we shall consider the reduction only when such a harmonic minorant exists, it is not necessary to dwell on the subject.

LEMMA **5.22** If u_1, u_2 are superharmonic on the open set R and possess harmonic minorants on R, the greatest harmonic minorant of $u_1 + u_2$ is the sum of the greatest harmonic minorants of u_1 and u_2.

Proof. Let $\{B_j\}$ be a sequence of balls defining the reduction of a superharmonic function u over R and let $\{u_j\}$ be the sequence of functions defining u_∞. Since the Poisson integral is linear in the boundary function, $(u_1 + u_2)_j = u_{1j} + u_{2j}$. Therefore $(u_1 + u_2)_\infty = u_{1\infty} + u_{2\infty}$ and the lemma follows from Theorem 5.21.

THEOREM **5.23** If the open set R has a Green function G, then the greatest harmonic minorant of $G(x, \cdot)$, $x \in R$, is the zero function.

Proof. We know that $G(x, \cdot) \geq 0$ on R and that $G(x, \cdot)$ is superharmonic on R. By Theorem 5.21 $G(x, \cdot)$ has a greatest harmonic minorant since the zero function is a harmonic function. Suppose h_x is a harmonic minorant of $G(x, \cdot)$. Then $G(x, \cdot) - h_x \geq 0$ and can be represented as the sum of u_x and a harmonic function; that is, $G(x, \cdot) - h_x \in \mathfrak{B}_x$. Therefore $G(x, \cdot) \leq G(x, \cdot) - h_x$ by the minimality of the Green function. This shows that $h_x \leq 0$ and that the zero function is the greatest harmonic minorant of $G(x, \cdot)$.

To prove the symmetry of the Green function, we shall use the method used to define the reduction of a superharmonic function. This will require a modification of the sequence $\{B_j\}$ of balls. Let R be an open set. We can assume that the set $C = \bigcup_{j=1}^\infty \partial B_j$ is closed relative to R. Such a sequence can always be constructed as follows. Let $\{U_j\}$ be a sequence of open sets with compact closures such that $U_j \uparrow R$ and $\overline{U}_j \subset R$. It is easily seen using compactness arguments that there is a sequence of balls $\{B_j\}$ such that only a finite number of balls intersect any $\overline{U}_j$. If the sequence is modified so that each ball appears infinitely often in a new sequence, this will still be true. Suppose now that $\xi_k \in C = \bigcup_{j=1}^\infty \partial B_j$ and $\xi_k \to \xi \in R$. Since $U_j \uparrow R$, $\xi \in U_{j_0}$ for some j_0. Since at most a finite number of balls intersect $\overline{U}_{j_0}$, there is some k_0 such that $\xi_k \in \bigcup_{j=1}^{k_0} \partial B_j$ for large k and it easily follows that $\xi \in \partial B_j$ for some j. This shows that $C = \bigcup_{j=1}^\infty \partial B_j$ is closed relative to R. We can therefore assume a sequence $\{B_j\}$ of balls such that (i) $R = \bigcup_{j=1}^\infty B_j$, (ii) $\overline{B}_j \subset R$, (iii) $\bigcup_{j=1}^\infty \partial B_j$ is closed relative to R, and (iv) each B_j occurs infinitely often in the sequence.

THEOREM **5.24** If the open set R has a Green function G, then $G(x, y) = G(y, x)$ for all $x, y \in R$. In particular, $G(\cdot, y)$ is harmonic on $R \sim \{y\}$ for each $y \in R$.

Proof. Let $\{B_j\}$ be a sequence of balls satisfying the above four properties. Then $R_0 = R \sim \bigcup_{j=1}^\infty \partial B_j$ is open. For each $x \in R$, $G(x, \cdot) = u_x + h_x$, where h_x is harmonic on R. To simplify the notation we write $u(x, y)$ for $u_x(y)$ and $h(x, y)$ for $h_x(y)$. Now

(d) $$G_j(x, \cdot) = u_j(x, \cdot) + h(x, \cdot),$$

where $G_j(x, \cdot)$ and $u_j(x, \cdot)$ denote the jth terms of the "sweeping out" method of defining the reduction of $G(x, \cdot)$ and $u(x, \cdot)$, respectively, over R. The second term on the right side of (d) does not have a subscript j because $h(x, \cdot)$ is harmonic and therefore invariant under the smoothing operation. As $j \to \infty$, $G_j(x, \cdot)$ approaches the greatest harmonic minorant of $G(x, \cdot)$ which is the zero function by Theorem 5.23. Likewise $u_j(x, \cdot) \to u_\infty(x, \cdot)$, the greatest harmonic minorant of $u(x, \cdot)$ over R. Letting $j \to \infty$ in (d) we get

(e) $$0 = u_\infty(x, \cdot) + h(x, \cdot).$$

Therefore $G(x, \cdot) = u(x, \cdot) - u_\infty(x, \cdot)$. We now show that $u_\infty(\cdot, y)$ is harmonic for each $y \in R$. To do this it is necessary to reexamine the smoothing process defining u_∞. Let y be a fixed point of R. If $y \notin B_1$, then $u_1(x, y) = u(x, y)$ for all x and $u_1(\cdot, y)$ is harmonic on $R \sim \{y\}$. Suppose then that $y \in B_1 = B_{z_1, \rho_1}$. Then $u_1(x, y)$ is given for all x by (in the $n \geq 3$ case)

$$u_1(x, y) = \frac{1}{\sigma_n \rho_1} \int_{\partial B_1} \frac{1}{\|x - z\|^{n-2}} \frac{\rho_1^2 - \|y - z_1\|^2}{\|y - z\|^n} \, d\sigma(z).$$

If we apply the Laplacian $\Delta_{(x)}$ to both sides of this equation, then the Laplacian of the right side may be taken under the integral at least for points $x \notin \partial B_1$. Since the Laplacian of the first factor under the integral is zero, $u_1(\cdot, y)$ is harmonic on $R \sim \partial B_1$. Therefore $u_1(\cdot, y)$ is harmonic on $R \sim \{y\} \sim \partial B_1$. Consider now $u_2(\cdot, y)$. If $y \notin B_2$, then $u_2(x, y) = u_1(x, y)$ for all x and $u_2(\cdot, y)$ is harmonic on $R \sim \{y\} \sim \partial B_1$. If $y \in B_2 = B_{z_2, \rho_2}$, then

$$u_2(x, y) = \frac{1}{\sigma_n \rho_2} \int_{\partial B_2} u_1(x, z) \frac{\rho_2^2 - \|y - z_2\|^2}{\|y - z\|^n} \, d\sigma(z).$$

Applying the Laplacian $\Delta_{(x)}$ to both sides of this equation, $u_2(\cdot, y)$ is harmonic on $R \sim \partial B_1 \sim \partial B_2$. Therefore, $u_2(\cdot, y)$ is harmonic on $R \sim \{y\} \sim \bigcup_{j=1}^2 \partial B_j$. By an induction argument, it follows that each $u_j(\cdot, y)$ is harmonic on $R \sim \{y\} \sim \bigcup_{j=1}^\infty \partial B_j$. But more is true; for some subsequence $\{j_k\}$, $y \in B_{j_k}$ for all $k \geq 1$. As in the case of $u_2(\cdot, y)$, $u_{j_k}(\cdot, y)$ is actually harmonic on $R \sim \bigcup_{j=1}^{j_k} \partial B_j$. That is, the sequence of functions $\{u_{j_k}(\cdot, y)\}$ is harmonic on $R \sim \bigcup_{j=1}^\infty \partial B_j$. Since the sequence decreases to $u_\infty(\cdot, y)$, $u_\infty(\cdot, y)$ is harmonic on $R \sim \bigcup_{j=1}^\infty \partial B_j$. Actually more is true since $u_\infty(\cdot, y)$ is independent of the sequence $\{B_j\}$ of balls used; that is, $u_\infty(\cdot, y)$ is harmonic on R for each $y \in R$. Since $G(x, y) = u(x, y) - u_\infty(x, y)$ and $u(x, y) = u_x(y) = u_y(x) = u(y, x)$, for each $y \in R$

(i) $u(\cdot, y) - u_\infty(\cdot, y) \geq 0$ on R,
(ii) $u(\cdot, y) - u_\infty(\cdot, y)$ is the sum of u_y and a harmonic function.

By the minimality of the Green function, $u(\cdot, y) - u_\infty(\cdot, y) \geq G(y, \cdot)$; that is, $G(x, y) \geq G(y, x)$ and therefore equality.

COROLLARY **5.25** Let R be an open set having a Green function G and let R_0 be a component of R. If $x \in R_0$, then $G_R(x, y) = 0$ for $y \in R \sim R_0$. Moreover, the Green function G_{R_0} for R_0 is given by $G_{R_0} = G \mid_{R_0 \times R_0}$.

Proof. Consider a fixed $x \in R_0$. Let $G(x, \cdot) = u_x + h_x$ where h_x is harmonic on R. Since $G(x, \cdot) \geq 0$, $u_x \geq -h_x$ and u_x has a harmonic minorant on R. Then u_x has a harmonic minorant on R_0 and therefore a greatest harmonic minorant h_0 on R_0. Let

$$h_x^* = \begin{cases} h_0 & \text{on} \quad R_0, \\ u_x & \text{on} \quad R \sim R_0. \end{cases}$$

Then h_x^* is harmonic on R and $u_x \geq h_x^*$ on R, but, since $-h_x$ is the greatest harmonic minorant of u_x, $-h_x \geq h_x^*$ and $G(x, \cdot) = u_x + h_x \leq u_x - h_x^* = 0$ on $R \sim R_0$. Since we always have $G(x, \cdot) \geq 0$, it follows that $G(x, \cdot) = 0$ on $R \sim R_0$. A similar argument proves the second assertion.

THEOREM **5.26** If R has a Green function G, then G is continuous on $R \times R$ (in the extended sense at points of the diagonal of $R \times R$). Moreover, if $\bar{B}_{x,\delta} \subset R$, then $L(G(z, \cdot) : x, \delta)$ is a continuous function of $z \in R$.

Proof. G is obviously continuous in the extended sense at points of the diagonal of $R \times R$. If $(x_0, y_0) \in R \times R$ with $x_0 \neq y_0$, let U and V be balls with centers x_0 and y_0, respectively, with disjoint closures in R. If x_0 and y_0 are in different components of R, then $G = 0$ on $U \times V$ and G is trivially continuous at (x_0, y_0). We can therefore assume that x_0 and y_0 belong to the same component of R. By Lemma 5.8, there is a constant $m > 0$ such that $G(x_0, \cdot) \leq m$ on V. Since $G(\cdot, y)$ is positive and harmonic on U for each $y \in V$, there is a constant $k > 0$ such that $G(x, y)/G(x_0, y) \leq k$ for all $x \in U$ and $y \in V$ by Theorem 2.16. Therefore $G(x, y) = [G(x, y)/G(x_0, y)]G(x_0, y) \leq km$ for all $x \in U, y \in V$. This shows that family $\{G(x, \cdot) : x \in U\}$ is uniformly bounded on V. Since $G(x, y) = u_x(y) + h_x(y)$ and $u_x(y)$ is bounded on $U \times V$, the family $\{h_x : x \in U\}$ of harmonic functions is uniformly bounded on V and equicontinuous on any compact subset of V according to Theorem 2.18. Since $h_x(y)$ is a symmetric function,

$$\begin{aligned} |G(x, y) - G(x_0, y_0)| &\leq |u_x(y) - u_{x_0}(y_0)| + |h_x(y) - h_{x_0}(y_0)| \\ &\leq |u_x(y) - u_{x_0}(y_0)| + |h_x(y) - h_x(y_0)| + |h_{y_0}(x) - h_{y_0}(x_0)|. \end{aligned}$$

The first term on the right can be made arbitrarily small by the joint continuity of $u_x(y)$ at (x_0, y_0), the second term can be made arbitrarily small by the equicontinuity of the family $\{h_x : x \in U\}$, and the last term can be made arbitrarily small by the continuity of h_{y_0} at x_0. This completes the proof that

G is continuous at (x_0, y_0). Now let $\bar{B}_{x,r} \subset R$ and let $\nu_{x,r}$ be a unit mass uniformly distributed on $\partial B_{x,r}$. Using the continuity of G and the Lebesgue dominated convergence theorem, one can show that

$$L(G(z, \cdot) : x, r) = \int G(z, y)\, d\nu_{x,r}(y)$$

is continuous at $z_0 \in R \sim \partial B_{x,r}$. In particular, $L(G(z, \cdot) : x, \delta)$ is continuous on $R \sim \partial B_{x,\delta}$. Suppose $z_0 \in \partial B_{x,\delta}$. Since the functions $L(G(z, \cdot) : x, \delta + r)$ are continuous at z_0 and increase to $L(G(z, \cdot) : x, \delta)$ as $r \downarrow 0$, $L(G(z, \cdot) : x, \delta)$ is l.s.c. at z_0; likewise the functions $L(G(z, \cdot) : x, \delta - r)$ are continuous at z_0 and decrease to $L(G(z, \cdot) : x, \delta)$ as $r \downarrow 0$ and $L(G(z, \cdot) : x, \delta)$ is u.s.c. at z_0.

Only connected everywhere dense open sets present a problem in so far as the existence of a Green function is concerned.

THEOREM **5.27** If $R \subset E^2$ is a non-connected open set, then R has a Green function.

Proof. Let $\{R_j\}$ be the components of R of which there are more than one. Suppose $x \in R_1$. By Theorem 5.16 R_1 has a Green function; i.e., there is a non-negative superharmonic function v_x on R_1 of the form $v_x = u_x + h_x$ where h_x is harmonic on R_1. Now u_x is harmonic on the remaining components of R. Extend h_x to R by letting $h_x = -u_x$ on $R \sim R_1$. Then $u_x + h_x \in \mathfrak{V}_x$ and $\mathfrak{V}_x$ is non-empty. Since this can be done on each component of R, $\mathfrak{V}_x \neq \varnothing$ for all $x \in R$ and R has a Green function.

CHAPTER 6

Green Potentials

Physically, a potential function represents the potential energy of a unit mass or charge at a point of E^n due to a distribution of mass or charge on the space. As such, potential functions are of interest in their own right. In this chapter, however, such functions are studied as a means of obtaining a representation theorem for superharmonic functions. This representation theorem has already been proved for smooth superharmonic functions. In fact, if u is superharmonic and has continuous second partials on a neighborhood of the closure of a ball $B = B_{y,\rho} \subset E^n$, $n \geq 3$, then for $x \in B$

$$\text{(a)} \qquad u(x) = -\frac{1}{\sigma_n(n-2)} \int_{\partial B} u(z) D_{\mathbf{n}} G_B(x, z)\, d\sigma(z) - \frac{1}{\sigma_n(n-2)} \int_B G_B(x, z)\, \Delta u(z)\, dz$$

according to Theorem 5.1. The first term on the right side of (a) is just $\mathbf{PI}(u, B)$ and represents a function harmonic on B. Recalling that $G_B(x, \cdot) = u_x + h_x$, where each h_x is harmonic on B, we can put the second term on the right side of (a) into the form

$$\text{(b)} \qquad -\frac{1}{\sigma_n(n-2)} \int_B u_x(z)\, \Delta u(z)\, dz - \frac{1}{\sigma_n(n-2)} \int_B h_x(z)\, \Delta u(z)\, dz.$$

We show in this chapter that the second term of (b) is harmonic on B. The first term of (b) is just the potential of a unit mass at x due to a mass distribution of density Δu over B and is the ordinary potential function that one computes in an elementary physics course. Equation (a) therefore shows that u has a representation as a sum of a harmonic function and a "potential" type function. The reader should not consider the foregoing remarks as a definition of a potential since we shall use the Green function $G(x, \cdot)$ rather than u_x in the definition. The above representation is the calculus version of the important Riesz representation theorem.

1. Definitions and Properties

In view of the introductory remarks, we can immediately give the real variable definition of potential.

Definition. Let R be an open subset of E^n having a Green function G. If μ is a signed measure on R, then

$$G\mu(x) = \int G(x, y)\, d\mu(y),$$

if defined everywhere on R, is called the **Green potential** of μ. If μ is a measure on R and $G\mu$ is superharmonic on R, then $G\mu$ is called the **potential** of μ.

With suitable restrictions on μ, $G\mu$ will be defined and superharmonic on R or a difference of two functions superharmonic on R.

LEMMA **6.1** If μ is a measure on the open set R having a Green function G, then $G\mu$ is either superharmonic or identically $+\infty$ on each component of R.

Proof. Let $\{U_j\}$ be an increasing sequence of open subsets of R with compact closures $\overline{U}_j \subset R$ such that $R = \bigcup_{j=1}^{\infty} U_j$. For each $x \in R$ define $G_j^*(x, \cdot) = \min[G(x, \cdot), j]$ and

$$u_j(x) = \int_{U_j} G_j^*(x, y)\, d\mu(y).$$

Since μ is a Borel measure, $\mu(U_j) \leq \mu(\overline{U}_j) < +\infty$. Using the symmetry of the Green function G and the fact that $G(x, \cdot)$ is continuous (in the extended sense) on R, we see that $G(x_i, \cdot) \to G(x, \cdot)$ as $x_i \to x$ and that the same is true of G_j^*. Since G_j^* is bounded by j, we obtain

$$u_j(x_i) = \int_{U_j} G_j^*(x_i, y)\, d\mu(y)$$

$$\to \int_{U_j} G_j^*(x, y)\, d\mu(y) = u_j(x)$$

as $x_i \to x$ by the Lebesgue dominated convergence theorem. This shows that each u_j is continuous on R. Since $U_j \uparrow R$ and $G_j^*(x, \cdot) \uparrow G(x, \cdot)$ as $j \to \infty$, $u_j \uparrow G\mu$ as $j \to \infty$ by the Lebesgue monotone convergence theorem. $G\mu$, as the limit of an increasing sequence of continuous functions, is l.s.c. on R. Using the symmetry of G again, $G(\cdot, y)$ is superharmonic on R for each y. Now for $\overline{B}_{x,\delta} \subset R$

$$L(G\mu : x, \delta) = L\left(\int_R G(\cdot\,, y)\, d\mu(y) : x, \delta\right)$$
$$= \frac{1}{\sigma_n \delta^{n-1}} \int_{\partial B_{x,\delta}} \left(\int_R G(z, y)\, d\mu(y)\right) d\sigma(z).$$

Since G is continuous in the extended sense on $R \times R$ by Theorem 5.26, Fubini's theorem may be applied to obtain

$$L(G\mu : x, \delta) = \int_R L(G(\cdot\,, y) : x, \delta)\, d\mu(y)$$
$$\leq \int_R G(x, y)\, d\mu(y) = G\mu(x).$$

Therefore $G\mu$ is non-negative on R, l.s.c. on R, super-mean-valued on R, and therefore superharmonic on any component on which it is not identically $+\infty$.

LEMMA **6.2** If μ is a measure on the open set R having a Green function G such that $\mu(R) < +\infty$ and $\mu(B) = 0$ for some closed ball $\bar{B} \subset R$, then $G\mu$ is superharmonic on the component of R containing B.

Proof. Let x be the center of the ball B. Then $G(x, \cdot)$ is bounded outside B by Lemma 5.8. Since $\mu(B) = 0$,

$$G\mu(x) = \int_R G(x, y)\, d\mu(y) = \int_{R \sim B} G(x, y)\, d\mu(y) < +\infty.$$

Therefore $G\mu$ is not identically $+\infty$ on the component containing B and superharmonic on this component by Lemma 6.1.

THEOREM **6.3** If μ is a measure on the open set R having a Green function G and $\mu(R) < +\infty$, then $G\mu$ is a potential.

Proof. Assume first that R is connected. Consider any $x \in R$ and $B = B_{x,\delta}$ with $\bar{B} \subset R$. Then $\mu = \mu|_B + \mu|_{R \sim B}$ where $\mu|_B$ and $\mu|_{R\sim B}$ are measures vanishing on $R \sim B$ and B, respectively. It is clear from the definition of a potential that $G\mu = G\mu|_B + G\mu|_{R\sim B}$. Since $\mu|_{R\sim B}(B) = 0$ and $\mu|_{R\sim B}(R) < +\infty$, $G\mu|_{R\sim B}$ is superharmonic on R by Lemma 6.2. Since $\bar{B} \subset R$, there is a closed ball $\bar{B}_1 \subset R \sim \bar{B}$ such that $\mu|_B(B_1) = 0$. Since $\mu|_B(R) < +\infty$, $G\mu|_B$ is superharmonic on R by Lemma 6.2. It follows from Theorem 4.12 that $G\mu = G\mu|_B + G\mu|_{R\sim B}$ is superharmonic on R, assuming that R is connected. Suppose now that R is not connected and let $R = \bigcup_j R_j$, where $\{R_j\}$ is the countable (possibly finite) collection of components of R. Let $\mu_j = \mu|_{R_j}$. Then

$\mu_j(R_j) < +\infty$. By Corollary 5.25, the Green function G_{R_j} for R_j is just the restriction of G to $R_j \times R_j$. Since $G(x, \cdot)$ vanishes outside the component containing x by Corollary 5.25, for $x \in R_j$

$$G\mu(x) = \int G(x, y)\, d\mu(y) = \int_{R_j} G(x, y)\, d\mu(y) = G_{R_j}\mu_j(x).$$

By the first part of the proof $G_{R_j}\mu_j$ is superharmonic on R_j and therefore $G\mu$ is superharmonic on each R_j.

Historically, the Green potential was preceded by the logarithmic potential in the $n = 2$ case and the Newtonian potential in the $n \geq 3$ case. The logarithmic potential U^μ of a measure μ on E^2 having compact support is defined by

$$U^\mu(x) = -\int \log\|x - y\|\, d\mu(y), \qquad x \in E^2.$$

If μ is any measure on E^n, $n \geq 3$, the Newtonian potential U^μ is defined by

$$U^\mu(x) = \int \frac{1}{\|x - y\|^{n-2}}\, d\mu(y), \qquad x \in E^n.$$

Note that the logarithmic potential need not be non-negative. Since the Newtonian potential U^μ of a measure is just the potential $G\mu$ where G is the Green function for E^n, the preceding results apply directly to U^μ; for example, if $\mu(E^n) < +\infty$, then U^μ is superharmonic. Suppose now that μ is a measure on E^2 having compact support and consider the logarithmic potential U^μ. The arguments used in the proof of Lemma 6.1 to show that $G\mu$ is super-mean-valued and l.s.c. cannot be used for U^μ since the logarithmic function may be negative and Fubini's theorem is not directly applicable. Consider any ball $B_{x_0,\delta}$ and let C be the support of μ. Since $-\log\|x - y\|$ is continuous in the extended sense on $\bar{B}_{x_0,\delta} \times C$, there is a constant c such that $-\log\|x - y\| + c \geq 0$ for $x \in \bar{B}_{x_0,\delta}$, $y \in C$. For $x \in \bar{B}_{x_0,\delta}$

$$U^\mu(x) + c\mu(C) = \int [-\log\|x - y\| + c]\, d\mu(y).$$

Applying the argument in the proof of Lemma 6.1, we can show from this equation that U^μ is l.s.c. on $B_{x_0,\delta}$ and therefore on E^2. Computing spherical averages and applying Fubini's theorem,

$$\begin{aligned} L(U^\mu : x_0, \delta) + c\mu(C) &= \int L(-\log\|x - y\| + c : x_0, \delta)\, d\mu(y) \\ &\leq \int (-\log\|x_0 - y\| + c)\, d\mu(y) \\ &= U^\mu(x_0) + c\mu(C). \end{aligned}$$

Therefore $U^\mu(x_0) \geq L(U^\mu : x_0, \delta)$ for any ball $B_{x_0,\delta}$ and U^μ is super-mean-valued. It is also easily seen that U^μ cannot take on the value $-\infty$. If $x_0 \notin C$ then $-\log\|x_0 - y\|$ is bounded on C and $U^\mu(x_0) < +\infty$. This shows that the logarithmic potential U^μ is superharmonic whenever μ has compact support. Moreover, if R is an open set having a Green function G_R and the measure μ has compact support in R, then U^μ and $G_R\,\mu$ differ on R by a harmonic function. This can be seen from the equation

$$G_R\,\mu(x) = \int u_x(y)\,d\mu(y) + \int h_x(y)\,d\mu(y),$$

where $G_R(x, \cdot) = u_x + h_x$ on R, and the fact that the second integral is a harmonic function of $x \in R$ (see Lemma 6.7).

THEOREM **6.4** Let R be an open set having a Green function G and let $G\mu$ be the potential of a measure μ. Then the greatest harmonic minorant of $G\mu$ is zero.

Proof. Let $u = G\mu$ and let $\{B_j\}$ be a sequence of balls satisfying the requirements for defining the reduction of a superharmonic function over R. Let $\{u_j\}$ be the sequence of functions defining u_∞, the reduction of u over R. By Theorem 5.21, the assertion to be proved is that $u_\infty = 0$ on R. Likewise let $G_j(\cdot, y)$ denote the sequence of functions defining the reduction of $G(\cdot, y)$ over R. By Theorem 5.23 the sequence $G_j(\cdot, y)$ decreases to the zero function since the greatest harmonic minorant of $G(\cdot, y)$ is the zero function. Actually u_j and G_j are related as follows:

$$\text{(c)} \qquad u_j(x) = \int_R G_j(x, y)\,d\mu(y), \qquad x \in R.$$

The proof of (c) is by induction on j. The usual first step of the inductive proof will be omitted because of its similarity with the general case. By definition,

$$u_j(x) = \begin{cases} u_{j-1}(x), & x \in R \sim B_j \\ \mathbf{PI}(u_{j-1}, B_j)(x), & x \in B_j \end{cases}$$

with a similar definition of $G_j(x, y)$ for fixed y. If $x \notin B_j$, then $u_j(x) = u_{j-1}(x)$ and $G_j(x, y) = G_{j-1}(x, y)$. Assuming that (c) is true with j replaced by $j - 1$, we have for $x \notin B_j$

$$u_j(x) = u_{j-1}(x) = \int G_{j-1}(x, y)\,d\mu(y) = \int G_j(x, y)\,d\mu(y).$$

Now for $x \in B_j$

$$u_j(x) = \int_{\partial B_j} u_{j-1}(y) P(x, y)\,d\sigma(y),$$

where $P(x, y)$ is the kernel of the Poisson integral over ∂B_j. Again under the assumption that (c) is true with j replaced by $j-1$, for $x \in B_j$

$$u_j(x) = \int_{\partial B_j} u_{j-1}(y)P(x, y)\, d\sigma(y)$$

$$= \int_{\partial B_j} \left(\int_R G_{j-1}(y, z)\, d\mu(z) \right) P(x, y)\, d\sigma(y).$$

By Fubini's theorem, for $x \in B_j$,

$$u_j(x) = \int_R \left(\int_{\partial B_j} G_{j-1}(y, z) P(x, y)\, d\sigma(y) \right) d\mu(z)$$

$$= \int_R G_j(x, z)\, d\mu(z).$$

Therefore (c) is true for j whenever it is true for $j-1$ and (c) is true for all $j \geq 1$. The assertion of the theorem can be obtained from (c) as follows. Since $G\mu$ is superharmonic, it is finite-valued a.e. on R by Theorem 4.10. Choose x_0 such that $G\mu(x_0) < +\infty$. Then

$$G\mu(x_0) = \int_R G(x_0, y)\, d\mu(y) < +\infty;$$

that is, $G(x_0, \cdot)$ is integrable relative to μ. Since $0 \leq G_j(x_0, \cdot) \leq G(x_0, \cdot)$ and $\lim_{j \to \infty} G_j(x_0, \cdot) = 0$, we may apply the Lebesgue dominated convergence theorem to (c) to obtain

$$u_\infty(x_0) = \lim_{j \to \infty} \int_R G_j(x_0, y)\, d\mu(y) = 0.$$

Since $u_\infty \geq 0$ and is harmonic on R, $u_\infty = 0$ on R by the minimum principle.

COROLLARY **6.5** If R is an open set having a Green function G and the potential $G\mu$ of a measure μ is harmonic on R, then μ is the zero measure.

Proof. Let $h = G\mu$ be harmonic on R. Since h is a harmonic minorant of itself, $h = 0$ on R by the preceding theorem. Thus for any $x \in R$

$$0 = \int_R G(x, y)\, d\mu(y).$$

Since $G(x, \cdot) > 0$ on the component of R containing x, each component of R has μ-measure zero.

Although the preceding corollary excludes harmonic potentials, it is possible for a potential to be harmonic on a proper subset of R.

THEOREM **6.6** If $G\mu$ is the potential of a measure μ on the open set R having a Green function G, then $G\mu$ is harmonic on any open set of μ-measure zero.

Proof. Let U be a nonempty open subset of R with $\mu(U) = 0$. Then for any $x \in R$,

$$G\mu(x) = \int_R G(x, y)\, d\mu(y) = \int_{R \sim U} G(x, y)\, d\mu(y).$$

Suppose $\bar{B}_{x,\delta} \subset U$. By Fubini's theorem

$$A(G\mu : x, \delta) = A\left(\int_{R \sim U} G(\cdot\,, y)\, d\mu(y) : x, \delta\right)$$

$$= \int_{R \sim U} A(G(\cdot\,, y) : x, \delta)\, d\mu(y).$$

For $y \in R \sim U$, $G(\cdot\,, y)$ is harmonic on U. Since $\bar{B}_{x,\delta} \subset U$, $A(G(\cdot\,, y) : x, \delta) = G(x, y)$ and

$$A(G\mu : x, \delta) = \int_{R \sim U} G(x, y)\, d\mu(y) = G\mu(x).$$

Since $G\mu$ is locally integrable on R, this shows that $G\mu$ is continuous on U. It follows that $G\mu$ is harmonic on U by Theorem 4.9.

We have seen that if $h_1, \ldots, h_p$ are functions harmonic on an open set R and $\alpha_1, \ldots, \alpha_p$ are any real numbers, then $\sum_{i=1}^{p} \alpha_i h_i$ is also harmonic on R. The following lemma is an extension of this elementary fact.

LEMMA **6.7** Let U and V be open subsets of E^n, μ a measure on U, and H a non-negative function on $U \times V$. If (i) for each $y \in V$, $H(\cdot\,, y)$ is continuous on U, (ii) for each $x \in U$, $H(x, \cdot)$ is harmonic on V, and (iii) $h(y) = \int_U H(x, y)\, d\mu(x) < +\infty$ for each $y \in V$, then h is harmonic on V.

Proof. By hypothesis, $H(x, y)$ is continuous in each variable separately. This implies that $H(x, y)$ is measurable in both variables jointly. Rather than proving this in the present context of $2n$-dimensional space, we shall discuss a special case from which a proof for the more general case can be constructed. Suppose $U = V = (-\infty, +\infty)$ and $H(x, y)$ is defined on $U \times V$ and continuous in each variable separately. Define a sequence $\{\phi_k\}$ of functions on $(-\infty, +\infty)$ as follows:

$$\phi_k(x) = \frac{j+1}{2^k} \quad \text{if} \quad \frac{j}{2^k} \leq x < \frac{j+1}{2^k}.$$

For each x, $\lim_{k\to\infty} \phi_k(x) = x$. Now define $H_k(x, y) = H(\phi_k(x), y)$. It is easy to see that $H_k(x, y)$ is measurable in both variables jointly and that $\lim_{k\to\infty} H_k(x, y) = H(x, y)$, so that $H(x, y)$ is measurable in both variables jointly. Let this suffice as justification for the assertion of joint measurability. Therefore $H(x, y)$ is a non-negative jointly measurable function to which Fubini's theorem can be applied. Suppose $\bar{B}_{y,\delta} \subset V$. Then

$$\begin{aligned} A(h : y, \delta) &= A\left(\int_U H(x, \cdot)\, d\mu(x) : y, \delta\right) \\ &= \int_U A(H(x, \cdot) : y, \delta)\, d\mu(x) \\ &= \int_U H(x, y)\, d\mu(x) \\ &= h(y) < +\infty. \end{aligned}$$

This shows that h is locally integrable and also continuous on V. It follows from Theorem 4.9 that h is harmonic on V.

We shall frequently need to compare Green potentials for different regions, especially when one is a subset of the other.

THEOREM **6.8** Let R be an open set having a Green function G_R, let S be an open subset of R with Green function G_S, and let μ be a measure on R such that $\mu(R \sim S) = 0$ and $G_R\,\mu$ is a potential. Then there is a non-negative harmonic function h on S such that $G_R\mu = G_S\,\mu|_S + h$ on S.

Proof. Since $S \subset R$, $G_S \le G_R$ on $S \times S$ by Theorem 5.14. Then for each $x \in S$

$$G_S\,\mu|_S(x) = \int_S G_S(x, y)\, d\mu(y) \le \int_S G_R(x, y)\, d\mu(y) = G_R\,\mu(x).$$

This also shows that $G_S\,\mu|_S$ is a potential. Since $G_R\,\mu$ is superharmonic, it is finite-valued a.e. on S. Then except possibly for a set of x in S of Lebesgue measure zero

$$\text{(d)} \qquad G_R\,\mu(x) - G_S\,\mu|_S(x) = \int_S [G_R(x, y) - G_S(x, y)]\, d\mu(y).$$

Now $G_R(x, y) = u_x(y) + h_x(y)$ and $G_S(x, y) = u_x(y) + h_x^*(y)$ where h_x and h_x^* are harmonic on R and S, respectively. Since $G_R(x, y)$ and $u_x(y)$ are symmetric functions, $h_x(y)$ is a symmetric function along with $h_x^*(y)$. For $x \ne y$, $G_R(x, y) - G_S(x, y) = H_y(x) \ge 0$, where $H_x(y)$ is the symmetric function $h_x(y) - h_x^*(y)$. Note that for each $y \in S$, $H_y(\cdot)$ is harmonic on S and that for each $x \in S$,

$H_y(x)$ is harmonic and therefore continuous on S. Supppse $\bar{B}_{x,\delta} \subset S$. Note that

$$A[G_R(\cdot\,, y) - G_S(\cdot\,, y) : x, \delta] = A(H_y : x, \delta) = H_y(x).$$

Applying Fubini's theorem to (d),

$$\begin{aligned} +\infty > A[G_R\mu - G_S\mu|_S : x, \delta] &= \int_S A[G_R(\cdot\,, y) - G_S(\cdot\,, y) : x, \delta]\, d\mu(y) \\ &= \int_S A[H_y : x, \delta]\, d\mu(y) \\ &= \int_S H_y(x)\, d\mu(y). \end{aligned}$$

The last integral on the right is a harmonic function of x by Lemma 6.7. Writing the last equation as

$$A(G_R\mu : x, \delta) = A(G_S\mu|_S : x, \delta) + \int_S H_y(x)\, d\mu(y)$$

and noting that the volume averages increase to $G_R\mu(x)$ and $G_S\mu|_S(x)$, respectively, as $\delta \downarrow 0$, we obtain

$$G_R\mu(x) = G_S\mu|_S(x) + \int_S H_y(x)\, d\mu(x)$$

on S.

We can now give the converse to Theorem 6.6.

THEOREM **6.9** Let R be an open set having a Green function G. If $G\mu$ is the potential of a measure μ and is harmonic on the open set $W \subset R$, then $\mu(W) = 0$.

Proof. Since $G\mu = G\mu|_W + G\mu|_{R\sim W}$ and $G\mu|_{R\sim W}$ is harmonic on W by Theorem 6.6, $G\mu|_W$ is harmonic on W. By the preceding theorem $G\mu|_W = G_W\mu|_W + h$, where G_W is the Green function for W and h is harmonic on W. This implies that $G_W\mu|_W$ is harmonic on W. By Corollary 6.5 $\mu|_W(W) = \mu(W) = 0$.

2. Total Sets of Continuous Functions

We shall digress in this section to develop some notions needed in the next section to show that if two potentials are equal then they come from the same measure.

Let R be an open subset of E^n. We shall let $\mathscr{K}^+ = \mathscr{K}_R^+$ be the set of functions f defined on E^n such that

(i) f is continuous,
(ii) $f \geq 0$,
(iii) f vanishes outside a compact subset of R.

For $f \in \mathscr{K}^+$, the closure of $\{x : f(x) \neq 0\}$ is called the **compact support** of f.

Definition. A subset $\mathscr{K}_0^+ \subset \mathscr{K}^+$ is called **total** if to each $v \in \mathscr{K}^+$ with compact support C, any neighborhood W of C, and any $\varepsilon > 0$ there corresponds a finite linear combination $f = \sum_{j=1}^p \alpha_j f_j$ of elements of $\mathscr{K}_0^+$, with $\alpha_j > 0$, $f_j \in \mathscr{K}_0^+$, and f_j having support in W for each j, such that

$$\|v - f\| = \sup_{x \in R} |v(x) - f(x)| < \varepsilon.$$

THEOREM **6.10** (*Cartan*) Let $\mathscr{K}_0^+$ be a class of functions in $\mathscr{K}^+$ with the following properties:

(i) If $f \in \mathscr{K}_0^+$, then every translate of f belongs to $\mathscr{K}_0^+$ provided it belongs to $\mathscr{K}^+$.
(ii) If $x \in R$ and $\delta > 0$, then there is an $f \in \mathscr{K}_0^+$ which vanishes outside $B_{x,\delta}$ but does not vanish identically.

Then $\mathscr{K}_0^+$ is a total subset of $\mathscr{K}^+$.

Proof. Without loss of generality, we can assume that $0 \in R$. Let v be an element of $\mathscr{K}^+$ with compact support C. We can assume that v is not the zero function. Let W be a neighborhood of C with $\overline{W} \subset R$ and let

$$\omega = d(C, R \sim W) > 0.$$

Let $\varepsilon > 0$. Since v is continuous and has compact support, there is a $\delta > 0$ (which we can assume to be less than $\omega/2$) such that $|v(x) - v(y)| < \varepsilon/2$ whenever $\|x - y\| < \delta$. Then $d(C, R \sim W) = \omega > 2\delta$. By (ii) there is an $f_\delta \in \mathscr{K}_0^+$ which vanishes outside $B_{0,\delta}$ but does not vanish identically. Since $f_\delta \geq 0$,

$$\alpha = \frac{1}{\int f_\delta(x)\, dz} > 0.$$

We can assume that $\int f_\delta(z)\, dz = 1$ by multiplying f_δ by the constant α if necessary and absorb α into the coefficients of the linear combination later. Since f_δ is continuous and has compact support, there is a $\rho > 0$ such that

$$|f_\delta(y) - f_\delta(x)| < \frac{\varepsilon}{2 \int v(z)\, dz}$$

whenever $\|y - x\| < \rho$. We can also assume that $\delta + \rho < \omega$. Moreover, there is a sequence $\{M_j\}$ of disjoint measurable sets such that $E^n = \bigcup_j M_j$, diam $M_j < \rho$, and thus $\delta + \text{diam } M_j < \delta + \rho < \omega$. We can assume that only finitely many M_j's have a nonempty intersection with C. Such a sequence of M_j's can be constructed by covering E^n by a sequence of balls each of diameter less than ρ with only finitely many intersecting C and then forming a disjoint sequence of sets from the sequence of balls. Define

$$h(x) = \int f_\delta(x - y)v(y)\, dy, \qquad x \in E^n.$$

The integral is always finite since the integrand vanishes outside a compact set and is continuous. We can also write

$$h(x) = \int f_\delta(y)v(x - y)\, dy = \int_{B_{0,\delta}} f_\delta(y)v(x - y)\, dy.$$

Since

$$v(x) = \int_{B_{0,\delta}} f_\delta(y)v(x)\, dy,$$

$$|h(x) - v(x)| \le \int_{B_{0,\delta}} f_\delta(y)|v(x - y) - v(x)|\, dy$$

Since $\|(x - y) - x\| = \|y\| < \delta$ for $y \in B_{0,\delta}$, $|v(x - y) - v(x)| < \varepsilon/2$ for $y \in B_{0,\delta}$. Therefore

$$|h(x) - v(x)| \le \frac{\varepsilon}{2} \int f_\delta(y)\, dy = \frac{\varepsilon}{2}$$

for all $x \in E^n$; that is, h approximates v uniformly within $\varepsilon/2$. We now show that h can be approximated uniformly within $\varepsilon/2$ by a finite linear combination of elements of $\mathscr{K}_0^+$. For each j, choose a fixed $y_j \in M_j$. Then for any $x \in E^n$

$$\begin{aligned}
&\left| h(x) - \sum_j f_\delta(x - y_j) \int_{M_j} v(y)\, dy \right| \\
&\quad = \left| \int f_\delta(x - y)v(y)\, dy - \sum_j f_\delta(x - y_j) \int_{M_j} v(y)\, dy \right| \\
&\quad = \left| \sum_j \int_{M_j} [f_\delta(x - y) - f_\delta(x - y_j)]v(y)\, dy \right|.
\end{aligned}$$

Since diam $M_j < \rho$ and $\|(x - y) - (x - y_j)\| = \|y_j - y\| < \rho$ for $y \in M_j$, the first factor in the integrand is bounded in absolute value by $\varepsilon/(2 \int v(y)\, dy)$. Therefore

$$\left| h(x) - \sum_j f_\delta(x - y_j) \int_{M_j} v(y)\, dy \right| \le \frac{\varepsilon}{2\int v(y)\, dy} \sum_j \int_{M_j} v(y)\, dy = \frac{\varepsilon}{2}.$$

This shows that h is approximated uniformly within $\varepsilon/2$ by

$$f(x) = \sum_j f_\delta(x - y_j) \int_{M_j} v(y)\, dy.$$

It follows that f approximates v uniformly within ε. Note that the sum defining f is really a finite sum since only a finite number of the M_j's intersect the support C of v. For the remaining M_j's, $\int_{M_j} v(y)\, dy = 0$. We need only show that each function

$$g_j(x) = f_\delta(x - y_j) \int_{M_j} v(y)\, dy$$

has support in W provided it is not identically zero. Suppose $M_j \cap C \neq \varnothing$ and $f_\delta(x - y_j) \neq 0$. Then $\|x - y_j\| < \delta$ and $d(x, C) \leq d(x, y_j) + d(y_j, C) < \delta + \rho < \omega$. Since $\omega = d(C, R \sim W)$, $x \notin R \sim W$. This shows that the support of those g_j's which are not identically zero is contained in W.

THEOREM **6.11** If R is an open subset of E^n, μ_1 and μ_2 are measures on R, and

$$\int_R f\, d\mu_1 = \int_R f\, d\mu_2$$

for all $f \in \mathscr{K}_0^+$, where $\mathscr{K}_0^+$ is a total subset of $\mathscr{K}^+ = \mathscr{K}_R^+$, then $\mu_1 = \mu_2$.

Proof. Since $g \in \mathscr{K}^+$ can be approximated uniformly by a finite linear combination of elements of $\mathscr{K}_0^+$, the above equation must hold for all $f \in \mathscr{K}^+$. Let C be a compact subset of R and let χ be its indicator function. There is then a sequence $\{\phi_j\}$ in $\mathscr{K}^+$ such that $\phi_j \downarrow \chi$; for example, let

$$\phi_j(x) = \max(2e^{-jd(x, C)} - 1, 0).$$

Since

$$\int_R \phi_j\, d\mu_1 = \int_R \phi_j\, d\mu_2,$$

$$\mu_1(C) = \int_R \chi\, d\mu_1 = \int_R \chi\, d\mu_2 = \mu_2(C)$$

for all compact sets $C \subset R$. Hence $\mu_1 = \mu_2$, since both are regular Borel measures.

3. Total Sets of Continuous Potentials

We shall look at some examples in order to construct a particular total set of functions. The first example is not a potential.

EXAMPLE **1** ($n = 2$) Consider a ball $B_{x,\delta} \subset E^2$, let $\nu_{x,\delta}$ be a unit measure uniformly distributed on $\partial B_{x,\delta}$, and let

$$u_{x,\delta}(y) = \int \log \frac{1}{\|y - z\|}\, d\nu_{x,\delta}(z), \qquad y \in E^2.$$

The measure $\nu_{x,\delta}$ is just surface area normalized by the total surface area of $\partial B_{x,\delta}$. From this point of view, $u_{x,\delta}(y)$ is just the average of u_y over $\partial B_{x,\delta}$. We shall now show that

$$u_{x,\delta}(y) = \begin{cases} \log \dfrac{1}{\delta} & \text{for} \quad \|y - x\| \le \delta \\[2ex] \log \dfrac{1}{\|x - y\|} & \text{for} \quad \|y - x\| \ge \delta. \end{cases}$$

As in the proof of Theorem 5.26, it is easily seen that $u_{x,\delta}$ is continuous. It is also easily seen that $u_{x,\delta}$ is a function only of $r = \|y - x\|$. If one takes the Laplacian of both sides of the equation defining $u_{x,\delta}$, then one can take the Laplacian under the integral and show that $u_{x,\delta}$ is harmonic on $B_{x,\delta}$. Since $u_{x,\delta}$ is a function only of $r = \|x - y\|$, it must be of the form $\alpha \log r + \beta$ on $B_{x,\delta}$; but since it is continuous on $B_{x,\delta}$, $\alpha = 0$ and $u_{x,\delta} = \beta$ on $B_{x,\delta}$. Since

$$u_{x,\delta}(x) = \int \log \frac{1}{\|x - z\|}\, d\nu_{x,\delta}(z) = \log \frac{1}{\delta},$$

$u_{x,\delta} = \log 1/\delta$ on $B_{x,\delta}$. Suppose $y \notin \bar{B}_{x,\delta}$. Then $u_{x,\delta}(y)$ is just the average over $\partial B_{x,\delta}$ of a function which is harmonic on a neighborhood of $\bar{B}_{x,\delta}$ and is therefore equal to its value at the center of the ball; that is, $u_{x,\delta}(y) = \log(\|y - x\|)^{-1}$ for $y \notin \bar{B}_{x,\delta}$.

EXAMPLE **2** ($n \ge 3$) Consider $B_{x,\delta} \subset E^n$, let $\nu_{x,\delta}$ be as in the preceding example, and let

$$u_{x,\delta}(y) = \int \frac{1}{\|y - z\|^{n-2}}\, d\nu_{x,\delta}(z).$$

As in the preceding example, we can show that

$$u_{x,\delta}(y) = \begin{cases} \dfrac{1}{\delta^{n-2}} & \text{if} \quad \|y - x\| \le \delta \\[2ex] \dfrac{1}{\|y - x\|^{n-2}} & \text{if} \quad \|y - x\| \ge \delta. \end{cases}$$

In this case $u_{x,\delta}$ is a potential and, in fact, $u_{x,\delta} = G\nu_{x,\delta}$ where G is the Green function for E^n.

The above computations make it easy to prove the following theorem for logarithmic and Newtonian potentials.

THEOREM **6.12** (*Gauss*) Let μ be a measure on E^n with compact support S_μ and let $B = B_{x,\delta}$. If $S_\mu \cap \bar{B} = \emptyset$, then the average of U^μ over ∂B is just $U^\mu(x)$; if $S_\mu \subset B$, the average of U^μ over ∂B depends only upon the total mass of μ and is equal to $\mu(S_\mu)\log 1/\delta$ in the $n = 2$ case and is equal to $\mu(S_\mu)/\delta^{n-2}$ in the $n \geq 3$ case.

Proof. Since μ has compact support,

$$U^\mu(y) = \int_{S_\mu} u_y(z)\, d\mu(z) = \int_{S_\mu} u_z(y)\, d\mu(z)$$

is superharmonic. Now

$$L(U^\mu : x, \delta) = \int_{S_\mu} L(u_z : x, \delta)\, d\mu(z).$$

If $S_\mu \cap \bar{B}_{x,\delta} = \emptyset$, then $L(u_z : x, \delta) = u_z(x) = u_x(z)$ for $z \in S_\mu$ and

$$L(U^\mu : x, \delta) = \int_{S_\mu} u_x(z)\, d\mu(z) = U^\mu(x).$$

If, however, $S_\mu \subset B_{x,\delta}$, then

$$L(U^\mu : x, \delta) = \int_{S_\mu} u_{x,\delta}(z)\, d\mu(z);$$

but for $z \in S_\mu \subset B_{x,\delta}$ we have seen in Examples 1 and 2 that $u_{x,\delta}(z)$ is equal to $\log 1/\delta$ or $1/\delta^{n-2}$ according as $n = 2$ or $n \geq 3$. Therefore $L(U^\mu : x, \delta)$ is equal to $\mu(S_\mu)(\log 1/\delta)$ or $\mu(S_\mu)/\delta^{n-2}$ according as $n = 2$ or $n \geq 3$.

EXAMPLE **3** ($n = 2$) Let R be an open set having a Green function G, let $\bar{B}_{x,\delta} \subset R$, and let $v_{x,\delta/2}$, $v_{x,\delta}$ be defined as before. Put $\tau_{x,\delta} = v_{x,\delta/2} - v_{x,\delta}$. Then $G\tau_{x,\delta}$ is a continuous Green potential with

$$G\tau_{x,\delta} \begin{cases} \geq & \text{on} \quad R \\ =0 & \text{on} \quad R \sim B_{x,\delta} \\ \not\equiv 0. & \end{cases}$$

To see this, consider $u_{x,\delta/2} - u_{x,\delta}$. Then

(i) for $y \in B_{x,\delta/2}$

$$u_{x,\delta/2}(y) - u_{x,\delta}(y) = \log \frac{1}{\delta/2} - \log \frac{1}{\delta} = \log 2 > 0;$$

(ii) for $y \in \bar{B}_{x,\delta} \sim B_{x,\delta/2}$

$$u_{x,\delta/2}(y) - u_{x,\delta}(y) = \log \frac{1}{\|x-y\|} - \log \frac{1}{\delta} = -\log \frac{\|x-y\|}{\delta} \geq 0;$$

(iii) for $y \notin \bar{B}_{x,\delta}$,

$$u_{x,\delta/2}(y) - u_{x,\delta}(y) = \log \frac{1}{\|x-y\|} - \log \frac{1}{\|x-y\|} = 0.$$

Since $G(y, \cdot) = u_y + h_y$, where h_y is harmonic on R,

$$G\tau_{x,\delta}(y) = \int [u_y(z) + h_y(z)] \, dv_{x,\delta/2}(z) - \int [u_y(z) + h_y(z)] \, dv_{x,\delta}(z)$$

$$= u_{x,\delta/2}(y) - u_{x,\delta}(y) + \int h_y(z) \, dv_{x,\delta/2}(z) - \int h_y(z) \, dv_{x,\delta}(z).$$

Since the last two integrals are just averages of the harmonic function h_y, their difference is $h_y(x) - h_y(x) = 0$. Therefore $G\tau_{x,\delta} = u_{x,\delta/2} - u_{x/\delta}$ on R.

EXAMPLE 4 ($n \geq 3$) Let R be an open subset of E^n with Green function G, let $\bar{B}_{x,\delta} \subset R$, and let $v_{x,\delta/2}$, $v_{x,\delta}$ be defined as before. If $\tau_{x,\delta} = v_{x,\delta/2} - v_{x,\delta}$, then $G\tau_{x,\delta}$ is a Green potential having the same properties as in the preceding example.

THEOREM 6.13 (*Cartan*) Let R be an open subset of E^n having a Green function G. Then the set of Green potentials $\mathcal{K}_0^+ = \{G\tau_{x,\delta} : \bar{B}_{x,\delta} \subset R\}$ is a total subset of $\mathcal{K}_R^+$.

There is nothing to prove in view of the preceding examples and Theorem 6.10.

4. Riesz Decomposition of Superharmonic Functions

In this section we show that a non-negative superharmonic function can be represented as the sum of a potential and a harmonic function. We begin with the potential theory version of Fubini's theorem.

THEOREM 6.14 (*Reciprocity Theorem*) Let R be an open subset of E^n having a Green function G and let μ, ν be two measures on the Borel subsets of R. Then

$$\int_R G\mu \, d\nu = \int_R G\nu \, d\mu.$$

Proof. By Fubini's theorem and the symmetry of the Green function

$$\begin{aligned}\int_R G\mu \, d\nu &= \int_R \left(\int_R G(x, y) \, d\mu(y) \right) d\nu(x) \\ &= \int_R \left(\int_R G(x, y) \, d\nu(x) \right) d\mu(y) \\ &= \int_R \left(\int_R G(y, x) \, d\nu(x) \right) d\mu(y) \\ &= \int_R G\nu(y) \, d\mu(y).\end{aligned}$$

Note that the reciprocity theorem holds even if the integral of $G\mu$ relative to ν is infinite, in which case the same is true of $G\nu$ relative to μ.

THEOREM **6.15** Let μ and ν be measures on the open set R having a Green function G such that $G\mu$ and $G\nu$ are potentials. If $G\mu = G\nu$ at points at which both are finite-valued, then $\mu = \nu$. More generally, let W be an open subset of R such that $G\mu - G\nu = h$ on the subset of W of points of finiteness of both $G\mu$ and $G\nu$ where h is harmonic on W. Then μ and ν are identical on Borel subsets of W.

Proof. Let Σ be the set of points of W where $G\mu$ and $G\nu$ are both finite. For $x \in \Sigma$, $G\mu(x) = G\nu(x) + h(x)$. Since $G\mu = G\mu|_W + G\mu|_{R\sim W}$ and $G\mu|_{R\sim W}$ is harmonic on W, $G\mu(x) = G\mu|_W(x) + h'(x)$, $x \in \Sigma$, where h' is harmonic on W. Applying the same argument to $G\nu$, $G\nu(x) = G\nu|_W(x) + h''(x)$, $x \in \Sigma$, where h'' is harmonic on W. Letting $\tilde{\mu} = \mu|_W$, $\tilde{\nu} = \nu|_W$, then $G\tilde{\mu}(x) = G\tilde{\nu}(x) + h^*(x)$, $x \in \Sigma$, where h^* is harmonic on W. Since $W \sim \Sigma$ has Lebesgue measure zero,

$$A(G\tilde{\mu} : x, \delta) = A(G\tilde{\nu} : x, \delta) + h^*(x)$$

whenever $\bar{B}_{x,\delta} \subset W$. Letting $\delta \downarrow 0$, $G\tilde{\mu}(x) = G\tilde{\nu}(x) + h^*(x)$ for all $x \in W$. Let G_W be the Green function of W which exists since W is an open subset of a set with a Green function. Applying Theorem 6.8 to $G\tilde{\mu}$ and $G\tilde{\nu}$, there is a harmonic function h^{**} on W such that $G_W \tilde{\mu} = G_W \tilde{\nu} + h^{**}$ on W. For $\bar{B}_{x,\delta} \subset W$,

$$L(G_W \tilde{\mu} : x, \delta) = L(G_W \tilde{\nu} : x, \delta) + h^{**}(x).$$

Consider the surface average on the left. Letting $G_W(x, y) = u_x(y) + h_x(y)$, we have

$$G_W \tilde{\mu}(z) = \int_W u_z(y) \, d\tilde{\mu}(y) + \int_W h_z(y) \, d\tilde{\mu}(y).$$

By Fubini's theorem and the symmetry of the two integrands

$$L(G_W \tilde{\mu} : x, \delta) = \int_W L(u_y : x, \delta)\, d\tilde{\mu}(y) + \int_W L(h_y : x, \delta)\, d\tilde{\mu}(y)$$

$$= \int_W u_{x,\delta}(y)\, d\tilde{\mu}(y) + \int_W h_x(y)\, d\tilde{\mu}(y).$$

Applying the same argument to $L(G_W \tilde{\nu} : x, \delta)$, we obtain

$$\int_W u_{x,\delta}(y)\, d\tilde{\mu}(y) + \int_W h_x(y)\, d\tilde{\mu}(y) = \int_W u_{x,\delta}(y)\, d\tilde{\nu}(y) + \int_W h_x(y)\, d\tilde{\nu}(y) + h^{**}(x).$$

Replacing δ with $\delta/2$ and subtracting results in

$$\int_W [u_{x,\delta}(y) - u_{x,\delta/2}(y)]\, d\tilde{\mu}(y) = \int_W [u_{x,\delta}(y) - u_{x,\delta/2}(y)]\, d\tilde{\nu}(y).$$

Since the functions $u_{x,\delta/2} - u_{x,\delta}$ form a total subset of $\mathscr{K}_W^+$, $\tilde{\mu} = \tilde{\nu}$ by Theorem 6.11. Therefore μ and ν agree on the Borel subsets of W.

We now return to the Riesz decomposition theorem discussed in the introduction to this chapter.

LEMMA **6.16** Let u be superharmonic on the open set R and let B be a ball with $\bar{B} \subset R$. Then there is a unique measure μ on B such that $u = G_B \mu + h$ where h is the greatest harmonic minorant of u on B.

Proof. The proof is carried out only for the $n \geq 3$ case, the $n = 2$ case differing only by certain constants appearing before integrals. Since we are only concerned with a representation of u on B, we can assume that R is also a ball on which u is bounded below. We can also assume that u is harmonic on $R \sim \bar{B}$ for the following reason. Let v be the reduction of u over $R \sim \bar{B}$. Since u is bounded below on R, v is harmonic on $R \sim \bar{B}$ by Theorem 5.19. Moreover, $v = u$ on B by the same theorem. The function v itself need not be superharmonic on R but the lower regularization $\hat{v}$ of v is superharmonic on R by Theorem 4.16 since v is the limit of a decreasing sequence of superharmonic functions. Now $\hat{v}$ differs from v only on ∂B. Therefore $\hat{v}$ is superharmonic on R, $\hat{v} = u$ on B, and $\hat{v}$ is harmonic on $R \sim \bar{B}$. Insofar as the representation of u over B is concerned, we can assume that u has the same properties as $\hat{v}$; that is, u is harmonic on $R \sim \bar{B}$. Let B_0 be a ball such that $\bar{B} \subset B_0 \subset \bar{B}_0 \subset R$ and let V be a neighborhood of $\bar{B}$ such that $\bar{V} \subset B_0$. By Theorem 4.21 there is a sequence $\{u_j\}$ of functions superharmonic on R such that

(1) $u_j \uparrow u$ as $j \to \infty$,
(2) each u_j has continuous second partials on R,
(3) $u_j = u$ on $R \sim \bar{V}$ for all sufficiently large j,
(4) u_j is harmonic on $R \sim \bar{V}$ for all sufficiently large j.

By discarding a finite number of u_j's we can assume that (3) and (4) hold for all j. In particular, $u_j = u$ on ∂B_0 for all j. By Theorem 5.1 for $x \in B_0$

$$u_j(x) = -\frac{1}{\sigma_n(n-2)} \int_{\partial B_0} u_j(z) D_{\mathbf{n}} G_0(x, z)\, d\sigma(z)$$
$$- \frac{1}{\sigma_n(n-2)} \int_{B_0} G_0(x, z)\, \Delta u_j(z)\, dz,$$

where G_0 is the Green function for the ball B_0. Since $u_j = u$ on ∂B_0, the first integral on the right does not depend upon j and is, in fact, $\mathbf{PI}(u, B_0)$ which shall be denoted by h_0. If M is a Borel subset of $\bar{B}_0$, let

$$v_j(M) = -\frac{1}{\sigma_n(n-2)} \int_M \Delta u_j(z)\, dz.$$

Since u_j is superharmonic on R, $\Delta u_j \le 0$ and v_j is a measure on the Borel subsets of $\bar{B}_0$ with $v_j(\partial B_0) = 0$. The above representation of u_j can then be written $u_j = G_0 v_j + h_0$ on B_0. Since $u_j \uparrow u$, $w = \lim_{j \to \infty} G_0 v_j$ is defined on B_0 and is, in fact, equal to $u - h_0$. We shall now show that w is the potential of a measure on B_0. Consider $\|v_j\| = v_j(B_0)$. By Green's identity we have

$$v_j(B_0) = -\frac{1}{\sigma_n(n-2)} \int_{B_0} \Delta u_j(z)\, dz = -\frac{1}{\sigma_n(n-2)} \int_{\partial B_0} D_{\mathbf{n}} u_j(z)\, d\sigma(z);$$

but, since $u_j = u$ outside $\bar{V}$, $D_{\mathbf{n}} u_j(z) = D_{\mathbf{n}} u(z)$ on ∂B_0 and the latter integral is independent of j. Therefore $\|v_j\|$ is independent of j and there is a subsequence of the sequence $\{v_j\}$ which converges to a measure v on $\bar{B}_0$ in the w^*-topology. Since we are going to let $j \to \infty$, we can assume that the sequence itself has the property. We cannot simply let $j \to \infty$ in the equation

$$u_j(x) = \int_{B_0} G_0(x, z)\, dv_j(z) + h_0(x), \qquad x \in B_0,$$

since the integrand is not a bounded continuous function. This difficulty can be circumvented as follows. Suppose $\bar{B}_{x,\delta} \subset B_0$. By Fubini's theorem

$$L(G_0 v_j : x, \delta) = \int_{B_0} L(G_0(\cdot, z) : x, \delta)\, dv_j(z).$$

By Theorem 5.26, $L(G_0(\cdot, z) : x, \delta)$ is a continuous function of $z \in B_0$ and, in fact, has a continuous extension to $\bar{B}_0$ since $L(G_0(\cdot, z) : x, \delta) = G_0(x, z)$ for $z \in B_0 \sim \bar{B}_{x,\delta}$. Since

$$L(u_j : x, \delta) = \int_{B_0} L(G_0(\cdot, z) : x, \delta)\, dv_j(z) + h_0(x),$$

$\nu_j \xrightarrow{w^*} \nu$, and $u_j \uparrow u$,

$$L(u : x, \delta) = \int_{B_0} L(G_0(\cdot, z) : x, \delta)\, d\nu(z) + h_0(x).$$

Since u and $G_0(\cdot, z)$ are superharmonic, the averages increase to $u(x)$ and $G_0(x, z)$, respectively, as $\delta \downarrow 0$ and $u = G_0 \nu + h_0$ on B_0. Since $G_0 \nu = G_0 \nu|_B + G_0 \nu|_{B_0 \sim B}$ and $G_0 \nu|_{B_0 \sim B}$ is harmonic on B, $u = G_0 \nu|_B + h_1$ on B where h_1 is harmonic on B. By Theorem 6.8, $G_0 \nu|_B$ differs from $G_B \nu|_B$ by a function which is harmonic on B. Letting $\mu = \nu|_B$, it follows that $u = G_B \mu + h$ where h is harmonic on B. Since the greatest harmonic minorant of $G_B \mu$ on B is the zero function according to Theorem 6.4, the greatest harmonic minorant of u on B is h by Lemma 5.22. Let $u = G_B \tilde{\mu} + \tilde{h}$ be a second such representation. Since the greatest harmonic minorant of u on B is unique, $h = \tilde{h}$ and $G_B \mu = G_B \tilde{\mu}$ on B. It follows from Theorem 6.15 that $\mu = \tilde{\mu}$.

LEMMA **6.17** Let u be superharmonic on the open set R and let W be an open subset of R with compact closure $\overline{W} \subset R$. Then there is a unique measure μ on W such that $u = G_W \mu + h$, where h is the greatest harmonic minorant of u on W.

Proof. Suppose first that W is a subset of a ball B for which $\overline{B} \subset R$. According to the preceding lemma, there is a measure ν on B such that $u = G_B \nu + h_0$, where h_0 is the greatest harmonic minorant of u on B. Since $G_B \nu = G_B \nu|_W + G_B \nu|_{B \sim W}$ and the latter is harmonic on W, $u = G_B \nu|_W + h_1$, where h_1 is harmonic on W. Using the fact that $G_B \nu|_W$ differs from $G_W \nu|_W$ on W by a harmonic function, $u = G_W \nu|_W + h_2$, where h_2 is harmonic on W. Letting $\mu = \nu|_W$ and $h = h_2$, the assertion is true for this particular case. Returning to the general case, suppose only that $\overline{W}$ is a compact subset of R. Then $W = \bigcup_{i=1}^p W_i$, where each W_i is an open subset of a ball with closure in R. We have just shown the assertion to be true if $p = 1$. Since the proof is by induction on p, it suffices to consider the case $W = U_1 \cup U_2$, where U_1 and U_2 are open sets with compact closures in R for which the assertion is true. There are then measures μ_i on U_i and harmonic functions h_i on U_i, $i = 1, 2$, such that

$$\text{(e)} \qquad \begin{aligned} u &= G_{U_1}\mu_1 + h_1 \quad \text{on} \quad U_1, \\ u &= G_{U_2}\mu_2 + h_2 \quad \text{on} \quad U_2. \end{aligned}$$

Let $U = U_1 \cap U_2$. Comparing potentials as above, there is a harmonic function h_1^* on U such that $u = G_U \mu_1|_U + h_1^*$ on U. Similarly, $u = G_U \mu_2|_U + h_2^*$ with h_2^* harmonic on U. It follows from Theorem 6.15 that $\mu_1|_U = \mu_2|_U$. There is therefore a measure μ defined on the Borel subsets of $W = U_1 \cup U_2$ such that $\mu_1 = \mu|_{U_1}$ and $\mu_2 = \mu|_{U_2}$. Since $G_W \mu = G_W \mu|_{U_1} + G_W \mu|_{W\sim U_1}$ and the latter is harmonic on U_1, $G_W \mu = G_W \mu_1 + H_1$ on U_1 with H_1 harmonic. Comparing $G_W \mu_1$ with $G_{U_1} \mu_1$, there is a harmonic function H_1^* on U_1 such that $G_W \mu = G_{U_1} \mu_1 + H_1^*$ on U. Similarly, $G_W \mu = G_{U_2} \mu_2 + H_2^*$ on U_2 with H_2 harmonic. It follows from (e) that

$$u = G_W \mu + H_1^{**} \quad \text{on} \quad U_1,$$

$$u = G_W \mu + H_2^{**} \quad \text{on} \quad U_2,$$

where H_i^{**} is harmonic on U_i, $i = 1, 2$. It follows that $H_1^{**} = H_2^{**}$ on $U_1 \cap U_2$ and we can define a harmonic function h on $W = U_1 \cup U_2$ such that $h|_{U_i} = H_i^{**}$, $i = 1, 2$. The above equations then become simply $u = G_W \mu + h$. As in the preceding proof, h is the greatest harmonic minorant of u on W and μ is unique.

THEOREM **6.18** (*Riesz Decomposition Theorem*) Let R be an open subset of E^n having a Green function G and let u be superharmonic on R. Then there is a unique measure ν on R such that if W is an open subset with compact closure in R, then $u = G_W \nu|_W + h_W$, where h_W is the greatest harmonic minorant of u on W; if, in addition, $u \geq 0$ on R, then $u = G\nu + h$, where h is the greatest harmonic minorant of u on R.

Proof. Let $\{R_j\}$ be an increasing sequence of open sets with compact closures in R such that $R = \bigcup_j R_j$. Let ν_j and h_j be the measure and harmonic function on R_j, respectively, of the preceding lemma. Then $u = G_{R_j} \nu_j + h_j$ on R_j. Now ν_j and ν_{j+k} must agree on the Borel subsets of R_j for every $k \geq 1$. If M is any Borel subset of R, then $\{\nu_j(R_j \cap M)\}$ is a nondecreasing sequence and we can define $\nu(M) = \lim_{j\to\infty} \nu_j(R_j \cap M)$. If we extend ν_j to be zero on $\sim R_j$, then ν is the limit of an increasing sequence of measures and is therefore a measure. If M is a Borel subset of R_j, then $\nu_j(R_j \cap M) = \nu_j(M) = \nu_{j+k}(M) = \nu_{j+k}(R_{j+k} \cap M)$ for $k \geq 1$ and therefore $\nu(M) = \nu_j(M)$; that is, $\nu_j = \nu|_{R_j}$. Now let W be any open set with compact closure $\overline{W} \subset R$ and let $u = G_W \nu_W + h_W$ be the decomposition of the preceding lemma. Let j_0 be such that $W \subset R_j$ for all $j \geq j_0$ and consider only such j. Since $u = G_{R_j} \nu_j + h_j$ on R_j with h_j harmonic on R_j and $G_{R_j} \nu_j = G_{R_j} \nu_j|_W + G_{R_j} \nu_j|_{R_j\sim W}$ with the latter potential harmonic on W, $u = G_{R_j} \nu_j|_W + h_j^*$ with h_j^* harmonic on W. Comparing $G_{R_j} \nu_j|_W$ with $G_W \nu_j|_W$, $u = G_W \nu_j|_W + h_j^{**}$ where h_j^{**} is harmonic on W. It follows from the uniqueness of the representation of u on W that $\nu_W = \nu_j|_W = \nu|_W$. This completes the proof of the first assertion. Lastly, suppose $u \geq 0$ on R. Then $u = G_{R_j} \nu|_{R_j} + h_j$ on R_j with h_j harmonic thereon. Since $G_{R_j} \leq G_{R_{j+1}}$ on $R_j \times R_j$, for $x \in R_j$

$$G_{R_j}\nu|_{R_j}(x) = \int_{R_j} G_{R_j}(x, y)\, d\nu(y) \le \int_{R_j} G_{R_{j+1}}(x, y)\, d\nu(y)$$

$$\le \int_{R_{j+1}} G_{R_{j+1}}(x, y)\, d\nu(y) = G_{R_{j+1}}\nu|_{R_{j+1}}(x)$$

and the sequence $\{G_{R_{j+k}}\nu|_{R_{j+k}}\}_{k\ge 1}$ increases on R_j. In fact, for $x \in R_j$

$$G_{R_{j+k}}\nu|_{R_{j+k}}(x) = \int \chi_{R_{j+k}}(y) G_{R_{j+k}}(x, y)\, d\nu(y) \uparrow \int G(x, y)\, d\nu(y) = G\nu(x)$$

by Theorem 5.15. Since $0 \le u = G_{R_j}\nu|_{R_j} + h_j$ on R_j, it follows by taking greatest harmonic minorants that $h_j \ge 0$. The sequence $\{h_{j+k}\}_{k\ge 1}$ is a decreasing sequence of non-negative harmonic functions on R_j which has a limit which is harmonic on R_j. We can therefore define a harmonic function $h = \lim_{k\to\infty} h_{j+k}$ on R_j, for all j. It follows that $u = G\nu + h$ on R. Since the greatest harmonic minorant of $G\nu$ on R is the zero function, h is the greatest harmonic minorant of u.

COROLLARY **6.19** A positive superharmonic function defined on an open set R having a Green function is a potential of a measure if and only if its greatest harmonic minorant is zero.

Proof. The necessity follows from Theorem 6.4. The sufficiency follows from the Riesz decomposition theorem.

5. Continuity Properties of Potentials

We have seen that it is possible to approximate a superharmonic function by a finite-valued smooth superharmonic function. We shall now show that the discontinuities of the potential of a measure can be removed by giving up a small part of the measure. This is the point of the second theorem of this section. The first theorem is used to prove the second but is of interest in its own right.

THEOREM **6.20** (*Evans and Vasilesco*) Let R be an open set having a Green function G, let μ be a measure with compact support $K \subset R$, and let $v = G\mu$. If $v|_K$ is continuous at $x_0 \in K$, then v is continuous at x_0.

Proof. We shall carry out the proof for the $n \ge 3$ case. If $v(x_0) = +\infty$, then the result is trivially true by the l.s.c. of v. We can therefore assume that $v(x_0) < +\infty$. The Newtonian potential U^μ of the measure μ is superharmonic by Theorem 6.3. By Theorem 6.8 U^μ and v differ by a function which is harmonic on R. Since $v|_K$ is continuous at $x_0 \in K$, the same is true of $U^\mu|_K$. If we can show that U^μ is continuous at x_0, the same will be true of v. We first define a mapping of R onto K in the following way. If $x \in R$, let z_x be a point of K nearest x. The point z_x need not be unique, but there are various ways of making the map single-valued. We shall suppose that this has been done. Since $x_0 \in K$,

$$\|z_x - x_0\| \leq \|z_x - x\| + \|x - x_0\| \leq 2\|x - x_0\|$$

for any $x \in R$. As $x \to x_0$, it follows that $z_x \to x_0$. For $y \in K$, $\|x - z_x\| \leq \|x - y\|$ and

$$\|z_x - y\| \leq \|z_x - x\| + \|x - y\| \leq 2\|x - y\|.$$

Therefore

$$\frac{1}{\|x - y\|^{n-2}} \leq 2^{n-2} \frac{1}{\|z_x - y\|^{n-2}}.$$

If ν is any measure with support in K, then by integrating

$$U^\nu(x) = \int \frac{1}{\|x - y\|^{n-2}} \, d\nu(y) \leq 2^{n-2} \int \frac{1}{\|z_x - y\|^{n-2}} \, d\nu(y) = 2^{n-2} \, U^\nu(z_x).$$

Suppose $\varepsilon > 0$. Since $U^\mu(x_0) < +\infty$, there is a ball $B \subset R$ with center at x_0 such that

$$\int_B \frac{1}{\|x_0 - y\|^{n-2}} \, d\mu(y) < \varepsilon$$

by the continuity of the indefinite integral with respect to μ. Let $\mu' = \mu\,|_B$ and $\mu'' = \mu\,|_{R \sim B}$. By Theorem 6.6 $U^{\mu''}$ is harmonic on B and, in particular, continuous at x_0. Now $U^\mu = U^{\mu'} + U^{\mu''}$ and

$$\begin{aligned} |U^\mu(x) - U^\mu(x_0)| &\leq |U^{\mu''}(x) - U^{\mu''}(x_0)| + U^{\mu'}(x) + U^{\mu'}(x_0) \\ &\leq |U^{\mu''}(x) - U^{\mu''}(x_0)| + 2^{n-2} U^{\mu'}(z_x) + \varepsilon. \end{aligned}$$

Since $U^\mu = U^{\mu'} + U^{\mu''}$ with $U^\mu|_K$ continuous at x_0 and $U^{\mu''}$ continuous at x_0, $U^{\mu'}$ is continuous at x_0 when restricted to K. Since $z_x \in K$ and $z_x \to x_0$ as $x \to x_0$, $\lim_{x \to x_0} U^{\mu'}(z_x) = U^{\mu'}(x_0) < \varepsilon$. Therefore, for all x sufficiently close to x_0,

$$|U^\mu(x) - U^\mu(x_0)| \leq |U^{\mu''}(x) - U^{\mu''}(x_0)| + 2^{n-2}\varepsilon + \varepsilon.$$

By the continuity of $U^{\mu''}$ at x_0, the first term on the right can be made arbitrarily small by taking x sufficiently close to x_0. Since ε is arbitrary, it follows that U^μ is continuous at x_0. Since U^μ differs from v by a function which is harmonic on R, v is continuous at x_0.

THEOREM **6.21** Let R be an open set having a Green function G and let μ be a measure with compact support $K \subset R$ such that $G\mu < +\infty$ on K. Then given any $\varepsilon > 0$, there is a compact set $C \subset K$ such that (i) $\mu(K \sim C) < \varepsilon$, (ii) $G\mu\,|_C < +\infty$ on C, and (iii) $G\mu\,|_C$ is continuous on R.

Proof. Consider $G\mu$ which is finite on K. By Lusin's theorem, given $\varepsilon > 0$ there is a compact set $C \subset K$ such that $\mu(K \sim C) < \varepsilon$ and such that the restriction of $G\mu$ to C is continuous on C. Now $G\mu = G\mu\,|_C + G\mu\,|_{K \sim C}$. Since

$G\mu|_{K\sim C}$ is l.s.c., $G\mu|_C = G\mu - G\mu|_{K\sim C}$ is u.s.c. on C when restricted to C. Since $G\mu|_C$ is also l.s.c. on C, $G\mu|_C$ restricted to C is continuous on C. By Theorem 6.20 $G\mu|_C$ is continuous on C. Since $G\mu|_C$ is harmonic on $R \sim C$, $G\mu|_C$ is continuous on R.

Although little can be said about the continuity properties of Green potentials in general, something can be said if the measures are absolutely continuous with respect to Lebesgue measure. If R is an open subset of E^n having a Green function G and f is a measurable function on R, we shall let $Gf(y) = \int_R G(y, z)f(z)\,dz$ provided the integral is defined for all $y \in R$.

THEOREM **6.22** If R is an open set having a Green function G and f is a bounded Lebesgue integrable function on R, then Gf is continuous on R.

Proof. Since the positive and negative parts f^+ and f^- are integrable over R and $Gf = Gf^+ - Gf^-$, it can be assumed that $0 \leq f \leq m$ for some constant m. By Theorem 6.3 Gf is a potential. Let B be a ball with $\bar{B} \subset R$ and radius less than 1/2. Then $Gf = Gf\chi_B + Gf\chi_{R\sim B}$. Since the density $f\chi_{R\sim B}$ assigns zero mass to B, $Gf\chi_{R\sim B}$ is harmonic on B by Theorem 6.6 and therefore continuous on B. It remains only to investigate $Gf\chi_B$. Let $B = B_{y,\rho}$. If we can show that $Gf\chi_B$ is continuous on $B \sim \{y\}$, it would follow that Gf is continuous on $B \sim \{y\}$; but, since B is an arbitrary ball in R of radius less than 1/2, this would imply that Gf is continuous on R. Since $Gf\chi_B$ differs from $G_B f$ by a function harmonic on B by Theorem 6.8, it suffices to prove that $G_B f$ is continuous on $B \sim \{y\}$. For the remainder of the proof it is necessary to consider the $n = 2$ and $n \geq 3$ cases separately. Suppose $n = 2$ and $x_0 \in B \sim \{y\}$ and consider any ball $B_{x_0,r}$ whose closure is in B. Since

$$G_B(x, z) = \log \frac{\|y - x\|}{\rho} \frac{\|z - x^\star\|}{\|z - x\|}$$

for $z \in B \sim \{x\}$, where $x^\star$ is the inverse of x relative to ∂B,

$$|G_B f(x) - G_B f(x_0)| \leq \left| \int_{B\sim B_{x_0,r}} \log \frac{\|y - x\|\,\|z - x^\star\|\,\|z - x_0\|}{\|y - x_0\|\,\|z - x\|\,\|z - x_0^\star\|} f(z)\,dz \right|$$

$$+ \int_{B_{x_0,r}} G_B(x, z)f(z)\,dz + \int_{B_{x_0,r}} G_B(x_0, z)f(z)\,dz.$$

Recall that $G_B(x, z)$ is minimal in the class of non-negative functions of the form $-\log\|x - z\| + w_x(z)$ where w_x is superharmonic on B. Since $\|x - z\| < 1$ for $x, z \in B$, $-\log\|x - z\|$ is positive and majorizes $G_B(x, z)$. Then for any $x \in B$

$$\int_{B_{x_0,r}} G_B(x, z)f(z)\,dz \leq m \int_{B_{x_0,r}} -\log\|x - z\|\,dz.$$

The integral on the right will be increased if we integrate over $B_{x,r}$ rather than $B_{x_0,r}$. Thus

$$\int_{B_{x_0,r}} G_B(x,z)f(z)\,dz \le m\int_{B_{x,r}} -\log\|x-z\|\,dz = 2\pi m\int_0^r -\rho\log\rho\,d\rho.$$

Since the integral on the right approaches zero as $r\to 0$, given $\varepsilon>0$ there is an r' such that

$$\int_{B_{x_0,r'}} G_B(x,z)f(z)\,dz < \frac{\varepsilon}{2}$$

for any $x\in B$. It follows that

$$|G_Bf(x)-G_Bf(x_0)| \le \left|\int_{B\sim B_{x_0,r'}} \log\frac{\|y-x\|\,\|z-x^\star\|\,\|z-x_0\|}{\|y-x_0\|\,\|z-x\|\,\|z-x_0^\star\|} f(z)\,dz\right| + \varepsilon.$$

Consider only those x for which $x\in B_{x_0,r'/2}$. Then the logarithmic factor in the integrand approaches zero boundedly as $x\to x_0$. By the Lebesgue dominated convergence theorem $\limsup_{x\to x_0}|G_Bf(x)-G_Bf(x_0)|\le\varepsilon$, but, since ε is arbitrary, $\lim_{x\to x_0} G_Bf(x)=G_Bf(x_0)$ and G_Bf is continuous on $B\sim\{y\}$. The proof for the $n\ge 3$ case is just the same except for using the fact that $G_B(x,z)\le\|x-z\|^{-n+2}$ and the estimate

$$\int_{B_{x_0,r}} G_B(x,z)f(z)\,dz \le \sigma_n m\int_0^r \rho\,d\rho.$$

6. Boundary Behavior of Potentials

Pending a study of the Dirichlet problem for general regions, we can prove only relatively simple results concerning the boundary behavior of potentials. It is not possible to get the same nice results for potentials as for harmonic functions, even on a ball, but there is an important case for the ball that can be handled.

THEOREM **6.23** If μ is a measure with compact support K in a ball B, then $\lim_{x\to x_0} G_B\mu(x)=0$ for all $x_0\in\partial B$.

Proof. Only the $n=2$ case is proved, since the $n\ge 3$ case is essentially the same. Let $B=B_{y,\rho}$ and let z_0 be a fixed point of K. Moreover, let V be a neighborhood of K such that $\overline{V}\subset B$. For $x\in B\sim\overline{V}$ the functions $G_B(x,\cdot)$ are strictly positive harmonic functions on V. By Harnack's inequality there is a constant k (depending only upon V and K) such that $G_B(x,z)\le kG_B(x,z_0)$ for all $x\in B\sim\overline{V}$ and $z\in K$. Since $G_B(x,z_0)\to 0$ as $x\to x_0\in\partial B$ by Lemma 5.3, $G_B(x,\cdot)\to 0$ as $x\to x_0\in\partial B$ uniformly on K. Since $G_B\mu(x)=\int_K G_B(x,z)\,d\mu(z)$, $G_B\mu(x)\to 0$ as $x\to x_0\in\partial B$.

The requirement of compact support can be relaxed provided the measure has a density.

LEMMA **6.24** Let R be a connected open set having a Green function G, finite Lebesgue measure if $n \geq 3$, and bounded if $n = 2$. If f is a bounded measurable function on R and $\lim_{x \to x_0 \in \partial R} G(x, z_0) = 0$ for some $z_0 \in R$, then $\lim_{x \to x_0 \in \partial R} Gf(x) = 0$.

Proof. Suppose $n \geq 3$. By consideration of the positive and negative parts of f we can assume that $0 \leq f \leq m$. Since f is Lebesgue integrable on R, Gf is a potential. Let K_j be an increasing sequence of compact sets in R such that $K_j \uparrow R$ and let

$$\phi_j(x) = \int_{K_j} G(x, z) f(z)\, dz.$$

Then $\phi_j \uparrow Gf$. Given $\varepsilon > 0$, choose j_0 such that the Lebesgue measure λ_0 of $R \sim K_{j_0}$ is less than ε and let α be the Lebesgue measure of R. Since $G(x, z)$ is majorized by $\|x - z\|^{-n+2}$,

$$\begin{aligned} |\phi_{j_0+j}(x) - \phi_{j_0}(x)| &\leq \int_{K_{j_0+j} \sim K_{j_0}} \frac{1}{\|x - z\|^{n-2}} f(z)\, dz \\ &\leq m \int_{R \sim K_{j_0}} \frac{1}{\|x - z\|^{n-2}}\, dz. \end{aligned}$$

Choose ρ such that the volume of $B_{x,\rho}$ is λ_0; that is, $\lambda_0 = v_n \rho^n$ or $\rho = (\lambda_0/v_n)^{1/n}$. Then

$$\begin{aligned} |\phi_{j_0+j}(x) - \phi_{j_0}(x)| &\leq m \int_{R \sim K_{j_0}} \frac{1}{\|x - z\|^{n-2}}\, dz \\ &\leq m \int_{B_{x,\rho}} \frac{1}{\|x - z\|^{n-2}}\, dz \\ &= \frac{1}{2} \sigma_n m \rho^2 \\ &= \frac{1}{2} \sigma_n m \left(\frac{\lambda_0}{v_n}\right)^{2/n} \\ &\leq \frac{1}{2} \sigma_n m \left(\frac{\varepsilon}{v_n}\right)^{2/n} \end{aligned}$$

uniformly with respect to $x \in R$ and $j \geq 1$. Since $\lim_{x \to x_0 \in \partial R} G(x, z) = 0$ uniformly with respect to $z \in K_{j_0}$, there is a $\delta > 0$ such that $G(x, z) < \varepsilon$

for all $z \in K_{j_0}$ and all x such that $\|x - x_0\| < \delta$; for such x, $|\phi_{j_0}(x)| \leq \int_{K_{j_0}} G(x, z)f(z)\,dz \leq \varepsilon\, m\alpha$. Thus, if $\|x - x_0\| < \delta$ and $j \geq 1$, then

$$|\phi_{j_0+j}(x)| \leq |\phi_{j_0+j}(x) - \phi_{j_0}(x)| + |\phi_{j_0}(x)|$$
$$\leq \frac{1}{2}\sigma_n m\left(\frac{\varepsilon}{v_n}\right)^{2/n} + \varepsilon m\alpha,$$

and by letting $j \to \infty$

$$|Gf(x)| \leq \frac{1}{2}\sigma_n m\left(\frac{\varepsilon}{v_n}\right)^{2/n} + \varepsilon m\alpha;$$

but, since ε is arbitrary and the constants on the right do not depend upon ε and x, $Gf(x)$ can be made arbitrarily small by taking x sufficiently close to $x_0 \in \partial R$. The proof in the $n = 2$ case is precisely the same except that $G(x, z)$ need not be majorized by $-\log\|x - z\|$. But if we assume that R is bounded, then there is a constant c such that $-\log\|x - z\| + \log c = \log(c/\|x - z\|)$ is positive and therefore majorizes $G(x, z)$.

7. Poisson's Equation

Consider the equation $\Delta u = -\kappa_n f$, where f is a given function on an open set R having a Green function G and κ_n is a constant depending upon the dimensionality of E^n. If $n = 2$, then $\kappa_n = 2\pi$; whereas, if $n \geq 3$, then $\kappa_n = \sigma_n(n - 2)$. This equation is known as Poisson's equation. With sufficient smoothness conditions on the given function f, one can show that there is at least one solution to the equation. But since such conditions are not in the mainstream of this book, we shall consider the equation $\tilde{\Delta} u = -\kappa_n f$ where $\tilde{\Delta}$ is the generalized Laplacian. If it can be shown that a solution u of this equation has continuous second partials on R, then u satisfies the equation $\Delta u = -\kappa_n f$ since in this case Δu and $\tilde{\Delta} u$ are the same.

THEOREM **6.25** If R is an open set having a Green function G and f is a bounded continuous Lebesgue integrable function on R, the equation $\tilde{\Delta} u = -\kappa_n f$ has the solution $u = Gf$.

Proof. We can assume $f \geq 0$ for if the theorem is true for non-negative functions, then

$$\tilde{\Delta} Gf = \tilde{\Delta} Gf^+ - \tilde{\Delta} Gf^- = -\kappa_n f^+ + \kappa_n f^- = -\kappa_n f.$$

Let $B = B_{y,\rho} \subset R$. Since $Gf = Gf\chi_B + Gf\chi_{R\sim B}$ and $Gf\chi_{R\sim B}$ is harmonic on B by Theorem 6.6, $\tilde{\Delta} Gf = \tilde{\Delta} Gf\chi_B$ on B. Moreover, since $Gf\chi_B$ differs from $G_B f$

by at most a harmonic function h for which $\tilde{\Delta}h = 0$, it suffices to prove that $\tilde{\Delta}G_B f = -\kappa_n f$ on B. Suppose $x_0 \in B_{y,\rho}$. For $\bar{B}_{x_0,\delta} \subset B$

$$L(G_B f : x_0, \delta) = \int_{B_{x_0,\delta}} L[G_B(\cdot, z) : x_0, \delta] f(z)\, dz$$

$$+ \int_{B \sim B_{x_0,\delta}} L[G_B(\cdot, z) : x_0, \delta] f(z)\, dz$$

by Fubini's theorem. If $n = 2$, then $G_B(x, z) = -\log\|x - z\| + h_x(z)$, where $h_x(z) = h_z(x)$. From Example 1 for $z \in B_{x_0,\delta}$

$$L[G_B(\cdot, z) : x_0, \delta] = L_{(x)}[-\log\|x - z\| : x_0, \delta] + L_{(x)}[h_z(x) : x_0, \delta]$$

$$= \log\left(\frac{1}{\delta}\right) + h_z(x_0);$$

but for $z \in B \sim \bar{B}_{x_0,\delta}$, $L[G_B(\cdot, z) : x_0, \delta] = G_B(x_0, z)$. Therefore

$$L(G_B f : x_0, \delta) = \int_{B_{x_0,\delta}} \left[\log\left(\frac{1}{\delta}\right) + h_{x_0}(z)\right] f(z)\, dz$$

$$+ \int_{B \sim B_{x_0,\delta}} G_B(x_0, z) f(z)\, dz$$

$$= \int_{B_{x_0,\delta}} \log\left(\frac{1}{\delta}\right) f(z)\, dz - \int_{B_{x_0,\delta}} \log \frac{1}{\|x_0 - z\|} f(z)\, dz$$

$$+ G_B f(x_0).$$

It follows that

$$4\,\frac{L(G_B f : x_0, \delta) - G_B f(x_0)}{\delta^2} = \frac{4}{\delta^2} \int_{B_{x_0,\delta}} \log \frac{\|x_0 - z\|}{\delta} f(z)\, dz.$$

Consider a measure ν_δ on $B_{x_0,\delta}$ having a density $-(1/\delta^2)\log(\|x_0 - z\|/\delta)$. By introducing spherical coordinates (r, θ) relative to x_0,

$$\nu_\delta(B_{x_0,\delta}) = -\frac{2\pi}{\delta^2} \int_0^\delta r \log \frac{r}{\delta}\, dr = \frac{\pi}{2},$$

independently of δ. The measures ν_δ obviously converge in the w^*-topology to $(\pi/2)\,\delta_{x_0}$ as $\delta \to 0$ where δ_{x_0} is a unit measure concentrated on $\{x_0\}$. Since f is continuous on R,

$$\frac{4}{\delta^2} \int_{B_{x_0,\delta}} \log \frac{\|x_0 - z\|}{\delta} f(z)\, dz \to -2\pi f(x_0)$$

as $\delta \downarrow 0$ and $\tilde{\Delta} G_B f(x_0) = -2\pi f(x_0)$. Now suppose $n \geq 3$. In this case, for $z \in B_{x_0, \delta}$, $L[G_B(\cdot, z) : x_0, \delta] = \delta^{-n+2} + h_z(x_0)$, whereas for $z \in B \sim \bar{B}_{x_0, \delta}$, $L[G_B(\cdot, z) : x_0, \delta] = G_B(x_0, z)$. Then

$$L(G_B f : x_0, \delta) = \int_{B_{x_0, \delta}} [\delta^{-n+2} + h_{x_0}(z)] f(z)\, dz + \int_{B \sim B_{x_0, \delta}} G_B(x_0, z) f(z)\, dz$$

$$= \int_{B_{x_0, \delta}} \frac{f(z)}{\delta^{n-2}}\, dz - \int_{B_{x_0, \delta}} \frac{1}{\|x_0 - z\|^{n-2}} f(z)\, dz + G_B f(x_0)$$

It follows that

$$\text{(f)} \qquad 2n \frac{L(G_B f : x_0, \delta) - G_B f(x_0)}{\delta^2} = 2n \left\{ \frac{1}{\delta^n} \int_{B_{x_0, \delta}} f(z)\, dz - \frac{1}{\delta^2} \int_{B_{x_0, \delta}} \frac{1}{\|x_0 - z\|^{n-2}} [f(z) - f(x_0)]\, dz - \frac{f(x_0)}{\delta^2} \int_{B_{x_0, \delta}} \frac{1}{\|x_0 - z\|^{n-2}}\, dz \right\}.$$

The first term in the braces is just the average of the continuous function f over $B_{x_0, \delta}$ except for a factor of v_n^{-1} and approaches $v_n f(x_0)$ as $\delta \downarrow 0$. Note that

$$\int_{B_{x_0, \delta}} \frac{1}{\|x_0 - z\|^{n-2}}\, dz = \sigma_n \int_0^\delta r\, dr = \frac{\sigma_n \delta^2}{2}.$$

Since f is continuous, if $\varepsilon > 0$, then

$$\left| \frac{1}{\delta^2} \int_{B_{x_0, \delta}} \frac{1}{\|x_0 - z\|^{n-2}} [f(z) - f(x_0)]\, dz \right| < \frac{\sigma_n \varepsilon}{2}$$

for small δ and therefore the second term in the braces in (f) approaches zero as $\delta \downarrow 0$. Therefore

$$\lim_{\delta \downarrow 0} 2n \frac{L(G_B f : x_0, \delta) - G_B f(x_0)}{\delta^2} = 2n \left[v_n f(x_0) - \frac{\sigma_n}{2} f(x_0) \right];$$

but, since $n v_n = \sigma_n$, the right side is just $-\sigma_n(n-2) f(x_0)$ and therefore $\tilde{\Delta} G_B f(x_0) = -\sigma_n(n-2) f(x_0)$.

Of course, the solution u of Poisson's equation $\tilde{\Delta} u = -\kappa_n f$ is not unique, for if h is any harmonic function on R then $u + h$ is also a solution. The solution $u = Gf$, however, is unique if one imposes the condition that the solution approaches zero at the boundary of R. Further discussion of this condition must await a study of the boundary behavior of the Green function.

We know that if h is harmonic on the open set R then $\tilde{\Delta} h = 0$ on R. The converse is true.

THEOREM **6.26** If u is continuous on the open set R and $\tilde{\Delta} u = 0$ on R, then u is harmonic on R.

Proof. Let B be a ball with closure in R and let $w = G_B 1$. Then $\tilde{\Delta} w = \kappa_n$ on B. Also let $v = \mathbf{PI}(u, B)$. Now $\tilde{\Delta}(u - v + tw) = -\kappa_n t < 0$ and $\tilde{\Delta}(v - u + tw) = -\kappa_n t < 0$ on B for all $t > 0$. Moreover, w is continuous on B by Theorem 6.22 and approaches zero at the boundary of B by Lemma 6.24. Since $u = v$ on ∂B, $u - v + tw$ and $v - u + tw$ have continuous extensions to $\bar{B}$ which vanish on ∂B for all $t > 0$. It follows from Theorem 4.2 that both functions satisfy the minimum principle. It follows that $u - v + tw \geq 0$ and $v - u + tw \geq 0$ on B for all $t > 0$. Letting $t \downarrow 0$, we get $u = v$ and, since v is harmonic on B, the same is true of u.

CHAPTER 7

Capacity

In this chapter we look at those sets that are negligible, in a sense, insofar as potential theory is concerned. Such sets are called polar sets. After defining them and establishing some properties, we develop the notion of capacity of a set. In doing so we define inner and outer capacities of a set in a manner quite similar to the construction of inner and outer measures. Lastly, we return to the study of decreasing sequences of superharmonic functions.

1. Polar Sets

We shall now study in detail the set of points at which a superharmonic function takes on the value $+\infty$. From one point of view these sets are negligible.

Definition. A set $Z \subset E^n$ is said to be a **polar set** if there is an open set $U \supset Z$ and a function u superharmonic on U such that $u = +\infty$ on Z (and perhaps elsewhere).

EXAMPLE **1** Let x_0 be a fixed point of E^n. Then $\{x_0\}$ is a polar set, since u_{x_0} is superharmonic on E^n and $+\infty$ on $\{x_0\}$.

EXAMPLE **2** ($n = 3$) Let $\{x_j\}$ be a sequence of distinct points in E^3. Then $Z = \{x_1, x_2, \ldots\}$ is a polar set. To see this let y be a fixed point in $\sim Z$. Choose constants $c_j > 0$ such that

$$\sum_j \frac{c_j}{\|y - x_j\|} < +\infty.$$

Defining $u(x) = \sum c_j \|x - x_j\|^{-1}$, u is superharmonic on E^3, since it is the limit of an increasing sequence of superharmonic functions and $u(y) < +\infty$. Clearly, $u = +\infty$ on Z.

THEOREM **7.1** If u is superharmonic on the open set R, the set of points at which $u = +\infty$ is a G_δ set. If Z is a polar set, it is a subset of a G_δ polar set. Moreover, any subset of a polar set is a polar set.

Proof. Since $\{x : u(x) = +\infty\} \cap R = \bigcap_j \{x : u(x) > j\} \cap R$ and each of the latter sets is open by the l.s.c. of u, $\{x : u(x) = +\infty\} \cap R$ is a G_δ set. The latter two statements of the theorem are immediate from the definition of polar set and the first statement.

THEOREM **7.2** If $Z \subset E^n$ is a polar set, then Z has n-dimensional Lebesgue measure zero.

Proof. Let u be superharmonic on an open set $U \supset Z$ and $+\infty$ on Z (and perhaps elsewhere). Since u is finite a.e. by Theorem 4.10, Z has Lebesgue measure zero.

The converse of Theorem 7.2 is not true as will be seen from the following example.

EXAMPLE **3** A line segment of E^2 is not a polar set. Let $I \subset E^2$ be a line segment and assume there is a function u that is superharmonic on a neighborhood of I and $u = +\infty$ on I. We use the fact that superharmonicity is invariant under rotations about a point. By choosing a sufficiently small subinterval I_1 of I and rotating I_1 about one of its endpoints we can construct an interval I_2 perpendicular to I_1 with an endpoint in common with an endpoint of I_1 and a function u_2 which is superharmonic on a ball containing I_1 and I_2 and which is $+\infty$ on I_2. The function u_2 is defined as u under the above rotation. Continuing in this manner, we can construct a square and four superharmonic functions with at least one of the four taking on the value $+\infty$ on a given side of the square and all four superharmonic on a neighborhood of the square. By taking the sum of the four functions we obtain a function v which is superharmonic on a neighborhood of the square and $+\infty$ on the boundary of the square. By the minimum principle v must be $+\infty$ on the entire square but this contradicts the fact that a superharmonic function is finite a.e. on its domain. The assumption that a line segment in E^2 is a polar set is therefore incorrect.

EXAMPLE **4** A line segment in E^3 is a polar set. Consider, for example, the line segment I joining $(0, 0, 0)$ to $(1, 0, 0)$ and one-dimensional Lebesgue measure μ on this segment. Let

$$\begin{aligned} v(x) &= \int_I \frac{1}{\|x - z\|}\, d\mu(z), \qquad x \in E^3 \\ &= \int_0^1 \frac{1}{[(x_1 - z_1)^2 + x_2^2 + x_3^2]^{1/2}}\, dz_1. \end{aligned}$$

Since $\|x - z\|^{-1}$ is the Green functon for E^3, v is the potential of a measure. By Theorem 6.3 v is superharmonic on E^3. Let $x \in I$. Then $x = (x_1, 0, 0)$ and

$$v(x) = \int_0^1 \frac{1}{|x_1 - z_1|} dz_1 = +\infty.$$

Therefore $v = +\infty$ on I and this shows that I is a polar set.

In a sense the definition of polar set is local, since it requires only that a superharmonic function be defined on a neighborhood of the set and takes on the value $+\infty$ on the set. We shall show that this local aspect of the definition is not necessary.

THEOREM **7.3** If $Z \subset E^2$ is a polar set, then there is superharmonic function u on E^2 such that $u = +\infty$ on Z.

Proof. Let U be a neighborhood of Z on which there is defined a superharmonic function v such that $v = +\infty$ on Z. We can assume that $v \geq 0$ on U by replacing U with the open set $U \cap \{y : v(y) > 0\}$, if necessary. Letting $U_j = U \cap B_{0,j}$, we have $Z = \bigcup_j (U_j \cap Z)$. By a simple change of notation, if necessary, we can assume that each U_j is nonempty. Since U_j is bounded, it has a Green function G_{U_j}. By the Riesz decomposition theorem there is a measure ν_j associated with v such that $v = G_{U_j}\nu_j + h_j$ where h_j is harmonic on U_j. Since $v = +\infty$ on $U_j \cap Z$, $G_{U_j}\nu_j = +\infty$ on $U_j \cap Z$. Since $U_j \subset B_{0,j}$, $G_{U_j}\nu_j \leq G_{B_{0,j}}\nu_j$ and the latter potential is $+\infty$ on $U_j \cap Z$. Define

(a) $$w_j(x) = \int_{B_{0,j}} \log \frac{2j}{\|x - y\|} d\nu_j(y), \qquad x \in E^2.$$

Then w_j is the logarithmic potential of ν_j, except for a constant, and as such is superharmonic on E^2. Moreover, $w_j = +\infty$ on $U_j \cap Z$, since

$$\log \frac{2j}{\|x - y\|} \geq G_{B_{0,j}}(x, y), \qquad x, y \in B_{0,j}$$

and $G_{B_{0,j}}\nu_j = +\infty$ on $U_j \cap Z$. In summary, (i) w_j is superharmonic on E^2, (ii) $w_j \geq 0$ on $B_{0,j}$, and (iii) $w_j = +\infty$ on $U_j \cap Z$. Since each w_j is finite a.e. on $B_{0,1}$, there is a point $x_0 \in B_{0,1}$ such that $-\infty < w_j(x_0) < +\infty$ for all j. Now choose a sequence of positive numbers $\{b_j\}$ such that the series $\sum b_j w_j(x_0)$ converges. Define $u = \sum b_j w_j$ on E^2. Obviously $u = +\infty$ on Z. We need show only that u is superharmonic on E^2 or, what is the same, that u is superharmonic on each $B_{0,k}$. Let k be a fixed positive integer. For $j \geq k$, $w_j \geq 0$ on $B_{0,j} \supset B_{0,k}$. Now

$$u(x) = \sum_{j=1}^{k-1} b_j w_j(x) + \sum_{j=k}^{\infty} b_j w_j(x), \qquad x \in B_{0,k}.$$

The finite sum is superharmonic since the b_j's are positive. The partial sums of the tail infinite series form an increasing sequence of non-negative superharmonic functions (on $B_{0,k}$) and the limit is either identically $+\infty$ or superharmonic on $B_{0,k}$ by Theorem 4.15. Since the sum is finite at $x_0 \in B_{0,1} \subset B_{0,k}$, the sum is superharmonic on $B_{0,k}$. This shows that u is superharmonic on E^2 and $u = +\infty$ on Z (and perhaps elsewhere).

THEOREM **7.4** If $Z \subset E^n$, $n \geq 3$, is a polar set and G is the Green function for E^n, there is a measure μ on E^n such that the potential $G\mu$ is $+\infty$ on Z.

Proof. By hypothesis there is an open set $U \supset Z$ and a superharmonic function v on U with $v = +\infty$ on Z. As in the preceding proof, we can assume that $v \geq 0$ on U. Let ν be the measure associated with v and U by the Riesz decomposition theorem. Extend ν to E^n by putting $\nu(\sim U) = 0$. Then $v = G_U \nu + h$, where G_U is the Green function for U and h is harmonic on U. Let $\{U_j\}$ be an increasing sequence of nonempty open sets having compact closures such that $E^n = \bigcup_j U_j$ and let $\{G_{U_j}\}$ be the associated sequence of Green functions. If $\nu_j = \nu|_{U_j}$, then $\nu_j(U_j) = \nu(U_j) < +\infty$ for each j. Also extend ν_j to E^n by putting $\nu_j(\sim U_j) = 0$. If $v_j = G\nu_j$, then $v_j \geq 0$ and is superharmonic, since $\nu_j(E^n) = \nu_j(U_j) < +\infty$. Since

$$v = G_U \nu + h = G_U \nu_j + G_U \nu|_{U \sim U_j} + h$$

and $\nu|_{U \sim U_j}$ assigns zero measure to U_j, $v = G_U \nu_j + h_j^*$ on U_j, where h_j^* is harmonic on U_j; but by Theorem 6.8 $G_U \nu_j = G_{U_j} \nu_j + h_j^{**}$ on U_j, where h_j^{**} is harmonic on U_j. Since $v = +\infty$ on $U_j \cap Z$, $G_U \nu_j = +\infty$ on $U_j \cap Z$ and therefore $G_{U_j} \nu_j = +\infty$ on $U_j \cap Z$. Since $v_j = G\nu_j \geq G_{U_j} \nu_j = +\infty$ on $U_j \cap Z$, $v_j = +\infty$ on $U_j \cap Z$. Let y be a fixed point of U_1 such that $v_j(y) < +\infty$ for all j and let $\{c_j)$ be a sequence of positive numbers such that $\sum c_j v_j(y)$ converges. Put $u = \sum c_j v_j$. Since the v_j's are non-negative superharmonic functions and $u(y) < +\infty$, u is a non-negative superharmonic function on E^n and $u = +\infty$ on Z. By the Riesz decomposition theorem there is a measure μ on E^n and a harmonic function h such that $u = G\mu + h$. Since $u = +\infty$ on Z, $G\mu = +\infty$ on Z.

The two preceding theorems permit a more familiar description of smallness of sets.

THEOREM **7.5** If Z is a polar set, its intersection with any sphere has surface area zero.

Proof. By the two preceding theorems, there is a superharmonic function u such that $u = +\infty$ on Z. Let B be any ball and consider $Z \cap \partial B$. Since u is integrable relative to surface area on ∂B, $Z \cap \partial B$ has surface area zero.

THEOREM **7.6** If $\{Z_j\}$ is a sequence of polar sets, then $\bigcup_j Z_j$ is a polar set.

Proof. Let u_j be superharmonic and $+\infty$ on Z_j. Since the u_j's are finite a.e., there is a point $y \in B_{0,1}$ such that $u_j(y) < +\infty$ for all j. Consider u_j just on $B_{0,j}$. Since u_j is l.s.c., $\inf_{\|x\| \le j} u_j(x)$ is finite. Thus we can choose a sequence $\{b_j\}$ of positive numbers such that

$$\sum_j b_j\Big[u_j(y) - \inf_{\|x\| \le j} u_j(x)\Big]$$

converges. Let

$$u = \sum_j b_j\Big[u_j - \inf_{\|x\| \le j} u_j(x)\Big].$$

Consider a fixed ball $B_{0,k}$. All terms except the first $k-1$ defining u are non-negative on $B_{0,k}$. The sum of the first $k-1$ is superharmonic and the sum of the tail series is superharmonic on $B_{0,k}$ by choice of the b_j's. It follows that u is superharmonic on $B_{0,k}$ for every positive integer k and therefore superharmonic. Since $u_j = +\infty$ on Z_j, $u = +\infty$ on $\bigcup_j Z_j$.

2. Superharmonic Extensions

In this section we take up the problem of continuation of superharmonic functions across polar sets.

THEOREM **7.7** Let R be an open set and let Z be a relatively closed polar subset of R. If u is superharmonic on $R \sim Z$ and locally bounded below on R, then it has a unique superharmonic extension on R.

Proof. We first note that $R \sim Z$ is dense in R for if this were not the case then Z would have a non-empty interior and therefore positive Lebesgue measure, a contradiction. Since $R \sim Z$ is dense in R, we can define $\hat{u}(x) = \liminf_{y \to x,\, y \in R \sim Z} u(y)$ for $x \in R$. Since u is locally bounded below on R, $\hat{u}$ cannot take on the value $-\infty$. Clearly, $\hat{u}$ is l.s.c. on R. Since $\hat{u} = u$ on the open set $R \sim Z$, $\hat{u}$ is superharmonic on $R \sim Z$ and therefore not identically $+\infty$ on any component of R. To show that $\hat{u}$ is superharmonic on R it remains only to show that $\hat{u}$ is locally super-mean-valued on R. This is true at points of $R \sim Z$ since $\hat{u}$ is superharmonic on this open set. Consider any $x \in Z$ and $\bar{B}_{x,\delta} \subset R$. Let $\{x_j\}$ be a sequence of distinct points in $R \sim Z$ such that $\lim_{j \to \infty} x_j = x$ and $\lim_{j \to \infty} \hat{u}(x_j) = \hat{u}(x)$. For each j we can choose a number $\delta_j > 0$ such that $\bar{B}_{x_j, \delta_j} \subset R \sim Z$ and $\bar{B}_{x_i, \delta_i} \cap \bar{B}_{x_j, \delta_j} = \varnothing$ for $i \neq j$. Since Z is a polar set, there is a superharmonic function v such that $v = +\infty$ on Z.

We can assume that $v(x_j) < +\infty$ for each j for the following reason. By Lemma 4.17, the function

$$v_1 = \begin{cases} \mathbf{PI}(v, B_1) & \text{on} \quad B_1, \\ v & \text{on} \quad R \sim B_1, \end{cases}$$

where $B_j = B_{x_j, \delta_j}$, is superharmonic on R. By choice of B_1, $v = v_1 = +\infty$ on Z and $v_1(x_1) < +\infty$. Inductively define a sequence $\{v_j\}$ by

$$v_j = \begin{cases} \mathbf{PI}(v_{j-1}, B_j) & \text{on} \quad B_j, \\ v_{j-1} & \text{on} \quad R \sim B_j. \end{cases}$$

The sequence $\{v_j\}$ is a decreasing sequence of superharmonic functions on R, $v_j = +\infty$ on Z, and $v_j(x_j) < +\infty$. Consider $v^* = \lim_{j\to\infty} v_j$. Given any neighborhood of x, the v_j's agree outside the neighborhood except possibly for a finite number of terms of the sequence. This means that v^* is superharmonic on $R \sim \{x\}$. To show that v^* is superharmonic on R we need only prove that v^* is l.s.c. at x, v^* being locally super-mean-valued at x since $v^*(x) = +\infty$. Since $v(x) = +\infty$ and v is l.s.c. at x, for each integer p there is a neighborhood U_p of x such that $v \geq p$ on U_p. Moreover, we can assume that there is an integer j_p such that $\bar{B}_j \subset U_p$ for all $j \geq j_p$, whereas $\bar{B}_j \subset R \sim U_p$ for $j < j_p$. It follows that $v_j \geq p$ on U_p for all $j \geq j_p$ and therefore $v^* \geq p$ on U_p. This shows that $+\infty = v^*(x) = \lim_{y\to x} v^*(y)$; that is, v^* is continuous at x. Therefore we can assume that $v = +\infty$ on Z and $v(x_j)$ is finite for each j. Since $\hat{u}$ cannot take on the value $-\infty$ and is l.s.c. on R, $\hat{u} + \varepsilon v$ is defined and l.s.c. on R for every $\varepsilon > 0$. Since $\hat{u} + \varepsilon v = u + \varepsilon v$ on $R \sim Z$, $\hat{u} + \varepsilon v$ cannot be identically $+\infty$ on any component of R. For $y \in Z$ and $\bar{B}_{y,\rho} \subset R$, $+\infty = (\hat{u} + \varepsilon v)(y) \geq L(\hat{u} + \varepsilon v : y, \rho)$. This shows that $\hat{u} + \varepsilon v$ is superharmonic on R. Now $\hat{u}(x_j) + \varepsilon v(x_j) \geq A(\hat{u} : x_j, \delta) + \varepsilon A(v : x_j, \delta)$. Since $v(x_j)$ is finite, we can let $\varepsilon \to 0$ to obtain

$$\hat{u}(x_j) \geq A(\hat{u} : x_j, \delta) = \frac{1}{v_n \delta^n} \int \chi_{B_{x_j,\delta}}(y)\hat{u}(y)\, dy.$$

Since $\hat{u}$ is bounded below on the compact closure of a neighborhood of x, we can choose δ so that all but a finite number of the integrands on the right are bounded below. With this in mind we can apply Fatou's lemma to obtain $\hat{u}(x) = \lim_{j\to\infty} \hat{u}(x_j) \geq A(\hat{u} : x, \delta)$ for all sufficiently small δ. This concludes the proof that $\hat{u}$ is superharmonic on R. To prove the uniqueness let u^* be a second such extension. Since Z has Lebesgue measure zero, $\hat{u} = u^*$ a.e. and $A(\hat{u} : x, \delta) = A(u^* : x, \delta)$ whenever $B_{x,\delta} \subset R$. Letting $\delta \downarrow 0$, we get $\hat{u}(x) = u^*(x)$ by Lemma 4.18.

COROLLARY **7.8** Let R be an open set, Z, a relatively closed polar subset of R, and h, a function that is harmonic on $R \sim Z$ and locally bounded on R. Then h has a unique harmonic extension h^* on R.

Proof. By Theorem 7.7 h has a superharmonic extension u' on R. Applying the same theorem to $-h$, it has a subharmonic extension u'' on R. Now $h = u' = u''$ on $R \sim Z$. Since Z has Lebesgue measure zero, $A(u' : x, \delta) = A(u'' : x, \delta)$, whenever $B_{x,\delta} \subset R$. Letting $\delta \downarrow 0$, we obtain $u'(x) = u''(x)$ for all $x \in R$ and u' is therefore harmonic on R.

The two preceding results attest to the negligibility of polar sets. We now apply the concept of polar set to the problem of existence of a Green function in the two-dimensional case.

LEMMA **7.9** If R is an open subset of E^2 having a Green function, then ∂R is not a polar set.

Proof. Suppose first that R is not dense in E^2 and that ∂R is polar. There is then a superharmonic function v such that $v = +\infty$ on ∂R. Let R_0 be a component of R and let

$$v^* = \begin{cases} +\infty & \text{on } R_0, \\ v & \text{otherwise.} \end{cases}$$

The function v^* is easily seen to be l.s.c., locally super-mean-valued, and does not take on the value $-\infty$; since v^* is finite at some point of $\sim \bar{R}$, v^* is superharmonic; but this is a contradiction, since v^* is identically $+\infty$ on a nonempty open set, namely R_0. Now let R be a dense open subset of E^2 having a Green function G. For $x_0 \in R$, $G(x_0, \cdot)$ is bounded and harmonic on $R \sim \bar{B}$, where $\bar{B} = \bar{B}_{x_0,\delta} \subset R$. Now $\partial R \subset E^2 \sim \bar{B}$. If ∂R were a polar set, then $G(x_0, \cdot)$ would have a harmonic extension on $E^2 \sim \bar{B}$ by Corollary 7.8. Since $G(x_0, \cdot) = u_{x_0} + h_{x_0}$, where h_{x_0} is harmonic on R and u_{x_0} is harmonic on $E^2 \sim \bar{B}$, this would mean that h_{x_0} would have a harmonic extension to E^2, which is also denoted by h_{x_0}. Since $G(x_0, \cdot)$ would have a bounded extension to $R \sim \bar{B}$ and $\lim_{\|x\| \to +\infty} u_{x_0}(x) = \lim_{\|x\| \to +\infty}(-\log\|x - x_0\|) = -\infty$, we would have $\lim_{\|x\| \to +\infty} h_{x_0}(x) = +\infty$. By the minimum principle h_{x_0} would be identically $+\infty$ in contradiction to the finiteness of harmonic functions.

In E^n, $n \geq 3$, it is possible for an open set R to have a Green function even though ∂R is a polar set. This is true for example if R is the complement of a singleton set or, as Example 4 shows, even a line segment.

The following theorem generalizes Corollary 4.3 considerably and supports the notion of negligibility of polar sets. It also exhibits an essential difference between the two-dimensional case and the higher dimensional case. In the two-dimensional case the point at infinity is a negligible point but in the higher dimensional case the point at infinity cannot be ignored.

THEOREM **7.10** Let R be an open subset of E^n which has a Green function, let u be superharmonic and bounded below on R, and suppose there is a number α such that $\liminf_{y \to z \in \partial R} u(y) \geq \alpha$ except possibly for z in a polar subset Z of ∂R. If (i) $n = 2$, or (ii) $n \geq 3$ and R is bounded, or (iii) $n \geq 3$, R is unbounded, and $\liminf_{\|y\| \to +\infty} u(y) \geq \alpha$, then $u \geq \alpha$ on R.

Proof. Suppose first that R is a bounded subset of E^n, $n \geq 2$. Let v be a superharmonic function which takes on the value $+\infty$ on $Z \subset \partial R$. Since $\bar{R}$ is compact and v is bounded below on $\bar{R}$, we can assume that $v \geq 0$ on $\bar{R}$ by adding a constant to v if necessary. Consider the function $u + \varepsilon v$ which is superharmonic on R for $\varepsilon > 0$. It is easily seen that

$$\liminf_{\substack{y \to z \in \partial R \\ y \in R}} (u + \varepsilon v)(y) \geq \begin{cases} \alpha & \text{if } z \in \partial R \sim Z, \\ +\infty & \text{if } z \in Z. \end{cases}$$

By Corollary 4.3 $u + \varepsilon v \geq \alpha$ on R. Suppose $\bar{B}_{x,\delta} \subset R$. Then $A(u : x, \delta) + \varepsilon A(v : x, \delta) \geq \alpha$. Letting $\varepsilon \to 0$ first and then letting $\delta \downarrow 0$, we obtain $u(x) \geq \alpha$ for all $x \in R$. Suppose now that R is unbounded. The rest of the proof requires separate consideration of the $n = 2$ and $n \geq 3$ cases. Consider first the $n = 2$ case. Since R has a Green function, ∂R is not a polar set by Lemma 7.9. It follows that there is a point $z_0 \in \partial R$ such that $\liminf_{y \to z_0} u(y) \geq \alpha$. If $\eta > 0$, then there is a ball $B = B_{z_0, \delta}$ such that $u(y) > \alpha - \eta$ for $y \in R \cap \bar{B}$. Consider the function $w(y) = \log\|y - z_0\| = -u_{z_0}(y)$, which is harmonic, bounded below by $\log \delta$ outside $B_{z_0, \delta}$, and approaches $+\infty$ as $\|y\| \to +\infty$. By adding a constant to w we can say that there is a function v' which is harmonic and nonnegative on $U = R \sim \bar{B}$ for which $\lim_{\substack{\|y\| \to +\infty \\ y \in U}} v'(y) = +\infty$. For each $\varepsilon > 0$ the function $u + \varepsilon v'$ has the properties that $\liminf_{\substack{y \to z \in \partial U \\ y \in U}} (u + \varepsilon v')(y) \geq \alpha - \eta$ except possibly for z in a polar subset of ∂U and $\lim_{\substack{\|y\| \to +\infty \\ y \in U}} (u + \varepsilon v')(y) = +\infty$. Thus for all sufficiently large positive integers j, $(u + \varepsilon v') \geq \alpha - \eta$ on $\partial B_{z_0, j} \cap U$. It follows that $\liminf_{\substack{y \to z \\ y \in U \cap B_{z_0, j}}} (u + \varepsilon v')(y) \geq \alpha - \eta$ for all z in $\partial(U \cap B_{z_0, j})$ except possibly for z in a polar set. It follows from the first part of the proof (for bounded sets) that $u + \varepsilon v' \geq \alpha - \eta$ on $U \cap B_{z_0, j}$ for every j. Therefore, $u + \varepsilon v' \geq \alpha - \eta$ on U. Letting $\varepsilon \to 0$, $u \geq \alpha - \eta$ on U. But since $u > \alpha - \eta$ on $R \cap \bar{B}$ and $R = (R \cap \bar{B}) \cup U$, $u \geq \alpha - \eta$ on R. Since η was an arbitrary positive number, $u \geq \alpha$ on R. Now consider the $n \geq 3$ case. Since $\liminf_{\|y\| \to +\infty} u(y) \geq \alpha$, for each $\eta > 0$ there corresponds a positive integer j_0 such that $u(y) \geq \alpha - \eta$ for all $y \in R \sim B_{0,j}$, $j \geq j_0$. Then $\liminf_{\substack{y \to z \\ y \in R \cap B_{0,j}}} u(y) \geq \alpha - \eta$ for $z \in \partial(R \cap B_{0,j})$ except possibly for z in a polar set. By the first part of the proof (for bounded sets), $u \geq \alpha - \eta$ on $R \cap B_{0,j}$ for every $j \geq j_0$. It follows that $u \geq \alpha - \eta$ on R and since η is an arbitrary positive number, $u \geq \alpha$ on R.

3. Balayage

In this section we develop the notion of "balayage," which is due to Poincaré. As conceived by Poincaré, "balayage," or "sweeping out," is just the smoothing process used to define the reduction of a superharmonic function u. Balayage, as defined below, amounts to the same thing.

Let R be an open subset of E^n having a Green function G and let u be a non-negative superharmonic function on R. All definitions are relative to R. If E is any subset of R, let

$$\Phi_E^u = \{v \in \mathscr{S}_R : v \geq 0 \quad \text{on} \quad R, v \geq u \quad \text{on} \quad E\}$$

and let

$$\mathbf{R}_E^u = \inf\{v : v \in \Phi_E^u\}.$$

(Recall that $\mathscr{S}_R$ is the class of functions superharmonic on R.)

Definition. $\mathbf{R}_E^u$ is called the **réduite** (or reduced function) of u relative to E in R.

$\mathbf{R}_E^u$, as the infimum of a family of superharmonic functions, need not be superharmonic. Consider, for example, the fundamental harmonic function $u = u_0$ for E^3 with pole at 0. Let $R = E^3$ and $E = \{0\}$. Then $\mathbf{R}_E^u$ is $+\infty$ on E, zero elsewhere, and therefore not l.s.c.

In the general case $\mathbf{R}_E^u$ fails to be superharmonic on R only because of the l.s.c. requirement. Since $u \in \Phi_E^u$ and the elements of Φ_E^u are non-negative, $0 \leq \mathbf{R}_E^u \leq u$. Thus $\mathbf{R}_E^u > -\infty$ and $\mathbf{R}_E^u$ cannot be $+\infty$ on any component of R since this is true of u. $\mathbf{R}_E^u$ is easily seen to be super-mean-valued on R as follows. If $v \in \Phi_E^u$, then

$$\begin{aligned} v(x) &\geq L(v : x, \delta) \\ &\geq L(\mathbf{R}_E^u : x, \delta) \end{aligned}$$

whenever $\bar{B}_{x,\delta} \subset R$ (assuming $\mathbf{R}_E^u$ to be measurable). Since the set

$$\{v(x) : v \in \Phi_E^u\}$$

is bounded below by $L(\mathbf{R}_E^u : x, \delta)$,

$$\mathbf{R}_E^u(x) = \inf_{v \in \Phi_E^u} v(x) \geq L(\mathbf{R}_E^u : x, \delta)$$

whenever $\bar{B}_{x,\delta} \subset R$ (assuming $\mathbf{R}_E^u$ to be measurable). This shows that only the l.s.c. requirement stands in the way of $\mathbf{R}_E^u$ being superharmonic.

LEMMA **7.11** The lower regularization $\hat{\mathbf{R}}_E^u$ of $\mathbf{R}_E^u$ is superharmonic on R. Moreover,

(i) $u \geq \mathbf{R}_E^u \geq \hat{\mathbf{R}}_E^u \geq 0$ on R,
(ii) $u = \mathbf{R}_E^u$ on E,
(iii) $u = \mathbf{R}_E^u = \hat{\mathbf{R}}_E^u$ on the interior of E,
(iv) $\mathbf{R}_E^u = \hat{\mathbf{R}}_E^u$ on $R \sim \bar{E}$ and both are harmonic on $R \sim \bar{E}$.

Proof. Properties (i), (ii), and (iii) are obvious. We need verify only that $\hat{\mathbf{R}}_E^u$ is superharmonic on R; but this follows from the fact that Φ_E^u is left-directed and Theorem 4.16. To prove (iv) we need only prove that $\mathbf{R}_E^u$ is harmonic on $R \sim \bar{E}$, for then $\mathbf{R}_E^u$ is continuous on $R \sim \bar{E}$ and $\mathbf{R}_E^u = \hat{\mathbf{R}}_E^u$ on $R \sim \bar{E}$. To show that $\mathbf{R}_E^u$ is harmonic on $R \sim \bar{E}$ we need show only that Φ_E^u is a saturated family of superharmonic functions on $R \sim \bar{E}$ according to Theorem 5.9. This is easily seen to be the case.

According to the preceding lemma, $\hat{\mathbf{R}}_E^u$ differs from $\mathbf{R}_E^u$ only on the boundary of E. We may actually have $\hat{\mathbf{R}}_E^u < \mathbf{R}_E^u$ at such points. If u is the fundamental harmonic function for E^3 with pole 0 and $E = \{0\}$, then

$$\mathbf{R}_E^u(y) = \begin{cases} +\infty & \text{if } \ y = 0, \\ 0 & \text{if } \ y \neq 0. \end{cases}$$

Clearly $\hat{\mathbf{R}}_E^u = 0$ and $\hat{\mathbf{R}}_E^u < \mathbf{R}_E^u$ on $E = \partial E$.

Definition. $\hat{\mathbf{R}}_E^u$ is called the **balayage** (or **regularized reduced function**) of u relative to E in R.

We show later that $\hat{\mathbf{R}}_E^u$ differs from $\mathbf{R}_E^u$ only on a polar set. Before taking up the next topic, we compare reduced functions.

THEOREM **7.12** Let u and v be non-negative superharmonic functions on R. Then

(i) $E \subset F \subset R \Rightarrow \hat{\mathbf{R}}_E^u \leq \hat{\mathbf{R}}_F^u$,
(ii) $u \leq v \Rightarrow \hat{\mathbf{R}}_E^u \leq \hat{\mathbf{R}}_E^v$,
(iii) $\lambda > 0 \Rightarrow \hat{\mathbf{R}}_E^{\lambda u} = \lambda \hat{\mathbf{R}}_E^u$,
(iv) $\hat{\mathbf{R}}_E^{u+v} \leq \hat{\mathbf{R}}_E^u + \hat{\mathbf{R}}_E^v$,
(v) K a compact subset of $R \Rightarrow \hat{\mathbf{R}}_K^u$ is a potential.

Proof. If $E \subset F$, then $\Phi_F^u \subset \Phi_E^u$ and $\mathbf{R}_E^u \leq \mathbf{R}_F^u$. The first assertion follows by taking the lower regularization of each side of this inequality. Parts (ii) and (iii) are proved similarly. If $f \in \Phi_E^u$ and $g \in \Phi_E^v$, then $f + g \in \Phi_E^{u+v}$ and $\hat{\mathbf{R}}_E^{u+v} \leq \mathbf{R}_E^{u+v} \leq f + g$, where $\hat{\mathbf{R}}_E^{u+v}$, f, and g are superharmonic on R. For the time being fix g. Then $A(\hat{\mathbf{R}}_E^{u+v} : x, \delta) \leq A(f : x, \delta) + A(g : x, \delta) \leq f(x) + A(g : x, \delta)$ whenever $\bar{B}_{x,\delta} \subset R$. Taking the infimum over $f \in \Phi_E^u$,

$$A(\hat{\mathbf{R}}_E^{u+v} : x, \delta) \leq \mathbf{R}_E^u + A(g : x, \delta).$$

Since the averages are continuous on $R_\delta = \{y : d(y, \sim R) > \delta\}$ by Theorem 4.19, $A(\hat{\mathbf{R}}_E^{u+v} : x, \delta) \leq \hat{\mathbf{R}}_E^u(x) + A(g : x, \delta)$ for $x \in R_\delta$. Letting $\delta \downarrow 0$ we obtain $\hat{\mathbf{R}}_E^{u+v}(x) \leq \hat{\mathbf{R}}_E^u(x) + g(x)$ on R by Theorem 4.19. Repeating this argument, $\hat{\mathbf{R}}_E^{u+v} \leq \hat{\mathbf{R}}_E^u + \hat{\mathbf{R}}_E^v$. To show that $\hat{\mathbf{R}}_K^u$ is a potential whenever K is a compact subset of R, we must show that its greatest harmonic minorant is zero. Assume first that u is bounded. Let R_0 be any component of R and let $x_0 \in R_0$. By Theorem 5.23 and Corollary 5.25, the greatest harmonic minorant of $G(x_0, \cdot)|_{R_0}$ is the zero function. Since $G(x_0, \cdot)$ is superharmonic on R_0 and strictly positive thereon, it has a strictly positive infimum on $R_0 \cap K$. Then for some $\lambda > 0$, $\lambda G(x_0, \cdot)$ majorizes u over $R_0 \cap K$ and the function w which is equal to $\lambda G(x_0, \cdot)$ on R_0 and equal to u on $R \sim R_0$ belongs to Φ_K^u; therefore, $0 \leq \hat{\mathbf{R}}_K^u \leq \mathbf{R}_K^u \leq \lambda G(x_0, \cdot)$ on R_0. It follows that the greatest harmonic minorant of $\hat{\mathbf{R}}_K^u|_{R_0}$ is the zero function for any component R_0 of R. It is easily seen that the greatest harmonic minorant of $\hat{\mathbf{R}}_K^u$ is the zero function. $\hat{\mathbf{R}}_K^u$ is therefore a potential by the Riesz decomposition theorem. Suppose now that u is unbounded on K. By the Riesz decomposition theorem, $u = v + h$ where v is a potential and h is harmonic on R. Since $0 \leq \hat{\mathbf{R}}_K^v \leq \mathbf{R}_K^v \leq v$ and v is a potential, $\hat{\mathbf{R}}_K^v$ is a potential. Since h is bounded on K, $\hat{\mathbf{R}}_K^h$ is a potential by the preceding part of the proof. By (iv) $\hat{\mathbf{R}}_K^u \leq \hat{\mathbf{R}}_K^v + \hat{\mathbf{R}}_K^h$. Since the sum of two potentials is again a potential, $\hat{\mathbf{R}}_K^u$ is a potential.

It is now possible to prove slightly improved versions of Theorems 7.3 and 7.4.

LEMMA **7.13** If u is superharmonic on $B_{y,\rho}$ and $0 < r < \rho$, then there is a superharmonic function u^* which agrees with u on $B_{y,r}$ and is finite on $\sim \bar{B}_{y,r}$.

Proof. By reducing ρ slightly, we can assume that u is bounded below on $\bar{B}_{y,\rho}$ and superharmonic on a neighborhood of $\bar{B}_{y,\rho}$. We can also assume that $u \geq 0$ on $\bar{B}_{y,\rho}$. Let $E = \bar{B}_{y,r}$ and consider $\hat{\mathbf{R}}_E^u$, the balayage of u over E relative to $B_{y,\rho}$. Now $\hat{\mathbf{R}}_E^u = u$ on $B_{y,r}$ and is harmonic on $B_{y,\rho} \sim \bar{B}_{y,r}$. By Theorem 7.12 $\hat{\mathbf{R}}_E^u$ is the potential of a measure having support in $\bar{B}_{y,r} \subset B_{y,\rho}$. By Theorem 6.23 $\lim_{z \to z_0} \hat{\mathbf{R}}_E^u(z) = 0$ for all $z_0 \in \partial B_{y,\rho}$. Define

$$w(z) = \begin{cases} \hat{\mathbf{R}}_E^u(z), & z \in B_{y,\rho}, \\ 0, & z \in \partial B_{y,\rho}. \end{cases}$$

By Theorem 5.5 w can be continued harmonically across $\partial B_{y,\rho}$; that is, there is a positive number $\delta < \rho - r$ and a function w^* defined on $B_{y,\rho+\delta}$ such that w^* agrees with w on $\bar{B}_{y,\rho}$ and is harmonic on $B_{y,\rho+\delta} \sim \bar{B}_{y,\rho-\delta}$. Now one can choose α and β such that $\alpha u_y + \beta$ agrees with w^* on $\partial B_{y,\rho}$ and $\alpha u_y + \beta \leq w^*$ on $B_{y,\rho+\delta/2} \sim \bar{B}_{y,\rho}$. Then

$$u^* = \begin{cases} w^* & \text{on} \quad \bar{B}_{y,\rho} \\ \alpha u_y + \beta & \text{on} \quad \sim\bar{B}_{y,\rho} \end{cases}$$

has the desired properties.

THEOREM **7.14** If $Z \subset E^n$ is a polar set and $x_0 \notin Z$, there is a superharmonic function v such that $v = +\infty$ on Z and $v(x_0) < +\infty$.

Proof. Let u be a superharmonic function which takes on the value $+\infty$ on Z. For each $y \in Z$ let B_{y,ρ_y} be a ball such that $x_0 \notin B_{y,\rho_y}$. A countable number of such balls corresponding to a sequence $\{y_j\}$ in Z suffice to cover Z. Let B_j be the ball corresponding to y_j. By the preceding lemma there is a superharmonic function u_j such that $u_j = +\infty$ on $B_j \cap Z$ and is finite outside $\bar{B}_j$. Since u_j is bounded below on $B_{x_0,j}$, there is a constant α_j such that $u_j + \alpha_j \geq 0$ on $B_{x_0,j}$. Now let $\{\beta_j\}$ be a sequence of positive numbers such that $\sum_j \beta_j(u_j(x_0) + \alpha_j)$ converges. The function $v = \sum_j \beta_j(u_j + \alpha_j)$ has the desired properties.

COROLLARY **7.15** There are discontinuous finite-valued superharmonic functions.

Proof. Let Z be a nonclosed polar set and let x be a limit point of Z which is not in Z. By the preceding theorem there is a superharmonic function v such that $v(x) < +\infty$ and $v = +\infty$ on Z. Now let $u = v \wedge (v(x) + 1)$. Then u is a finite-valued superharmonic function which takes on the value $v(x) + 1$ on Z.

4. Capacity and Capacitary Potential

In order to avoid duplication of most of the results of this section, we shall introduce a certain class of sets. Members of this class will be called "admissible," a term which is to be regarded as temporary. The necessity of doing this results from lack of knowledge at this point concerning the boundary behavior of the Green function.

Definition. An open set $R \subset E^n$ is **admissible** if (i) R has a Green function and (ii) there is a polar set $Z \subset \partial R$ such that $\lim_{y \to z \in \partial R \sim Z} G\mu(y) = 0$ for any measure μ having compact support in R.

Note that if R is an unbounded subset of E^n, $n \geq 3$, and μ is a measure having compact support in R, then $\lim_{\|y\| \to +\infty} G\mu(y) = 0$ since G is majorized by the Green function for E^n, namely $\|x - y\|^{-n+2}$. In Chapter 8 we prove that **any open set having a Green function is an admissible set**, but for the time being we know for certain only that each ball is admissible.

Throughout this section R is an admissible set and G denotes its Green function. If in the definition of the réduite we take E to be a compact set $K \subset R$ and $u = 1$, then $\mathbf{R}_K^1$ and $\hat{\mathbf{R}}_K^1$ will be denoted by W_K and V_K, respectively. It should be emphasized that W_K and V_K depend upon R and are actually functions on R. Then

(i) $1 \geq W_K \geq V_K \geq 0$ on R,
(ii) V_K is a potential on R,
(iii) $W_K = 1$ on K,
(iv) $1 = W_K = V_K$ on the interior of K,
(v) $W_K = V_K$ on $R \sim K$ and both are harmonic on $R \sim K$,
(vi) $\lim_{y \to z \in \partial R} V_K(y) = 0$ except possibly for z in a polar subset of ∂R and, in case R is an unbounded subset of E^n for $n \geq 3$, $\lim_{\|y\| \to +\infty} V_K(y) = 0$.

LEMMA **7.16** V_K is the potential of a measure with support in ∂K.

Proof. By (ii) V_K is the potential of a measure μ on R; that is, $V_K = G\mu$. Since V_K is harmonic on $R \sim \partial K$ by (iv) and (v), $\mu(R \sim \partial K) = 0$ by Theorem 6.9.

LEMMA **7.17** V_K is the greatest potential of measures μ with support in K such that $G\mu \leq 1$ on R.

Proof. Let μ be such a measure. Since R is an admissible set, $\lim_{y \to z \in \partial R} G\mu(y) = 0$ except possibly for z in a polar subset Z of ∂R and, in case R is an unbounded subset of E^n for $n \geq 3$, $\lim_{\|y\| \to +\infty} G\mu(y) = 0$. Let $u = G\mu$. Since $u \leq 1$ on R, $\limsup_{\substack{y \to z \in \partial K \\ y \in R \sim K}} u(y) \leq 1$. Consider any $v \in \Phi_K^1$ and the function $v - u$. Since $v \geq 1$ on K and l.s.c. on R,

$$\liminf_{\substack{y \to z \in \partial(R \sim K) \\ y \in R \sim K}} [v(y) - u(y)] \geq 0,$$

except possibly for $z \in Z$. This inequality holds also if R is an unbounded subset of E^n for $n \geq 3$ as $\|y\| \to +\infty$, $y \in R \sim K$. Since $v \geq 0$ and $u \leq 1$ on R, $v - u$ is bounded below on R. Since u is harmonic on $R \sim K$, $v - u$ is superharmonic on $R \sim K$. By Theorem 7.10 $v - u \geq 0$ on $R \sim K$. Since $v \geq 1 \geq u$ on K, $v \geq u$ on R for every $v \in \Phi_K^1$. Therefore $W_K = \mathbf{R}_K^1 \geq u$ on R and, since u is l.s.c. on R, $V_K = \hat{W}_K \geq \hat{u} = u$ on R.

Definition. For each compact subset K of the admissible set R, we call V_K the **capacitary potential** of K. The unique measure μ_K for which $V_K = G\mu_K$ is called the **capacitary distribution** for K. The **capacity** of K (relative to the admissible set R), denoted by $\mathscr{C}(K)$, is defined to be $\mathscr{C}(K) = \mu_K(K)$ with $\mathscr{C}(\varnothing) = 0$.

We now consider $V_K(x)$ and $W_K(x)$ as functions of the compact set $K \subset R$ with R and x fixed. Although the variable x is suppressed in the following, the reader should keep in mind that x is fixed. K, with or without subscripts, always denotes a compact subset of the admissible set R.

LEMMA **7.18** As a function of the compact set $K \subset R$, W_K has the following properties:

(i) Monotonicity: $K_1 \subset K_2 \Rightarrow W_{K_1} \leq W_{K_2}$.
(ii) Right-continuity: The mapping $K \rightsquigarrow W_K$ is right-continuous; that is, if K is a compact set and $\varepsilon > 0$, there is an open set U such that $W_{K_1} - W_K < \varepsilon$ whenever $K \subset K_1 \subset U$.
(iii) Strong subadditivity: $W_{K_1 \cup K_2} + W_{K_1 \cap K_2} \leq W_{K_1} + W_{K_2}$.

Proof. The monotonicity is immediate from the definition $W_K = \mathbf{R}_K^1$. As to (ii), it is easily seen that (ii) is equivalent to

(ii′) If $\{K_j\}$ is any decreasing sequence of compact sets, then $W_{K_j} \downarrow W_{\cap K_j}$.

Let $\{K_j\}$ be a decreasing sequence of compact sets and let $K = \cap K_j$. Since $\{W_{K_j}\}$ is a decreasing sequence, we can define $g = \lim_{j \to \infty} W_{K_j}$ on R. It is easily seen that $g = \inf\{v : v \in \bigcup_{j=1}^{\infty} \Phi_{K_j}^1\}$ and that the family $\bigcup_{j=1}^{\infty} \Phi_{K_j}^1$ is left-directed. By Theorem 4.16 the lower regularization $\hat{g}$ of g is superharmonic on R. By the Riesz decomposition theorem, $\hat{g} = G\nu + h$ where ν is a measure on R and h is the greatest harmonic minorant of $\hat{g}$. Since $0 \leq g \leq W_{K_j}$ on R, $0 \leq \hat{g} \leq V_{K_j}$ on R for each j. Since V_{K_j} is a potential, the greatest harmonic minorant of $\hat{g}$ is the zero function; that is, $h = 0$ and $\hat{g} = G\nu$ on R. For fixed j_0 consider the functions $W_{K_j}, j \geq j_0$, which are harmonic on $R \sim K_{j_0}$. By Theorem 2.19 $\hat{g}$ is harmonic on $R \sim K_{j_0}$. Since j_0 is arbitrary, $\hat{g}$ is harmonic on $R \sim K$. It follows from Theorem 6.9 that the support of ν is contained in K. Since $\hat{g} = G\nu \leq 1$ on R, $\hat{g} \leq V_K$ by Lemma 7.17. By (i) $W_{K_j} \geq W_K \geq V_K$ for all j and therefore $g \geq V_K$. Taking the lower regularization of both sides of this inequality, we get $\hat{g} \geq V_K$ and consequently $\hat{g} = V_K$. Since g is harmonic on $R \sim K$, $\hat{g} = g$ on $R \sim K$. Therefore $\hat{g} = g = \lim_{j \to \infty} W_{K_j} \geq W_K = V_K = \hat{g}$ on $R \sim K$ and $\lim_{j \to \infty} W_{K_j} = W_K$ on $R \sim K$. Since $W_{K_j} = 1$ on K and $W_K = 1$ on K, $\lim_{j \to \infty} W_{K_j} = W_K$ on K and (ii′) is established. It remains only to prove (iii). Let K_1 and K_2 be compact subsets of R. Suppose $z \in K_1 \cup K_2$ and, for example, $z \in K_1$. Then $W_{K_1 \cup K_2}(z) = 1$ and $W_{K_1}(z) = 1$. By (i) $W_{K_1 \cap K_2}(z) \leq W_{K_2}(z)$. Therefore $W_{K_1 \cup K_2}(z) + W_{K_1 \cap K_2} \leq 1 + W_{K_2}(z) = W_{K_1}(z) + W_{K_2}(z)$ and strong subadditivity holds for $z \in K_1 \cup K_2$. Let $U = R \sim (K_1 \cup K_2)$ and consider any $z \in \partial U$. Now $W_{K_1 \cup K_2}$, $W_{K_1 \cap K_2}$, W_{K_1}, and W_{K_2} all tend to zero at points of ∂R except possibly on a polar subset Z of ∂R and tend to zero as $\|y\| \to +\infty$ if R is an unbounded subset of E^n, $n \geq 3$, since these functions are equal to $V_{K_1 \cup K_2}$, $V_{K_1} \cap V_{K_2}$, V_{K_1} and V_{K_2}, respectively, on U. If $v_1 \in \Phi^1_{K_1}$, $v_2 \in \Phi^1_{K_2}$, and $z \in \partial R \sim Z$, then

$$\liminf_{\substack{y \to z \in \partial R \\ y \in U}} [v_1(y) + v_2(y) - W_{K_1 \cup K_2}(y) - W_{K_1 \cap K_2}(y)] \geq 0.$$

We now show that the same is true if $z \in \partial(K_1 \cup K_2)$. Consider any such z. Then $z \in K_1 \cup K_2$ and $W_{K_1 \cup K_2} \leq 1 = W_{K_1 \cup K_2}(z)$. This implies that

$$\limsup_{\substack{y \to z \\ y \in U}} [W_{K_1 \cup K_2}(y) + W_{K_1 \cap K_2}(y)] \leq W_{K_1 \cup K_2}(z) + \limsup_{\substack{y \to z \\ y \in U}} W_{K_1 \cap K_2}(y).$$

If $z \in K_1 \cap K_2$, then $W_{K_1 \cap K_2} \leq 1 = W_{K_1 \cap K_2}(z)$ and the second term on the right is bounded above by $W_{K_1 \cap K_2}(z)$. On the other hand, if $z \notin K_1 \cap K_2$, then $W_{K_1 \cap K_2}$ is harmonic and continuous at z and the second term on the right is $W_{K_1 \cap K_2}(z)$. It follows that for $z \in \partial(K_1 \cup K_2)$

$$\begin{aligned}\limsup_{\substack{y \to z \\ y \in U}} [W_{K_1 \cup K_2}(y) + W_{K_1 \cap K_2}(y)] &\leq W_{K_1 \cup K_2}(z) + W_{K_1 \cap K_2}(z) \\ &\leq W_{K_1}(z) + W_{K_2}(z) \\ &\leq v_1(z) + v_2(z)\end{aligned}$$

(where the strong subadditivity on $K_1 \cup K_2$ has been used). This shows that for $z \in \partial U \sim Z$

$$\liminf_{\substack{y \to z \\ y \in U}} [v_1(y) + v_2(y) - W_{K_1 \cup K_2}(y) - W_{K_1 \cap K_2}(y)] \geq 0$$

(or even if $\|y\| \to +\infty$, $y \in U$, in case R is an unbounded subset of E^n, $n \geq 3$). Since the function in the brackets is superharmonic and bounded below on U, it is non-negative on U by Theorem 7.10. This shows that $v_1 + v_2 \geq W_{K_1 \cup K_2} + W_{K_1 \cap K_2}$ on U and, by taking the infimum over $v_i \in \Phi^1_{K_i}$, $W_{K_1} + W_{K_2} \geq W_{K_1 \cup K_2} + W_{K_1 \cap K_2}$ on $R \sim (K_1 \cup K_2)$.

The following lemma gives an alternative definition of capacity which is sometimes easier to use than the original definition.

LEMMA **7.19** If K is a compact subset of the admissible set R, the capacity of K relative to R is given by $\mathscr{C}(K) = \sup\{\mu(K) : G\mu \leq 1$ on R, μ a measure with support in $K\}$.

Proof. Let U be an open set with compact closure such that $K \subset U \subset \overline{U} \subset R$ and let $v = V_U$. Then v is the potential of a measure ν and $v = G\nu = 1$ on K.

Let μ_1 and μ_2 be any two measures with support in K such that $G\mu_1 \le G\mu_2 \le 1$ on R. By Theorem 6.14

$$\mu_1(K) = \int_K 1\, d\mu_1 = \int G\nu\, d\mu_1 = \int G\mu_1\, d\nu \le \int G\mu_2\, d\nu = \mu_2(K).$$

If μ is any measure with support in K and $G\mu \le 1$ on R, then $G\mu \le V_K = G\mu_K$ by Lemma 7.17. We have just shown that $\mu(K) \le \mu_K(K) = \mathscr{C}(K)$.

5. Choquet Capacity

In this section we extend the notion of capacity to include noncompact sets. The extension applies to Borel sets and even analytic sets. The extension procedure is similar to that used in extending the domain of a measure. The reader should recall that we are limiting ourselves to subsets of an admissible set R having a Green function G.

THEOREM **7.20** The non-negative finite-valued set function $\mathscr{C}(K)$ defined on the class of compact subsets K of the admissible set R has the following properties:

(i) $\mathscr{C}(\varnothing) = 0$ and $K_1 \subset K_2 \Rightarrow \mathscr{C}(K_1) \le \mathscr{C}(K_2)$.
(ii) If $\{K_j\}$ is a decreasing sequence of compact sets and $K = \cap K_j$, then $\mathscr{C}(K) = \lim_{j\to\infty} \mathscr{C}(K_j)$.
(iii) $\mathscr{C}(K_1 \cup K_2) + \mathscr{C}(K_1 \cap K_2) \le \mathscr{C}(K_1) + \mathscr{C}(K_2)$.

Proof. Part (i) follows easily from Lemma 7.18. To prove (ii) let U be an open set with compact closure such that $K_1 \subset U \subset \overline{U} \subset R$. If ν denotes the capacitary distribution of $\overline{U}$, then $V_{\overline{U}} = G\nu$ and ν has support in ∂U. Moreover, $G\nu = 1$ on U and also on each K_j. Note that $V_{K_j} = W_{K_j}$ on $\partial U \subset R \sim K_j$. By Theorem 6.14 and the fact that ν has support in ∂U,

$$\begin{aligned}\mathscr{C}(K_j) &= \int G\nu\, d\mu_{K_j} = \int G\mu_{K_j}\, d\nu \\ &= \int V_{K_j}\, d\nu \\ &= \int W_{K_j}\, d\nu.\end{aligned}$$

Since $W_{K_j} \downarrow W_K$ by Lemma 7.18,

$$\mathscr{C}(K_j) \to \int W_K\, d\nu = \int V_K\, d\nu = \int G\mu_K\, d\nu = \int G\nu\, d\mu_K = \mathscr{C}(K).$$

To prove (iii), let U be an open set with compact closure $\overline{U} \subset R$ which contains $K_1 \cup K_2$ and let ν be the capacitary distribution for $\overline{U}$. Again by the reciprocity theorem and the fact that ν has support in $\partial U \subset R \sim (K_1 \cup K_2)$,

$$\begin{aligned}
\mathscr{C}(K_1 \cup K_2) + \mathscr{C}(K_1 \cap K_2) &= \int G\nu \, d\mu_{K_1 \cup K_2} + \int G\nu \, d\mu_{K_1 \cap K_2} \\
&= \int G\mu_{K_1 \cup K_2} \, d\nu + \int G\mu_{K_1 \cap K_2} \, d\nu \\
&= \int (V_{K_1 \cup K_2} + V_{K_1 \cap K_2}) \, d\nu \\
&= \int (W_{K_1 \cup K_2} + W_{K_1 \cap K_2}) \, d\nu \leq \int (W_{K_1} + W_{K_2}) \, d\nu \\
&= \int (V_{K_1} + V_{K_2}) \, d\nu.
\end{aligned}$$

Reverting to potentials and applying the reciprocity theorem in the same way, we get (iii).

If we are willing to assume a non-negative finite-valued set function $\mathscr{C}(\cdot)$ defined on the class of compact subsets of some space R which satisfies conditions (i), (ii), and (iii) of Theorem 7.20, the remainder of this section can be read quite independently of everything that preceded Theorem 7.20.

We now prove several lemmas which are self-explanatory.

LEMMA **7.21** If K, C_1, and C_2 are compact subsets of R, then

$$\mathscr{C}(K) + \mathscr{C}(K \cup C_1 \cup C_2) \leq \mathscr{C}(K \cup C_1) + \mathscr{C}(K \cup C_2).$$

Proof. If K_1 and K_2 are compact subsets of R, then

$$\mathscr{C}(K_1 \cup K_2) + \mathscr{C}(K_1 \cap K_2) \leq \mathscr{C}(K_1) + \mathscr{C}(K_2).$$

Replacing K_1 by $K \cup C_1$ and K_2 by $K \cup C_2$,

$$\begin{aligned}
\mathscr{C}(K \cup C_1 \cup C_2) &+ \mathscr{C}((K \cup C_1) \cap (K \cup C_2)) \\
&\leq \mathscr{C}(K \cup C_1) + \mathscr{C}(K \cup C_2);
\end{aligned}$$

but, since $\mathscr{C}((K \cup C_1) \cap (K \cup C_2)) = \mathscr{C}(K \cup (C_1 \cap C_2)) \geq \mathscr{C}(K)$, the inequality is established.

LEMMA **7.22** If C, K_1 and K_2 are compact subsets of R with $K_1 \subset K_2$, then $\mathscr{C}(C \cup K_2) - \mathscr{C}(K_2) \leq \mathscr{C}(C \cup K_1) - \mathscr{C}(K_1)$.

Proof. In the inequality of the preceding lemma, replace K by K_1, C_1 by K_2, and C_2 by C to obtain

$$\mathscr{C}(K_1) + \mathscr{C}(C \cup K_2) \leq \mathscr{C}(K_2) + \mathscr{C}(C \cup K_1).$$

LEMMA **7.23** Let $\{K_i : 1 \leq i \leq m\}$ and $\{C_i : 1 \leq i \leq m\}$ be finite collections of compact subsets of R such that $C_i \subset K_i$, $i = 1, \ldots, m$. Then

$$\mathscr{C}(\cup K_i) - \mathscr{C}(\cup C_i) \leq \sum_{i=1}^{m} [\mathscr{C}(K_i) - \mathscr{C}(C_i)].$$

Proof. The proof is by induction on m. Suppose first that $m = 2$. By the preceding lemma

$$\mathscr{C}(K_1 \cup K_2) + \mathscr{C}(C_2) \leq \mathscr{C}(K_1 \cup C_2) + \mathscr{C}(K_2)$$

and

$$\mathscr{C}(K_1 \cup C_2) + \mathscr{C}(C_1) \leq \mathscr{C}(C_1 \cup C_2) + \mathscr{C}(K_1).$$

By adding the last two inequalities we get the desired result for $m = 2$. Assume that the inequality holds for $m = 1, 2, \ldots, p$. Since $\bigcup_{i=1}^{p} C_i \subset \bigcup_{i=1}^{p} K_i$ and $C_{p+1} \subset K_{p+1}$, we can apply the $m = 2$ case and the induction hypothesis to obtain

$$\begin{aligned}
\mathscr{C}\left(\bigcup_{i=1}^{p+1} K_i\right) - \mathscr{C}\left(\bigcup_{i=1}^{p+1} C_i\right) &= \mathscr{C}\left(\bigcup_{i=1}^{p} K_i \cup K_{p+1}\right) - \mathscr{C}\left(\bigcup_{i=1}^{p} C_i \cup C_{p+1}\right) \\
&\leq \left[\mathscr{C}\left(\bigcup_{i=1}^{p} K_i\right) - \mathscr{C}\left(\bigcup_{i=1}^{p} C_i\right)\right] + [\mathscr{C}(K_{p+1}) - \mathscr{C}(C_{p+1})] \\
&\leq \sum_{i=1}^{p} [\mathscr{C}(K_i) - \mathscr{C}(C_i)] + [\mathscr{C}(K_{p+1}) - \mathscr{C}(C_{p+1})] \\
&= \sum_{i=1}^{p+1} [\mathscr{C}(K_i) - \mathscr{C}(C_i)].
\end{aligned}$$

Definition. If $E \subset R$, the **inner capacity** of E, denoted by $\mathscr{C}_*(E)$, is defined by

$$\mathscr{C}_*(E) = \sup\{\mathscr{C}(K) : K \text{ compact}, K \subset E\}.$$

If E is compact, then clearly $\mathscr{C}_*(E) = \mathscr{C}(E)$. It is also clear that $\mathscr{C}_*$ is a nonnegative monotone extended real-valued set function on the class of all subsets of R with $\mathscr{C}_*(\varnothing) = \mathscr{C}(\varnothing) = 0$.

LEMMA **7.24** If U_1 and U_2 are open subsets of R and K is a compact subset of $U_1 \cup U_2$, then there are compact sets $K_1 \subset U_1$ and $K_2 \subset U_2$ such that $K = K_1 \cup K_2$.

Proof. The sets $K \sim U_1$ and $K \sim U_2$ are closed disjoint sets and subsets of U_2 and U_1, respectively. There are disjoint open sets V_1 and V_2 such that $K \sim U_1 \subset V_1$ and $K \sim U_2 \subset V_2$. Let $K_1 = K \sim V_1$ and $K_2 = K \sim V_2$.

LEMMA **7.25** Let $\{U_i : 1 \le i \le m\}$ and $\{V_i : 1 \le i \le m\}$ be finite families of open subsets of R with $V_i \subset U_i$, $1 \le i \le m$. Then

$$\mathscr{C}_*(\cup U_i) + \sum_{i=1}^{m} \mathscr{C}_*(V_i) \le \sum_{i=1}^{m} \mathscr{C}_*(U_i) + \mathscr{C}_*(\cup V_i).$$

Proof. The proof is by induction on m. First consider the $m = 2$ case. For $i = 1, 2$, let C_1 and Σ be compact sets such that $C_i \subset V_i$ and $\Sigma \subset U_1 \cup U_2$. By the preceding lemma there are compact sets $\Sigma_i \subset U_i$, $i = 1, 2$, such that $\Sigma = \Sigma_1 \cup \Sigma_2$. By Lemma 7.23

$$\mathscr{C}((\Sigma_1 \cup C_1) \cup (\Sigma_2 \cup C_2)) - \mathscr{C}(C_1 \cup C_2) \le \mathscr{C}(\Sigma_1 \cup C_1) + \mathscr{C}(\Sigma_2 \cup C_2) - \mathscr{C}(C_1) - \mathscr{C}(C_2)$$

or

$$\mathscr{C}((\Sigma_1 \cup C_1) \cup (\Sigma_2 \cup C_2)) + \mathscr{C}(C_1) + \mathscr{C}(C_2) \le \mathscr{C}(\Sigma_1 \cup C_1) + \mathscr{C}(\Sigma_2 \cup C_2) + \mathscr{C}(C_1 \cup C_2).$$

Since $\Sigma \subset (\Sigma_1 \cup C_1) \cup (\Sigma_2 \cup C_2)$,

$$\mathscr{C}(\Sigma) + \mathscr{C}(C_1) + \mathscr{C}(C_2) \le \mathscr{C}_*(V_1 \cup V_2) + \mathscr{C}_*(U_1) + \mathscr{C}_*(U_2).$$

Taking the supremum over such Σ, C_1, and C_2,

$$\mathscr{C}_*(U_1 \cup U_2) + \mathscr{C}_*(V_1) + \mathscr{C}_*(V_2) \le \mathscr{C}_*(V_1 \cup V_2) + \mathscr{C}_*(U_1) + \mathscr{C}_*(U_2).$$

This proves the inequality for the $m = 2$ case. If the inequality is assumed to hold for $m = 2, \ldots, p$ and the $m = 2$ case is applied to the open sets $\bigcup_{i=1}^{p} U_i$, $\bigcup_{i=1}^{p} V_i$, U_{p+1}, and V_{p+1} and the induction hypothesis is applied, the inequality is seen to be true for $m = p + 1$.

Definition. If $E \subset R$, the **outer capacity** of E, denoted by $\mathscr{C}^*(E)$, is defined by

$$\mathscr{C}^*(E) = \inf\{\mathscr{C}_*(U) : U \text{ open}, U \supset E\}.$$

Definition. A subset E of R is **capacitable** if $\mathscr{C}_*(E) = \mathscr{C}^*(E)$ in which case the **capacity** $\mathscr{C}(E)$ is defined to be the common value.

LEMMA **7.26** The open subsets and compact subsets of R are capacitable.

Proof. That open sets are capacitable is obvious from the definition. Now consider any compact set K. We know $\mathscr{C}(K) = \mathscr{C}_*(K)$. Consider $\mathscr{C}^*(K)$. Given $\varepsilon > 0$ there is an open set U such that $K \subset K_1 \subset U$ implies $\mathscr{C}(K_1) - \mathscr{C}(K) < \varepsilon$, where K_1 is compact, by the right-continuity of capacity. Since

$$\mathscr{C}(K) = \mathscr{C}_*(K) \leq \mathscr{C}_*(U) = \sup_{K_1 \subset U} \mathscr{C}(K_1) \leq \mathscr{C}(K) + \varepsilon,$$

for any open set V such that $K \subset V \subset U$ we have $\mathscr{C}_*(K) \leq \mathscr{C}_*(V) \leq \mathscr{C}_*(U)$ and

$$\mathscr{C}(K) = \mathscr{C}_*(K) \leq \inf_{V \supset K} \mathscr{C}_*(V) \leq \mathscr{C}_*(U) \leq \mathscr{C}(K) + \varepsilon;$$

but, since the middle member is just $\mathscr{C}^*(K)$ and ε is arbitrary, $\mathscr{C}(K) = \mathscr{C}_*(K) = \mathscr{C}^*(K)$.

Obviously $\mathscr{C}^*$ is a non-negative monotone set function on the class of subsets of R. In order to show that a set $E \subset R$ is capacitable, we need show only that $\mathscr{C}^*(E) = \mathscr{C}_*(E)$; but, since $\mathscr{C}_*(E) = \sup_{K \subset E} \mathscr{C}(K)$, this amounts to showing that $\mathscr{C}^*(E) = \sup_{K \subset E} \mathscr{C}(K)$, where K is compact.

Definition. An extended real-valued non-negative set function ϕ^* defined on the subsets of R is called an **outer capacity** if

(i) $E \subset F \Rightarrow \phi^*(E) \leq \phi^*(F)$.

(ii) If $\{E_j\}$ is an increasing sequence of sets, then

$$\phi^*(\cup E_j) = \lim_{j \to \infty} \phi^*(E_j).$$

(iii) If $\{K_j\}$ is a decreasing sequence of compact sets, then

$$\phi^*(\cap K_j) = \lim_{j \to \infty} \phi^*(K_j).$$

A subset $E \subset R$ is **ϕ^*-capacitable** if

$$\phi^*(E) = \sup\{\phi^*(K) : K \text{ compact}, K \subset E\}.$$

Since $\mathscr{C}^*$ agrees with $\mathscr{C}$ on compact sets and (iii) in the above definition is known to hold for $\mathscr{C}$, $\mathscr{C}^*$ satisfies (iii) and, as we have remarked, (i) as well. We shall need the following lemma to show that $\mathscr{C}^*$ is an outer capacity.

LEMMA **7.27** Let $\{E_i : 1 \leq i \leq m\}$ and $\{F_i : 1 \leq i \leq m\}$ be finite families of subsets of R such that $F_i \subset E_i$, $1 \leq i \leq m$. Then

$$\mathscr{C}^*(\cup E_i) + \sum_{i=1}^{m} \mathscr{C}^*(F_i) \leq \mathscr{C}^*(\cup F_i) + \sum_{i=1}^{m} \mathscr{C}^*(E_i).$$

Proof. For each i let U_i be an open set such that $U_i \supset E_i$. Then $\mathscr{C}^*(\cup E_i) \leq \mathscr{C}_*(\cup U_i) \leq \sum_{i=1}^{m} \mathscr{C}_*(U_i)$ by Lemma 7.25. Taking the infimum over such U_i's containing E_i, we get $\mathscr{C}^*(\cup E_i) \le \sum_{i=1}^m \mathscr{C}^*(E_i)$, that is, $\mathscr{C}^*(\cdot)$ is finitely subadditive. If some $\mathscr{C}^*(E_i)$ is infinite, then there is nothing to prove. If each $\mathscr{C}^*(E_i)$ is finite, then all other members of the inequality are also finite. We assume finiteness. If $\varepsilon > 0$, there is an open set $U \supset \cup F_i$ and open sets $U_i \supset E_i$, $1 \le i \le m$, such that

$$\mathscr{C}_*(U) + \sum_{i=1}^m \mathscr{C}_*(U_i) \le \mathscr{C}^*(\cup F_i) + \sum_{i=1}^m \mathscr{C}^*(E_i) + \varepsilon.$$

Let $V_i = U_i \cap U \subset U_i$. By Lemma 7.25

$$\mathscr{C}_*(\cup U_i) + \sum_{i=1}^m \mathscr{C}_*(V_i) \le \mathscr{C}_*(\cup V_i) + \sum_{i=1}^m \mathscr{C}_*(U_i).$$

Since $U \supset \cup V_i$, $\mathscr{C}_*(\cup V_i) + \sum_{i=1}^m \mathscr{C}_*(U_i) \le \mathscr{C}_*(U) + \sum_{i=1}^m \mathscr{C}_*(U_i)$ and

$$\mathscr{C}_*(\cup U_i) + \sum_{i=1}^m \mathscr{C}_*(V_i) \le \mathscr{C}^*(\cup F_i) + \sum_{i=1}^m \mathscr{C}^*(E_i) + \varepsilon.$$

Since $E_i \subset U_i$, $\cup E_i \subset \cup U_i$ and $\mathscr{C}^*(\cup E_i) \le \mathscr{C}^*(\cup U_i) = \mathscr{C}_*(\cup U_i)$. Moreover, since $F_i \subset E_i \subset U_i$ and $F_i \subset U$, $F_i \subset V_i$ and $\mathscr{C}^*(F_i) \le \mathscr{C}^*(V_i) = \mathscr{C}_*(V_i)$. Therefore

$$\mathscr{C}^*(\cup E_i) + \sum_{i=1}^m \mathscr{C}^*(F_i) \le \mathscr{C}^*(\cup F_i) + \sum_{i=1}^m \mathscr{C}^*(E_i) + \varepsilon.$$

Since ε is arbitrary, the lemma is proved.

THEOREM **7.28** $\mathscr{C}^*$ is an outer capacity on the class of subsets of R.

Proof. We need show only that if $\{E_j\}$ is an increasing sequence of subsets of R then $\mathscr{C}^*(\cup E_j) = \lim_{j\to\infty} \mathscr{C}^*(E_j)$. Assume first that the E_j's are open. In this case $\cup E_j$ is open and capacitable along with the E_j's by Lemma 7.26. Let η be any number such that $\mathscr{C}^*(\cup E_j) > \eta$. Since $\cup E_j$ is capacitable, there is a compact set $K \subset \cup E_j$ such that $\mathscr{C}(K) > \eta$ (we have taken the liberty of dropping "$*$" since the sets are capacitable). Since $\{E_j\}$ is an open covering of the compact set K, there is a j_0 such that $K \subset E_j$ for all $j \ge j_0$. Thus $\mathscr{C}(K) \le \mathscr{C}(E_j) \le \mathscr{C}^*(\cup E_j)$ for all $j \ge j_0$ and $\mathscr{C}(E_j) > \eta$ for all $j \ge j_0$. Since η was any number less than $\mathscr{C}^*(\cup E_j)$, $\lim_{j\to\infty} \mathscr{C}^*(E_j) = \mathscr{C}^*(\cup E_j)$. Now suppose the E_j's are arbitrary subsets of R. If $\lim_{j\to\infty} \mathscr{C}^*(E_j) = +\infty$, then there is nothing to prove, since $\mathscr{C}^*(\cup E_j) \ge \mathscr{C}^*(E_j)$ for each j. We can therefore assume that the sequence $\{\mathscr{C}^*(E_j)\}$ is bounded and, in particular, each $\mathscr{C}^*(E_j)$ is finite. Let $\varepsilon > 0$. For each j there is an open set $U_j \supset E_j$ such that $\mathscr{C}(U_j) < \mathscr{C}^*(E_j) + \varepsilon 2^{-j}$. By the preceding lemma

$$\mathscr{C}^*\left(\bigcup_{j=1}^{p} U_j\right) - \mathscr{C}^*\left(\bigcup_{j=1}^{p} E_j\right) \le \sum_{j=1}^{p} [\mathscr{C}^*(U_j) - \mathscr{C}^*(E_j)]$$

$$< \sum_{j=1}^{p} \varepsilon 2^{-j} < \varepsilon.$$

Since the U_j's are open, the first part of the proof may be applied to give

$$\mathscr{C}^*(\cup E_j) \le \mathscr{C}^*(\cup U_j) = \lim_{p\to\infty} \mathscr{C}^*\left(\bigcup_{j=1}^{p} U_j\right) \le \lim_{p\to\infty} \mathscr{C}^*\left(\bigcup_{j=1}^{p} E_j\right) + \varepsilon.$$

Since ε is arbitrary, $\mathscr{C}^*(\cup E_j) \le \lim_{p\to\infty} \mathscr{C}^*(\bigcup_{j=1}^{p} E_j) = \lim_{p\to\infty} \mathscr{C}^*(E_p)$, and, because we always have $\lim_{p\to\infty} \mathscr{C}^*(E_p) \le \mathscr{C}^*(\cup E_j)$ by the monotonicity of $\mathscr{C}^*$, $\lim_{p\to\infty} \mathscr{C}^*(E_p) = \mathscr{C}^*(\cup E_j)$.

We need only the fact that $\mathscr{C}^*$ is an outer capacity to show that a certain class of sets is capacitable; that is, $\mathscr{C}^*$-capacitable. This class includes not only the open sets and compact sets, which we have already shown to be capacitable, but also the Borel sets. Even better, it contains the classical analytical sets as well.

Definition A subset of a Hausdorff space is a $\mathscr{K}_\sigma$ set if it is a countable union of compact sets. A subset of a Hausdorff space is a $\mathscr{K}_{\sigma\delta}$ set if it is a countable intersection of $\mathscr{K}_\sigma$ sets.

It is easily seen that the intersection of a closed set with a $\mathscr{K}_{\sigma\delta}$ set is again a $\mathscr{K}_{\sigma\delta}$ set and that a countable union of $\mathscr{K}_\sigma$ sets is again a $\mathscr{K}_\sigma$ set.

Definition. A subset $E \subset E^n$ is a $\mathscr{K}$-**analytic** set if there is a compact Hausdorff space X, a $\mathscr{K}_{\sigma\delta}$ subset D of X, and a continuous mapping f from D onto E. In particular, the compact subsets of E^n are $\mathscr{K}$-analytic for they are continuous images of themselves under the identity map.

THEOREM **7.29** The class $\mathscr{K}$ of $\mathscr{K}$-analytic subsets of the open set R is closed under countable unions and intersections and contains the Borel subsets of R.

Proof. We first show that $\mathscr{K}$ is closed under countable unions. Let $\{E_j\}$ be a sequence of $\mathscr{K}$-analytic subsets of R where E_j is the continuous image under f_j of some $\mathscr{K}_{\sigma\delta}$ set D_j in some compact Hausdorff space X_j. We can assume that the X_j's are disjoint. A topology can be assigned to $\cup X_j$ by taking as a base the collection of all open subsets of the X_j's. Let X be the one-point compactification of $\cup X_j$ under this topology. Then X is a compact Hausdorff space. Define a mapping $f\colon \cup D_j \rightsquigarrow R$ as follows. If $x \in D_j$, define $f(x) = f_j(x)$. This is clearly a continuous mapping onto $\cup E_j$. We need only show that $D = \cup D_j \subset \cup X_j$ is a $\mathscr{K}_{\sigma\delta}$ subset of $X = \cup X_j$. For each j let $D_j = \bigcap_{k=1}^{\infty} D_{jk}$, where each D_{jk} is a $\mathscr{K}_\sigma$ set. Consider $D = \cup_j D_j = \cup_j \cap_k D_{jk}$. Since the X_j's are disjoint, $D = \cup_j \cap_k D_{jk} = \cap_k \cup_j D_{jk}$. Since each D_{jk} is a $\mathscr{K}_\sigma$ set, $\cup_j D_{jk}$ is a $\mathscr{K}_\sigma$ set and D is a $\mathscr{K}_{\sigma\delta}$ subset of X. This shows that $\mathscr{K}$ is closed under countable unions. Note that any open subset of R is $\mathscr{K}$-analytic since each such set is a countable union of compact sets which are $\mathscr{K}$-analytic. We now show that $\mathscr{K}$ is closed under countable intersections. Let $\{E_j\}$ be a sequence of $\mathscr{K}$-analytic subsets of R and let $E_\delta = \cap E_j$. Let E_j be the continuous image under f_j of a $\mathscr{K}_{\sigma\delta}$ set D_j contained in the compact Hausdorff space X_j. Let $X = \Pi X_j$ and let $D \subset X$ be defined by $D = \Pi D_j$. As a product of compact spaces, X is compact in the usual product topology. For each j, consider the cylinder set $D_j^* = \Pi Y_i$ where $Y_i = X_i$ for $i \neq j$ and $Y_j = D_j$. Since D_j is a $\mathscr{K}_{\sigma\delta}$ subset of X_j, D_j^* is a $\mathscr{K}_{\sigma\delta}$ subset of X. Since $D = \cap_j D_j^*$, D is a $\mathscr{K}_{\sigma\delta}$ subset of X. Now let $\tilde{f}_j$ be the natural extension of f_j; that is, if $x = (,\ldots, x_j, \ldots) \in D$, where $x_j \in D_j$, then $\tilde{f}_j((\ldots, x_j, \ldots)) = f_j(x_j)$. Each $\tilde{f}_j$ maps D onto E_j. Since $\tilde{f}_j$ is the composition of the continuous projection map $\pi_j : (\ldots, x_j, \ldots) \rightsquigarrow x_j$ and the continuous map f_j, $\tilde{f}_j$ is a continuous map from D onto E_j. For each i and j, $\{x \in D : \tilde{f}_i(x) = \tilde{f}_j(x)\}$ is a closed subset of D by the continuity on D of $\tilde{f}_i$ and $\tilde{f}_j$. Since D is a $\mathscr{K}_{\sigma\delta}$ subset of X, $\{x \in D : \tilde{f}_i(x) = \tilde{f}_j(x)\}$ is again a $\mathscr{K}_{\sigma\delta}$ subset of X. Since a countable intersection of $\mathscr{K}_{\sigma\delta}$ sets is again a $\mathscr{K}_{\sigma\delta}$ set, $D_\delta = \{x \in D : \tilde{f}_i(x) = \tilde{f}_j(x)$ for all i and $j\} = \cap_{i,j}\{x \in D : \tilde{f}_i(x) = \tilde{f}_j(x)\}$ is also a $\mathscr{K}_{\sigma\delta}$ subset of X. Note that all of the mappings $\tilde{f}_i$ agree on D_δ; that is, $\tilde{f}_i|_{D_\delta}$ is independent of i. Fix i and let $\tilde{f} = \tilde{f}_i|_{D_\delta}$. Then $\tilde{f}$ is continuous on D_δ. Now $\tilde{f}(D_\delta) = \tilde{f}_i(D_\delta) \subset \tilde{f}_i(D) = f_i(D_i) = E_i$ for all i. This shows that $\tilde{f}(D_\delta) \subset \cap E_i = E_\delta$. Suppose now that $y \in E_\delta = \cap E_i$. Then for each i there is an $x_i \in D_i$ such that $f_i(x_i) = y$. Then $x = \{x_i\} \in D_\delta$ and $\tilde{f}(x) = y$; that is, $E_\delta = \tilde{f}(D_\delta)$ and E_δ is a $\mathscr{K}$-analytic subset of R. This concludes the proof that $\mathscr{K}$ is closed under countable intersections. It remains only to prove that $\mathscr{K}$ contains the Borel subsets of R. The class of Borel sets is the smallest class of sets which is closed under complementation and countable unions and which contains the compact subsets of R. Let $\mathscr{M}$ be the class of $\mathscr{K}$-analytic subsets of R having $\mathscr{K}$-analytic complements. Since an open set is a countable union of compact sets which are $\mathscr{K}$-analytic, any open set is $\mathscr{K}$-analytic. Since any compact set is $\mathscr{K}$-analytic along with its complement, $\mathscr{M}$ contains the class of compact subsets of R. Obviously $\mathscr{M}$ is closed under complementation and it is easily seen that $\mathscr{M}$ is closed under countable unions. Therefore, $\mathscr{M}$ contains the Borel subsets of R.

LEMMA **7.30** If ϕ^* is an outer capacity on the subsets of a Hausdorff space X, then each $\mathscr{K}_{\sigma\delta}$ subset of X is ϕ^*-capacitable.

Proof. Let E be a $\mathscr{K}_{\sigma\delta}$ subset of X. We must prove that $\phi^*(E) = \sup\{\phi^*(K) : K$ compact, $K \subset E\}$. Let $E = \bigcap_{j=1}^{\infty} E_j$, where each E_j is a $\mathscr{K}_\sigma$

set and let $E_j = \cup_k E_{jk}$, where each E_{jk} is a compact subset of X. We can assume that, for each j, $\{E_{jk}\}$ is an increasing sequence of compact sets. Let η be any number such that $\phi^*(E) > \eta$. Consider the sequence $\{E_{1k} \cap E\}$ which is increasing and has union $E_1 \cap E = E$. Since $\phi^*(E) = \phi^*(E_1 \cap E) = \sup_k \phi^*(E_{1k} \cap E)$, there is an integer p_1 such that if $F_1 = E_{1p_1} \cap E$ then $\phi^*(F_1) > \eta$. Suppose the integers $p_1, \ldots, p_{j-1}$ have been defined so that if $F_{j-1} = E \cap E_{1p_1} \cap \cdots \cap E_{(j-1)p_{j-1}}$ then $\phi^*(F_{j-1}) > \eta$. With j fixed, $\{E_{jk} \cap F_{j-1}\}$ is an increasing sequence of which the union is $E_j \cap F_{j-1}$. Now $E_j \cap F_{j-1} = F_{j-1}$ since $F_{j-1} \subset E \subset E_j$. Therefore $\phi^*(E_j \cap F_{j-1}) = \phi^*(F_{j-1}) > \eta$ and there is an integer p_j such that $\phi^*(E_{jp_j} \cap F_{j-1}) > \eta$. Define $F_j = E_{jp_j} \cap F_{j-1} = E \cap E_{1p_1} \cap \cdots \cap E_{jp_j}$. Let $F_j^* = E_{1p_1} \cap \cdots \cap E_{jp_j}$. Then $F_j = E \cap F_j^* \subset F_j^*$. Since $\phi^*(F_j) > \eta$, $\phi^*(F_j^*) > \eta$. Since $\{F_j^*\}$ is a decreasing sequence of compact sets, $\phi^*(\cap F_j^*) = \lim_{j\to\infty} \phi^*(F_j^*) \geq \eta$. Now $\cap F_j^* = \cap E_{jp_j} \subset \cap E_j = E$. Therefore $\cap F_j^*$ is a compact subset of E for which $\phi^*(\cap F_j^*) \geq \eta$. Since η is any number less than $\phi^*(E)$, $\phi^*(E) = \{\sup \phi^*(K) : K \text{ compact}, K \subset E\}$ and E is ϕ^*-capacitable.

LEMMA **7.31** Let X be a Hausdorff space, let f be a continuous map from X into the open set $R \subset E^n$, and define $\phi^*(E) = \mathscr{C}^*(f(E))$ for $E \subset X$. Then ϕ^* is an outer capacity of subsets of X and $f(E)$ is capacitable whenever E is ϕ^*-capacitable.

Proof. If $E \subset F$, then $\phi^*(E) = \mathscr{C}^*(f(E)) \leq \mathscr{C}^*(f(F)) = \phi^*(F)$ and ϕ^* is monotone. Let $\{E_j\}$ be an increasing sequence of subsets of X. Since $f(\cup E_j) = \cup f(E_j), \phi^*(\cup E_j) = \mathscr{C}^*(f(\cup E_j)) = \mathscr{C}^*(\cup f(E_j)) = \sup_j \mathscr{C}^*(f(E_j)) = \sup_j \phi^*(E_j)$. Now let $\{K_j\}$ be a decreasing sequence of compact subsets of X. We first show that $f(\cap K_j) = \cap f(K_j)$. Obviously $f(\cap K_j) \subset \cap f(K_j)$. To show the converse, let $y \in \cap f(K_j)$. Then for each j there is an $x_j \in K_j$ such that $f(x_j) = y$. Since the sequence $\{x_j\}$ lies in the compact set K_1, there is a subsequence $\{x_{j_k}\}$ with limit $x \in \cap K_j$. In view of the continuity of f, $y = f(x_{j_k}) \to f(x)$ as $k \to \infty$ so that $f(x) = y$ and $\cap f(K_j) \subset f(\cap K_j)$. This shows that $f(\cap K_j) = \cap f(K_j)$. Now $\phi^*(\cap K_j) = \mathscr{C}^*(f(\cap K_j)) = \mathscr{C}^*(\cap f(K_j)) = \lim_{j\to\infty} \mathscr{C}^*(f(K_j)) = \lim_{j\to\infty} \phi^*(K_j)$. This concludes the proof that ϕ^* is an outer capacity. Suppose now that $E \subset X$ is ϕ^*-capacitable and that λ is any real number with $\mathscr{C}^*(f(E)) = \phi^*(E) > \lambda$. There is then a compact set $K \subset E$ such that $\phi^*(K) > \lambda$. Therefore $\mathscr{C}^*(f(K)) > \lambda$ and, since f takes compact sets into compact sets, $f(K)$ is a compact subset of $f(E)$. This shows that $\mathscr{C}^*(f(E)) = \sup\{\mathscr{C}^*(K) : K \text{ compact}, K \subset f(E)\}$.

THEOREM **7.32** (*Choquet*) The $\mathscr{K}$-analytic subsets of the open set R are capacitable.

Proof. Let E be a $\mathscr{K}$-analytic subset of R contained in a compact set $C \subset R$. To show that E is capacitable, we must show that $\mathscr{C}^*(E) = \sup\{\mathscr{C}^*(K) : K$ compact, $K \subset E\}$. Since E is $\mathscr{K}$-analytic, there is a compact Hausdorff space X, a $\mathscr{K}_{\sigma\delta}$ subset D of X, and a continuous mapping $f : D \rightsquigarrow E$ such that $f(D) = E$. Let $\Gamma \subset X \times C$ denote the graph of f. Then $E = P_r\Gamma$, where P_r is the continuous projection map of $X \times C$ onto C. Since D is a $K_{\sigma\delta}$ subset of X, $D \times C$ is a $K_{\sigma\delta}$ subset of $X \times C$. Now $\Gamma \subset D \times C$ and we shall show that Γ is relatively closed in $D \times C$. Suppose $\{(x_a, f(x_a)) : a \in A\}$ is a net in Γ which converges to $(x, e) \in D \times C$. Since $x_a \to x$ on the directed set A, $f(x_a) \to f(x)$ in the same sense by the continuity of f. Since $f(x_a) \to e$ also and C is Hausdorff, $f(x) = e$ and $(x, e) \in \Gamma$. This shows that Γ is relatively closed in $D \times C$. It follows that there is a closed subset of $X \times C$ such that Γ is the intersection of this closed set with $D \times C$. For this closed subset of $X \times C$ we can take $\overline{\Gamma}$, the closure of Γ in $X \times C$; that is, $\Gamma = \overline{\Gamma} \cap (D \times C)$. Since $\overline{\Gamma}$ is closed and $D \times C$ is a $\mathscr{K}_{\sigma\delta}$ subset of $X \times C$, Γ is a $\mathscr{K}_{\sigma\delta}$ subset of $X \times C$. For $\Sigma \subset X \times C$, define $\phi^*(\Sigma) = \mathscr{C}^*(P_r\Sigma)$. Since the projection mapping $P_r : X \times C \rightsquigarrow C$ is continuous, ϕ^* is an outer capacity on the subsets of $X \times C$ by Lemma 7.31. By Lemma 7.30, Γ is ϕ^*-capacitable. It follows that $E = P_r\Gamma$ is capacitable by Lemma 7.31. Now let E be any $\mathscr{K}$-analytic subset of R. Then $E \subset \cup C_j$ where the C_j's are compact subsets of R. Let $E_j = E \cap C_j$. Then $E = \cup E_j$ and $\mathscr{C}^*(E) = \sup_j \mathscr{C}^*(E_j)$. Consider any number λ for which $\mathscr{C}^*(E) > \lambda$. There is then a j_0 such that $\mathscr{C}^*(E_{j_0}) > \lambda$. Since E_{j_0} is a $\mathscr{K}$-analytic set contained in the compact set $C_{j_0} \subset R$, E_{j_0} is capacitable and there is a compact set $K \subset E_{j_0} \subset E$ such that $\mathscr{C}^*(K) > \lambda$. This shows that $\mathscr{C}^*(E) = \sup\{\mathscr{C}^*(K) : K$ compact, $K \subset E\}$ and that E is capacitable.

6. Applications

We have seen that a polar set in E^n has n-dimensional Lebesgue measure zero. We now show that Lebesgue measure may be replaced by capacity.

THEOREM **7.33** (*Cartan*) A subset of an admissible set R is a polar set if and only if it has capacity zero relative to R.

Proof. Let $Z \subset R$ be a polar set. We can assume that $Z \subset W$, where W is an open subset of R with compact closure $\overline{W} \subset R$. This follows from the fact that $\mathscr{C}^*(Z) = \lim_{j \to \infty} \mathscr{C}^*(Z \cap W_j)$, where $\{W_j\}$ is an increasing sequence of open subsets of R with compact closure $\overline{W}_j \subset R$ and $R = \cup W_j$. By Theorems 7.3 and 7.4 there is a superharmonic function u such that $u = +\infty$ on Z. By the Riesz decomposition theorem, there is a measure μ on W such that $u = G_W\mu + h$ where h is harmonic on W. Since $u = +\infty$ on Z, $G_W\mu = +\infty$

on Z. Since $G_W \leq G$ on $W \times W$, $G\mu = +\infty$ on Z. Since $G\mu$ is l.s.c., $U_r = \{x \in R: G\mu(x) > r\}$ is an open subset of R for each $r > 0$. Consider any compact set $K \subset U_r$ and the capacitary distribution μ_K corresponding to K and R. By the reciprocity theorem and the fact that $V_K = G\mu_K \leq 1$ on R,

$$\mathscr{C}(K) = \int d\mu_K \leq \frac{1}{r}\int G\mu\, d\mu_K = \frac{1}{r}\int G\mu_K\, d\mu \leq \frac{1}{r}\mu(\overline{W}).$$

It follows that $\mathscr{C}_*(U_r) = \mathscr{C}(U_r) \leq (1/r)\,\mu(\overline{W})$. Since $\mathscr{C}^*(Z) \leq \inf_{r>0}\mathscr{C}(U_r)$, $\mathscr{C}^*(Z) = 0$. That Z has capacity zero relative to R follows from the fact that $0 \leq \mathscr{C}_*(Z) \leq \mathscr{C}^*(Z)$. As to the converse, suppose $\mathscr{C}^*(Z) = 0$. Since a countable union of polar sets is again polar, it suffices to prove that Z is a polar set in the case where Z is a subset of a connected open set W with compact closure $\overline{W} \subset R$. Let U be a nonempty open subset of W with $\overline{U} \subset W$. Then $Z \cap U$ and $Z \cap (W \sim U)$ both have outer capacity zero by the monotonicity of an outer capacity. If we can show that each is a polar set, then it would follow from Theorem 7.6 that Z is a polar set. In either case we can find a point not in the set of outer capacity zero which can be separated from the latter by means of neighborhoods. We might as well assume that Z has this property; that is, there is a point $y \in W \sim Z$ and open sets U and V such that (i) $y \in U \subset W$. (ii) $Z \subset V \subset W$, and (iii) $U \cap V = \varnothing$. Since $\mathscr{C}^*(Z) = 0$, for every positive integer j there is an open set V_j with $Z \subset V_j \subset V$ and $\mathscr{C}(V_j) < j^{-2}$. For the time being consider a fixed j. Now there is an increasing sequence $\{W_i\}$ of open sets such that $\overline{W}_i \subset W_{i+1} \subset V_j$ and $V_j = \cup W_i$. For each i let $K_i = \overline{W}_i$ and consider the corresponding capacitary potential $V_{K_i} = G\mu_{K_i}$ (relative to R), which is equal to 1 on W_i. Since $G(y, \cdot)$ is bounded outside any ball containing y, $\sup_{x \in V} G(y, \cdot) \leq m$ for some m. Using this along with the fact that the support of μ_{K_i} is contained in $K_i \subset V_j \subset V$,

$$V_{K_i}(y) = \int G(y, x)\, d\mu_{K_i}(x) \leq m\mu_{K_i}(K_i) \leq m\mathscr{C}(K_i) \leq m\mathscr{C}(V_j) \leq mj^{-2}.$$

Since $\{V_{K_i}\}$ is an increasing sequence, we can let $w_j = \lim_{i\to\infty} V_{K_i}$. Then $w_j(y) \leq mj^{-2}$. Since $\{V_{K_i}\}$ is an increasing sequence of potentials uniformly bounded by 1, w_j is a non-negative superharmonic function on R by Theorem 4.15. In addition, the fact that V_{K_i} is 1 on W_i implies that w_j is 1 on each W_i and therefore 1 on $V_j \supset Z$. Letting $w = \sum_{j=1}^{\infty} w_j$, $w(y) = \sum_{j=1}^{\infty} w_j(y) \leq \sum_{j=1}^{\infty} mj^{-2} < +\infty$. Since the partial sums of the series defining w form an increasing sequence of superharmonic functions on R, w is superharmonic on the component of R containing W by Theorem 4.15. Since $w_j = 1$ on Z, $w = +\infty$ on Z. This concludes the proof that Z is a polar set.

Using this theorem, one can show that the property of being of capacity zero is independent of the admissible set R used to define the capacity; that is, if $Z \subset R$ and has capacity zero, then Z is a polar set; if Z is a subset of some other admissible set R_1, then it has capacity zero relative to R_1.

Using some of the ideas developed in the preceding section, we now find it possible to improve Theorems 7.3, 7.4, and 7.14, at least for compact polar sets.

THEOREM **7.34** (*Evans*) If Z is a compact polar subset of the admissible set R having Green function G, then there is a measure μ such that $G\mu = +\infty$ on Z and $G\mu < +\infty$ on $R \sim Z$.

Proof. For each positive integer j let $U_j = \{y : d(y, Z) < j^{-1}\}$. Since there is a j_0 such that $\bar{U}_j \subset R$ for all $j \geq j_0$, we can assume that $\bar{U}_1 \subset R$ by a simple change of notation if necessary. Then $Z = \cap U_j$, the U_j's form a decreasing sequence, and the sequence of capacitary potentials V_{K_j}, where $K_j = \bar{U}_j$, forms a decreasing sequence of non-negative superharmonic functions. Moreover, $V_{K_j} = 1$ on Z and V_{K_j} is harmonic on $R \sim K_j$. Define v on $R \sim Z$ by $v = \lim_{j \to \infty} V_{K_j}$. Since for each positive integer p the functions V_{K_j}, $j \geq p$, are harmonic outside K_p, v is harmonic outside K_p by Theorem 2.19 and therefore harmonic on $R \sim Z$. Since $0 \leq v \leq 1$, v has a harmonic extension v^* to R by Corollary 7.8. Since $0 \leq v^* \leq V_{K_1}$ and V_{K_1} has greatest harmonic minorant 0, $v^* = 0$ on R and $v = 0$ on $R \sim Z$. Now let $\{R_j\}$ be an increasing sequence of open subsets of R with compact closures $\bar{R}_j \subset R$ such that $R = \cup R_j$. We can assume that $U_1 \subset R_1$. For each integer p the functions $V_{K_j}, j \geq p$, are continuous on the compact set $\bar{R}_p \sim U_p$ and decrease monotonically to the continuous function 0. By Dini's theorem, the convergence is uniform on $\bar{R}_p \sim U_p$. It follows that there is an integer $j_p \geq p$ such that $V_{K_j} < 2^{-p}$ on $\bar{R}_p \sim U_p$ for all $j \geq j_p$. Let $\phi_p = V_{K_{j_p}}$ and define $w = \sum_{p=1}^{\infty} \phi_p$. Now $w = +\infty$ on Z, since $\phi_p = V_{K_{j_p}} = 1$ on Z. If $y \in R \sim Z$, then $y \in \bar{R}_q \sim U_q$ for some q and

$$w(y) = \sum_{p=1}^{q-1} \phi_p(y) + \sum_{p=q}^{\infty} \phi_p(y) \leq (q-1) + \sum_{p=q}^{\infty} 2^{-p} < +\infty.$$

Therefore w is finite on $R \sim Z$. The function w is a non-negative superharmonic function on R. By the Riesz decomposition theorem there is a measure μ on R such that $w = G\mu + h$, where h is the greatest harmonic minorant of w. Therefore $G\mu$ is $+\infty$ on Z and finite on $R \sim Z$.

It has been pointed out that polar sets are negligible in several situations. On the other hand, if we try to put a nontrivial charge on a polar set, the polar set will be anything but negligible.

THEOREM **7.35** Let M be a Borel subset of the admissible set R having Green function G and let μ be a measure on R with $\mu(M) > 0$ such that $G\mu$ is bounded on R. Then M is not a polar set.

Proof. Assume that M is a polar subset of R. Then M has capacity zero relative to R. Now there is an open set W with compact closure $\overline{W} \subset R$ such that $\mu(W \cap M) > 0$. Since $W \cap M$ is a polar set, there is a superharmonic function v such that $v = +\infty$ on $W \cap M$. We can assume that $v \geq 0$ on $\overline{W}$ by adding a constant if necessary. Consider the measure ν associated with v and W by the Riesz decomposition theorem. We shall extend ν to R by letting $\nu(R \sim W) = 0$. By the decomposition theorem $v = G_W \nu + h \leq G\nu + h$ on W, where h is harmonic on W. Since $v = +\infty$ on $W \cap M$, $G\nu = +\infty$ on $W \cap M$. By the reciprocity theorem

$$+\infty = \int_R G\nu \, d\mu = \int_R G\mu \, d\nu;$$

but, since $G\mu$ is bounded on R and $\nu(R) = \nu(\overline{W}) < \infty$, we have a contradiction. It follows that M is not a polar set.

For a final application of some of the concepts developed in this chapter we return to decreasing sequences of superharmonic functions. We have seen in Theorem 4.16 that a decreasing sequence of superharmonic functions which is locally bounded below has a limit whose lower regularization is superharmonic. After obtaining some preliminary results we shall show that the limit differs from its lower regularization at most on a polar set. It should be pointed out that this result will apply to any open set R and not just to admissible sets.

LEMMA **7.36** Let R be an open set having Green function G and let $\{\mu_j\}$ be a sequence of measures on R which converges in the w^*-topology to μ. Then $G\mu \leq \liminf_{j\to\infty} G\mu_j$.

Proof. Consider any $x \in R$. Then $G(x, \cdot)$ is non-negative and l.s.c. on R. There is then a sequence $\{\phi_j\}$ of continuous functions with compact supports such that $\phi_j \uparrow G(x, \cdot)$. Then

$$\int \phi_i \, d\mu_j \leq \int G(x, \cdot) \, d\mu_j .$$

Since for fixed i, ϕ_i has compact support and $\mu_j \xrightarrow{w^*} \mu$,

$$\int \phi_i \, d\mu \leq \liminf_{j\to\infty} \int G(x, \cdot) \, d\mu_j .$$

Letting $i \to \infty$, we obtain $G\mu(x) \leq \liminf_{j\to\infty} G\mu_j(x)$.

LEMMA **7.37** If K is a nonpolar compact subset of the admissible set R, there is a nonzero measure μ with support K such that $G\mu$ is finite and continuous on R.

Proof. Consider the capacitary distribution μ_K of K relative to R. Since K is nonpolar, it has positive capacity; that is, $\mathscr{C}(K) = \mu_K(K) > 0$. Note that $0 \le V_K = G\mu_K \le 1$ on R. By Theorem 6.21 there is a compact set $C \subset K$ with $\mu_K(C) > 0$ such that $G\mu_K|_C$ is finite and continuous on R.

LEMMA **7.38** Let $\{\mu_i : i \in I\}$ be a family of measures on the ball B with the following properties:

(i) The support of each μ_i is contained in a compact set $K \subset B$.
(ii) The family $\{G_B\mu_i : i \in I\}$ is left-directed.
(iii) The set $\{\mu_i(B) : i \in I\}$ is bounded.

Then there is a measure μ such that $G_B\mu \le \inf_{i\in I} G_B\mu_i$ with equality except possibly on a polar set.

Proof. We can apply Lemma 2.20 to obtain a countable set $I_0 \subset I$ such that if g is l.s.c. on B and $g \le \inf_{i\in I_0} G_B\mu_i$ then $g \le \inf_{i \in I} G_B\mu_i$. Since the family $\{G_B\mu_i : i \in I\}$ is left-directed, we can assume that the sequence $\{G_B\mu_i : i \in I_0\}$ is monotone decreasing. Suppose $I_0 = \{i_j : j \ge 1\}$ and the sequence $\{G_B\mu_{i_j} : j \ge 1\}$ is decreasing. From the sequence $\{\mu_{i_j}\}$ we can extract a subsequence which converges in the w^*-topology to some measure μ. We can assume that the sequence itself has this property. By Lemma 7.36 $G_B\mu \le \lim_{j\to\infty} G_B\mu_{i_j} = \inf_{i\in I_0} G_B\mu_i$. But since $G_B\mu$ is l.s.c., $G_B\mu \le \inf_{i\in I} G_B\mu_i \le \lim_{j\to\infty} G_B\mu_{i_j}$. Consider the Borel set $E = \{x \in B : G_B\mu(x) < \lim_{j\to\infty} G_B\mu_{i_j}(x)\}$. Assume that E does not have inner capacity zero relative to B. There is then a compact set $C \subset E$ such that $G_B\mu < \lim_{j\to\infty} G_B\mu_{i_j}$ on C and C has positive capacity. By Lemma 7.37 there is a nonzero measure ν with support in C such that $G_B\nu$ is finite and continuous on $\bar{B}$ (if defined to be zero on ∂B). Since $G_B\nu$ is continuous, $\int G_B\nu\, d\mu_{i_j} \to \int G_B\nu\, d\mu$ as $j \to \infty$. By the reciprocity theorem and Fatou's lemma

$$\int \lim_{j\to\infty} G_B\mu_{i_j}\, d\nu \le \lim_{j\to\infty} \int G_B\mu_{i_j}\, d\nu = \lim_{j\to\infty} \int G_B\nu\, d\mu_{i_j} = \int G_B\nu\, d\mu = \int G_B\mu\, d\nu$$

or

$$\int \left(G_B\mu - \lim_{j\to\infty} G_B\mu_{i_j} \right) d\nu \ge 0;$$

but since the integrand is strictly negative on the support of ν, we have a contradiction. It follows that E has inner capacity zero and since Borel sets are capacitable, E has capacity zero. By Theorem 7.33 E is a polar set and $G_B\mu = \inf_{i\in I} G_B\mu_i = \lim_{j\to\infty} G_B\mu_{i_j}$ except possibly on a polar set.

Having proved that the limit of a decreasing sequence of potentials on a ball is equal to a potential except possibly on a polar set, the case of decreasing sequences of superharmonic functions is easily reduced to the first case.

THEOREM **7.39** (*Cartan*) Let $\mathscr{F}$ be a family of superharmonic functions which is locally bounded below on an open set R. Then $v = \inf\{u : u \in \mathscr{F}\}$ differs from the superharmonic lower regularization $\hat{v}$ of v at most on a polar set.

Proof. That $\hat{v}$ is superharmonic on R follows from Theorem 4.16. Since a countable union of polar sets is again polar, it suffices to prove that $\hat{v} = v$ except possibly on a polar subset of a ball $B \subset \bar{B} \subset R$. If fact, we can assume that R is a ball and that the family $\mathscr{F}$ is non-negative on R by replacing R with a ball containing $\bar{B}$ in its interior and having closure in R, if necessary. By replacing $\mathscr{F}$ with the collection of infimums of finitely many elements of $\mathscr{F}$, if necessary, we can assume that $\mathscr{F}$ is left-directed. Since each $u \in \mathscr{F}$ can be replaced with its balayage $\hat{\mathbf{R}}_B^u$ (relative to R) without affecting the values of u on B and the set $\{\hat{\mathbf{R}}_B^u : u \in \mathscr{F}\}$ is also left-directed, we can assume that each $u \in \mathscr{F}$ is the potential of a measure μ having support in the compact set $\bar{B} \subset R$. We can also assume that there is a $u_0 \in \mathscr{F}$ such that $u \le u_0$ for all $u \in \mathscr{F}$. By taking the balayage of 1 over a ball between B and R, we can find a measure σ such that $G_R \sigma = 1$ on $\bar{B}$. If $u \in \mathscr{F}$ and $u = G_R \mu$, then

$$\mu(\bar{B}) = \int G_R \sigma \, d\mu = \int G_R \mu \, d\sigma \le \int u_0 \, d\sigma \le \mu_0(\bar{B}),$$

where $u_0 = G_R \mu_0$. This shows that the set $\{\mu(\bar{B}) : u \in \mathscr{F}, u = G_R \mu\}$ is bounded and that Lemma 7.38 is applicable. Therefore $\hat{v} = v$ on B except possibly for a polar set.

COROLLARY **7.40** If E is any subset of an open set R and u is a non-negative superharmonic function on R, the réduite $\mathbf{R}_E^u$ differs from the balayage $\hat{\mathbf{R}}_E^u$ at most on a polar subset of ∂E.

CHAPTER 8

The Generalized Dirichlet Problem

In this chapter we shall study the Dirichlet problem for bounded open subsets of E^n. For a boundary function in a certain class, we show that an associated harmonic function can be constructed by a method which is known as the Perron-Wiener-Brelot method. Having constructed a formal solution to the Dirichlet problem, the relationship between the solution and the boundary function is investigated. This investigation brings in a classical concept of "regular boundary point". Finally, we apply these results to the Green function and potentials.

1. The Perron-Wiener-Brelot Method

Let R be any nonempty open subset of E^n and let f be any extended real-valued function defined on ∂R. The generalized Dirichlet problem is that of constructing a harmonic function h on R corresponding to the boundary function f. Of course, even in the case of a ball we cannot expect h to have the value of f at a point of ∂R as a limit upon approach to the boundary point, but we are able to show that this is the case at most boundary points for continuous boundary functions.

Definition. An extended real-valued function u is **hyperharmonic (hypoharmonic)** on R if it is superharmonic (subharmonic) or identically $+\infty$ $(-\infty)$ on each component of R.

Definition. The **upper class** of functions $\mathfrak{U}_f$ determined by the function f on ∂R is given by

$$\mathfrak{U}_f = \{u\colon u \text{ hyperharmonic on } R, \liminf_{y\to x} u(y) \geq f(x) \text{ for all } x \in \partial R,\ u \text{ bounded below on } R\}.$$

The **lower class** $\mathfrak{L}_f$ determined by f is given by

$$\mathfrak{L}_f = \{u : u \text{ hypoharmonic on } R, \limsup_{y \to x} u(y) \le f(x) \text{ for all } x \in \partial R, u \text{ bounded above on } R\}.$$

Note that $\mathfrak{U}_f$ contains the function that is identically $+\infty$ on R and that $\mathfrak{L}_f$ contains the function which is identically $-\infty$ on R.

Definition. $\overline{H}_f = \inf\{u : u \in \mathfrak{U}_f\}$ is the **upper solution** for the generalized Dirichlet problem for the boundary function f. $\underline{H}_f = \sup\{u : u \in \mathfrak{L}_f\}$ is the corresponding **lower solution.**

It is worth remarking that $\mathfrak{U}_f(\mathfrak{L}_f)$ may contain only the identically $+\infty$ $(-\infty)$ function. The requirement that elements of $\mathfrak{U}_f$ be bounded below may be impossible to satisfy by any nontrivial hyperharmonic function. If there is a superharmonic function in $\mathfrak{U}_f$, then R supports a non-negative superharmonic function. We shall see later that R supports a nonconstant positive superharmonic function if and only if R has a Green function. In spite of the generality concerning R at this point, we shall eventually have to limit ourselves to open sets having Green functions.

Before proving the next lemma a few words should be said about the upper and lower solutions relative to the components of R. Let W be a component and let $\mathfrak{U}_f^W$ and $\overline{H}_f^W$ be the upper class and upper solution, respectively, relative to the component W. We first remark that $\mathfrak{U}_{f|\partial W}^W = \mathfrak{U}_f|_W = \{u|_W : u \in \mathfrak{U}_f\}$. To verify this suppose $u \in \mathfrak{U}_{f|\partial W}^W$ and let $u^* = u$ on W and $u^* = +\infty$ on $R \sim W$. Clearly, $u^* \in \mathfrak{U}_f$ and $u = u^*|_W$. This shows that $\mathfrak{U}_{f|\partial W}^W \subset \mathfrak{U}_f|_W$. Suppose $u \in \mathfrak{U}_f$ and $x \in \partial W$. Then $x \in \partial R$ and

$$\liminf_{\substack{y \to x \\ y \in W}} u(y) \ge \liminf_{\substack{y \to x \\ y \in R}} u(y) \ge f(x).$$

Therefore $u|_W \in \mathfrak{U}_{f|\partial W}^W$ and $\mathfrak{U}_f|_W \subset \mathfrak{U}_{f|\partial W}^W$. It follows immediately that $\overline{H}_{f|\partial W}^W = \overline{H}_f|_W$. Because of this it usually suffices to consider components of R.

LEMMA **8.1** $\overline{H}_f$ and $\underline{H}_f$ are either identically $+\infty$, identically $-\infty$, or harmonic on each component of R.

Proof. We can assume that R is connected. If $\mathfrak{U}_f$ contains only the identically $+\infty$ function, then $\overline{H}_f = +\infty$ and we are through. Suppose $\mathfrak{U}_f$ contains a hyperharmonic function that is not identically $+\infty$ on R. Then $\overline{H}_f = \inf\{u : u \in \mathfrak{U}_f, u \text{ superharmonic on } R\}$. If we can show that the set of superharmonic members of $\mathfrak{U}_f$ forms a saturated family, it would follow from

Theorem 5.9 that $\overline{H}_f$ is either identically $-\infty$ or harmonic on R. Let u_1 and u_2 be two superharmonic members of $\mathfrak{U}_f$. If $x \in \partial R$, then $\liminf_{y \to x} u_i(y) \geq f(x)$ and it is easily seen that this implies that $\liminf_{y \to x}(u_1(y) \wedge u_2(y)) \geq f(x)$. Since u_1 and u_2 are bounded below, the same is true of the superharmonic function $u_1 \wedge u_2$, and we have shown that $u_1 \wedge u_2 \in \mathfrak{U}_f$. If u is a superharmonic member of $\mathfrak{U}_f$ and B is a ball with $\overline{B} \subset R$, then

$$u^* = \begin{cases} \mathbf{PI}(u, B) & \text{on} \quad B \\ u & \text{on} \quad R \sim B, \end{cases}$$

is a superharmonic member of $\mathfrak{U}_f$, since it is superharmonic on R,

$$\liminf_{y \to x} u^*(y) = \liminf_{y \to x} u(y) \geq f(x)$$

for all $x \in \partial R$, and u^* is bounded below by the same constant bounding u. Therefore the superharmonic members of $\mathfrak{U}_f$ constitute a saturated family.

Definition. If $\overline{H}_f = \underline{H}_f$ and both harmonic on R, then f is called a **resolutive boundary function** and $H_f = \overline{H}_f = \underline{H}_f$ is called the **Dirichlet solution** for f.

The above method of obtaining a harmonic function corresponding to the boundary function f is called the **Perron-Wiener-Brelot method.**

It was pointed out at the end of Chapter 2 that the Dirichlet problem does not always have a unique solution. We saw there that the function $h = \mathbf{PI}(c, f, R)$, the Poisson integral for the half-space $R = \{(x_1, \ldots, x_n) : x_n > 0\}$, solves the Dirichlet problem corresponding to the boundary function f and that $\lim_{y \to x} h(y) = f(x)$ for all $x \in \partial R$ if f is a bounded continuous function; but in view of the arbitrariness of the real number c, the solution is certainly not unique. To obtain a unique solution we must specify the behavior of the solution as $\|y\| \to +\infty$. This is sufficient reason to limit ourselves to bounded regions R for the time being; unbounded regions are taken up in Chapter 9.

LEMMA **8.2** Let R be a bounded open subset of E^n. If $u \in \mathfrak{L}_f$ and $v \in \mathfrak{U}_f$, then $u \leq v$ and $\underline{H}_f \leq \overline{H}_f$ on R.

Proof. We can assume that R is connected. If either v is not superharmonic or u is not subharmonic, then $u \leq v$ on R trivially. It suffices to consider the case where $v - u$ is superharmonic. If $x \in \partial R$ and $f(x)$ in finite, then

$$\liminf_{y \to x} (v - u)(y) \geq \liminf_{y \to x} v(y) - \limsup_{y \to x} u(y) \geq f(x) - f(x) = 0;$$

if $f(x) = +\infty$, $\liminf_{y \to x} (v - u)(y) \geq 0$, since $\liminf_{y \to x} v(y) = +\infty$ and $\limsup_{y \to x} u(y) < +\infty$ with a similar result holding if $f(x) = -\infty$. Therefore $\liminf_{y \to x}(v - u)(y) \geq 0$ for all $x \in \partial R$ and $v - u \geq 0$ on R by Corollary 4.3.

It is possible that the Perron-Wiener-Brelot method for constructing a solution of the Dirichlet problem for a given boundary function ignores certain boundary points. We give an example of a resolutive boundary function for which its value at one boundary point has nothing to do with the Dirichlet solution. Let $R = B_{0,1} \sim \{0\}$, $B = B_{0,1}$, and

$$f = \begin{cases} 1 & \text{at} \quad 0, \\ 0 & \text{on} \quad \partial B. \end{cases}$$

Then $\lambda G_B(0, \cdot) \in \mathfrak{U}_f$ for every $\lambda > 0$ and $-\lambda G_B(0, \cdot) \in \mathfrak{L}_f$ for every $\lambda > 0$. Clearly $H_f = \underline{H}_f = \overline{H}_f = 0$ on R and f is resolutive. No matter what value is assigned to f at 0, we always have $\underline{H}_f = \overline{H}_f = 0$ on R. Note also that $\lim_{y \to x} H_f(y) = f(x)$ except at 0. Remembering that a singleton set is a polar set, we can anticipate some of the results of this chapter. Generally speaking, the values of the boundary function on a polar set can be changed arbitrarily without affecting the Dirichlet solution.

LEMMA **8.3** Let R be a bounded open subset of E^n. If $f = g$ on ∂R except possibly on a polar subset of ∂R, then $\overline{H}_f = \overline{H}_g$ and $\underline{H}_f = \underline{H}_g$ on R.

Proof. Consider a fixed $x \in R$. By Theorem 7.14 there is a superharmonic function v such that $v = +\infty$ on the subset of ∂R where $f \neq g$ and $v(x) < +\infty$. Since R is bounded, we can assume that v is positive on R. If $u \in \mathfrak{U}_f$ and $\varepsilon > 0$, then $u + \varepsilon v$ belongs to both $\mathfrak{U}_f$ and $\mathfrak{U}_g$. Therefore $\overline{H}_g(x) \leq u(x) + \varepsilon v(x)$ for every $\varepsilon > 0$ and $\overline{H}_g(x) \leq u(x)$. Since u can be any function in $\mathfrak{U}_f$, $\overline{H}_g(x) \leq \overline{H}_f(x)$. Interchanging f and g, $\overline{H}_g(x) = \overline{H}_f(x)$,

COROLLARY **8.4** Let R be a bounded open subset of E^n and let Z be a polar subset of ∂R. Then $\overline{H}_f$ is the infimum of all lower bounded hyperharmonic functions u on R such that $\liminf_{y \to x} u(y) \geq f(x)$ for all $x \in \partial R \sim Z$.

Proof. Define $g = f$ on $\partial R \sim Z$ and $g = -\infty$ on Z. Then $f = g$ except on a polar subset of ∂R. Since the functions u of the hypothesis are just those in $\mathfrak{U}_g$ and $\overline{H}_f = \overline{H}_g$ by the preceding lemma, the assertion is proved.

Changing the boundary function f on a polar subset of ∂R in no way affects the solution H_f, if it exists. The same is true if a polar subset of R is deleted.

If there is any merit to the Perron-Wiener-Brelot method, it should give solutions that are consistent with certain natural solutions.

THEOREM **8.5** Let R be a bounded open subset of E^n. If f is bounded on ∂R and there is a harmonic function h on R such that $\lim_{y \to x} h(y) = f(x)$ for all $x \in \partial R$, then f is resolutive and $H_f = h$.

Proof. Since f is bounded, h is bounded by Corollary 4.3. Therefore h belongs to both $\mathfrak{U}_f$ and $\mathfrak{L}_f$ and $\overline{H}_f \le h \le \underline{H}_f$. Therefore $\underline{H}_f = h = \overline{H}_f$ and since h is harmonic f is a resolutive boundary function.

In particular, if h is harmonic on a neighborhood of the compact set R and the Perron-Wiener-Brelot method is applied to $h|_{\partial R}$, the solution is just what it should be.

Later we shall give necessary and sufficient conditions in order that a boundary function be resolutive. For the time being it is more convenient to study the class of resolutive boundary functions. Each of the assertions of the following lemma has either been proved already or is easily proved from the definitions.

LEMMA **8.6** Let R be a bounded open subset of E^n, let f and g be extended real-valued functions on ∂R, and let c be any real number.

(i) if $f = c$ on ∂R, then f is resolutive and $H_f = c$ on R.
(ii) $\overline{H}_{f+c} = \overline{H}_f + c$ and $\underline{H}_{f+c} = \underline{H}_f + c$. If f is resolutive, then $f + c$ is resolutive and $H_{f+c} = H_f + c$.
(iii) If $c > 0$, then $\overline{H}_{cf} = c\overline{H}_f$ and $\underline{H}_{cf} = c\underline{H}_f$. If f is resolutive, then cf is resolutive and $H_{cf} = cH_f$ for $c \ge 0$.
(iv) If $f \le g$, then $\overline{H}_f \le \overline{H}_g$ and $\underline{H}_f \le \underline{H}_g$.
(v) $\overline{H}_{-f} = -\underline{H}_f$. If f is resolutive, then $-f$ is resolutive with $H_{-f} = -H_f$.
(vi) $\overline{H}_{f+g} \le \overline{H}_f + \overline{H}_g$ and $\underline{H}_{f+g} \ge \underline{H}_f + \underline{H}_g$ whenever the sums are defined. If f and g are resolutive and $f + g$ is defined, then $f + g$ is resolutive with $H_{f+g} = H_f + H_g$.

LEMMA **8.7** Let R be a bounded open subset of E^n. The real-valued resolutive boundary functions form a linear class. Moreover, if $\{f_j\}$ is a sequence of such functions that converges uniformly to f, then f is resolutive and H_{f_j} converges uniformly to H_f.

Proof. The first assertion is contained in the preceding lemma. If $\|f_j - f\| < \varepsilon$, then $f < f_j + \varepsilon$ and $\overline{H}_f \le \overline{H}_{f_j} + \varepsilon$ by the preceding lemma. Likewise $\overline{H}_f \ge \overline{H}_{f_j} - \varepsilon$. Therefore $\|\overline{H}_f - \overline{H}_{f_j}\| = \|\overline{H}_f - H_{f_j}\| < \varepsilon$ and the sequence $\{H_{f_j}\}$ converges uniformly to $\overline{H}_f$. In the same way the sequence $\{H_{f_j}\}$ converges uniformly to $\underline{H}_f$. Since limits are unique, $\overline{H}_f = \underline{H}_f$. The above inequalities also show that $\overline{H}_f$ is finite and f is therefore resolutive.

LEMMA **8.8** Let R be a bounded open subset of E^n. If $\{f_j\}$ is an increasing sequence of boundary functions, $f = \lim_{j\to\infty} f_j$, and $\overline{H}_{f_1} > -\infty$ on R, then $\overline{H}_f = \lim_{j\to\infty} \overline{H}_{f_j}$; if, in addition, $\{f_j\}$ is a sequence of resolutive boundary functions, then $\overline{H}_f = \underline{H}_f$ and f is resolutive if either $\overline{H}_f$ or $\underline{H}_f$ is finite.

Proof. We can assume that R is connected. Since $\overline{H}_{f_j} \leq \overline{H}_f$, there is nothing to prove if $\lim_{j\to\infty} \overline{H}_{f_j} = +\infty$ on R. We can therefore assume $\lim_{j\to\infty} \overline{H}_{f_j}(x_0) < +\infty$ for some point $x_0 \in R$. It follows that $-\infty < H_{f_1}(x_0) \leq H_{f_j}(x_0) < +\infty$ and so each of the functions $\overline{H}_{f_j}$ is harmonic by Lemma 8.1. Since the limit of a increasing sequence of harmonic functions is either harmonic or identically $+\infty$ and $\lim_{j\to\infty}\overline{H}_{f_j}(x_0) < +\infty$, $\lim_{j\to\infty} \overline{H}_{f_j}$ is harmonic on R. Suppose $\varepsilon > 0$. For each j choose $v_j \in \mathfrak{U}_{f_j}$ such that

$$v_j(x_0) < \overline{H}_{f_j}(x_0) + \varepsilon 2^{-j}$$

and define

$$v = \lim_{i\to\infty} \overline{H}_{f_i} + \sum_{j=1}^{\infty} (v_j - \overline{H}_{f_j}).$$

For each j, $v_j - \overline{H}_{f_j}$ is superharmonic on R, non-negative, and smaller than $\varepsilon 2^{-j}$ at x_0. From the inequality

$$v \geq \lim_{i\to\infty} \overline{H}_{f_i} + (v_j - \overline{H}_{f_j}) = (\lim_{i\to\infty} \overline{H}_{f_i} - \overline{H}_{f_j}) + v_j$$

and the fact that $\lim_{i\to\infty} \overline{H}_{f_i} \geq \overline{H}_{f_j}$, we obtain $v \geq v_j$ for each j. Since v_j is bounded below, the same is true of v. It is easily seen that v is superharmonic on R. Moreover, for each $x \in \partial R$ and each j

$$\liminf_{y\to x} v(y) \geq \liminf_{y\to x} v_j(y) \geq f_j(x).$$

Therefore $\liminf_{y\to x} v(y) \geq f(x)$ for all $x \in \partial R$ and consequently $v \in \mathfrak{U}_f$. This means that $v(x_0) \geq \overline{H}_f(x_0)$. Since $\lim_{j\to\infty} \overline{H}_{f_j}(x_0) \leq \overline{H}_f(x_0) \leq v(x_0) \leq \lim_{j\to\infty} \overline{H}_{f_j}(x_0) + \varepsilon$ and ε is arbitrary, $\overline{H}_f(x_0) = \lim_{j\to\infty} \overline{H}_{f_j}(x_0)$. Since the latter quantity is finite at x_0, $\overline{H}_f(x_0)$ is finite and $\overline{H}_f$ is harmonic. Now $\overline{H}_f - \lim_{j\to\infty} \overline{H}_{f_j}$ is a non-negative harmonic function which vanishes at x_0 and must be identically zero by the minimum principle. If the f_j's are resolutive, then $\overline{H}_{f_j} = \underline{H}_{f_j}$ and $\overline{H}_f = \lim_{j\to\infty} \overline{H}_{f_j} = \lim_{j\to\infty} \underline{H}_{f_j} \leq \underline{H}_f$; it follows that $\overline{H}_f = \underline{H}_f$ and that f is resolutive if either is finite.

We are now ready to obtain significant results about resolutive functions.

LEMMA **8.9** If u is a bounded subharmonic function on the bounded open set R such that $f(x) = \lim_{y\to x} u(y)$ exists for all $x \in \partial R$, then f is a resolutive boundary function.

Proof. Note that f is bounded along with $\overline{H}_f$ and $\underline{H}_f$. $\overline{H}_f$ and $\underline{H}_f$ are therefore harmonic functions. Clearly $u \in \mathfrak{L}_f$ and $u \leq \underline{H}_f$. Then $\liminf_{y\to x} \underline{H}_f(y) \geq \liminf_{y\to x} u(y) = f(x)$ for all $x \in \partial R$ and therefore $\underline{H}_f \in \mathfrak{U}_f$. It follows that $\underline{H}_f \geq \overline{H}_f$; but since we always have $\underline{H}_f \leq \overline{H}_f$, $\underline{H}_f = \overline{H}_f$ and f is resolutive.

LEMMA **8.10** Let K be a compact subset of E^n and let f be a continuous function on K. Then, given $\varepsilon > 0$, there is a function u which is the difference of two continuous subharmonic functions defined on a ball containing K such that $\sup_{x \in K} |f(x) - u(x)| < \varepsilon$.

Proof. It follows from the Stone-Weierstrass theorem that for each $\varepsilon > 0$ there is a function u polynomial in the n coordinate variables such that $\sup_{x \in K} |f(x) - u(x)| < \varepsilon$. We need show only that any such u, restricted to a ball $B \supset K$, can be expressed as the difference of two continuous subharmonic functions. If $v(y) = \|y\|^2$, then $\Delta v = 2n$ on E^n and λv is a continuous subharmonic function on B for $\lambda \geq 0$. Choosing λ_0 such that $\Delta(u + \lambda_0 v) \geq 0$ on B, then $u = (u + \lambda_0 v) - \lambda_0 v$ with $u + \lambda_0 v$ and $\lambda_0 v$ continuous subharmonic functions on B.

THEOREM **8.11** (*Wiener*) If f is a continuous real-valued function on the boundary ∂R of the bounded open set R, then f is resolutive.

Proof. Since ∂R is compact we can apply the preceding lemma. Let B be a ball containing ∂R. There is then a function u which is the difference of two functions v and w which are continuous and subharmonic on B and which approximates f uniformly on ∂R. Since $v|_{\partial R}$ and $w|_{\partial R}$ are resolutive by Lemma 8.9, $u|_{\partial R} = v|_{\partial R} - w|_{\partial R}$ is resolutive. Since $u|_{\partial R}$ approximates f uniformly on ∂R, f is resolutive by Lemma 8.7.

According to this theorem, each continuous boundary function f determines a harmonic function H_f defined on the bounded open set R. Generally speaking, $\lim_{y \to x} H_f(y) = f(x)$ at points $x \in \partial R$ (but not always) for such boundary functions. We defer relating H_f to its boundary function f in order to generalize the preceding theorem.

2. Harmonic Measure

We continue to restrict R to being a bounded open set. Consider the Banach space $C(\partial R)$, the space of continuous functions on the compact set ∂R with the norm of $f \in C(\partial R)$ given by $\|f\| = \sup|f(x)|$. According to Theorem 8.11, each $f \in C(\partial R)$ determines a harmonic function H_f. In particular, each $f \in C(\partial R)$ determines a number $H_f(x)$, x a fixed point of R. Consider the mapping $L_x: C(\partial R) \rightsquigarrow E^1$ defined by $L_x(f) = H_f(x)$.

LEMMA **8.12** For each $x \in R$, L_x is a positive linear functional on $C(\partial R)$; moreover, there is a unique unit measure μ_x on the Borel subsets of ∂R such that $H_f(x) = L_x(f) = \int f \, d\mu_x$, $x \in R$, $f \in C(R)$.

Proof. Suppose $f, g \in C(\partial R)$. Then both are resolutive by Theorem 8.11 and $L_x(f+g) = H_{f+g}(x) = H_f(x) + H_g(x) = L_x(f) + L_x(g)$ by Lemma 8.6. It also follows from the same lemma that $L_x(af) = aL_x(f)$ for each real number a. This shows that L_x is linear on $C(\partial R)$. Suppose $f \geq 0$. Then $L_x(f) = H_f(x) \geq H_0(x) = 0$ by Lemma 8.6 and L_x is positive. Since $-\|f\| \leq f \leq \|f\|$, $-\|f\| = H_{-\|f\|}(x) \leq H_f(x) \leq H_{\|f\|}(x) = \|f\|$ by Lemma 8.6 and $|L_x(f)| = |H_f(x)| \leq \|f\|$. This proves that L_x is a bounded functional on $C(\partial R)$. The existence of the measure μ_x is just the assertion of the Riesz representation theorem for bounded linear functionals on $C(\partial R)$. Since $H_1(x) = 1$, $\mu_x(\partial R) = 1$ for all $x \in R$.

Before stating the next theorem, we shall digress into measure theory. Let $\mathscr{F}_0$ be the collection of compact subsets of ∂R (i.e., closed subsets since R is bounded). $\mathscr{F}_0$ is closed under finite intersections and contains ∂R. A system $\mathscr{G}$ of subsets of ∂R is called a ***d*-system** if (i) $\partial R \in \mathscr{G}$, (ii) $E, F \in \mathscr{G}$ with $E \subset F$ implies that $F \sim E \in \mathscr{G}$, and (iii) if $\{E_j\}$ is an increasing sequence in $\mathscr{G}$, then $\cup E_j \in \mathscr{G}$. Let $d(\mathscr{F}_0)$ denote the smallest d-system containing $\mathscr{F}_0$ (i.e., $d(\mathscr{F}_0)$ is the intersection of all d-systems containing $\mathscr{F}_0$ and is itself a d-system). Then $d(\mathscr{F}_0) \supset \mathscr{F}_0$. Also let $\sigma(\mathscr{F}_0)$ denote the smallest σ-algebra containing $\mathscr{F}_0$; that is, $\sigma(\mathscr{F}_0)$ is the class of Borel subsets of ∂R. Since $\sigma(\mathscr{F}_0)$ is a d-system containing $\mathscr{F}_0$, $d(\mathscr{F}_0) \subset \sigma(\mathscr{F}_0)$. We will now show that $d(\mathscr{F}_0) = \sigma(\mathscr{F}_0)$. Since a d-system is closed under complementation, it is closed under unions if closed under intersections. We first show that $d(\mathscr{F}_0)$ is closed under finite intersections. Let $\mathscr{H}_1 = \{E \in d(\mathscr{F}_0): E \cap F \in d(\mathscr{F}_0)$ for all $F \in \mathscr{F}_0\} \subset d(\mathscr{F}_0)$. Now $\mathscr{H}_1$ is a d-system which contains $\mathscr{F}_0$ and therefore $\mathscr{H}_1 \supset d(\mathscr{F}_0)$. It follows that $\mathscr{H}_1 = d(\mathscr{F}_0)$. Now let $\mathscr{H}_2 = \{E \in d(\mathscr{F}_0): E \cap F \in d(\mathscr{F}_0)$ for all $F \in d(\mathscr{F}_0)\}$. Again $\mathscr{H}_2$ is a d-system. If $F \in \mathscr{F}_0$, then $E \cap F \in d(\mathscr{F}_0)$ for all $E \in \mathscr{H}_1 = d(\mathscr{F}_0)$ by definition of $\mathscr{H}_1$ and consequently $\mathscr{F}_0 \subset \mathscr{H}_2$. Therefore $\mathscr{H}_2 \supset d(\mathscr{F}_0)$ and since the opposite relation holds, $\mathscr{H}_2 = d(\mathscr{F}_0)$. But this just shows that $d(\mathscr{F}_0)$ is closed under finite intersections and therefore closed under finite unions. Since $d(\mathscr{F}_0)$ is closed under monotone increasing limits and finite unions, $d(\mathscr{F}_0)$ is closed under countable unions; that is, $d(\mathscr{F}_0)$ is a σ-algebra and $d(\mathscr{F}_0) \supset \sigma(\mathscr{F}_0)$. Therefore, $d(\mathscr{F}_0) = \sigma(\mathscr{F}_0)$. We shall see an advantage in using d-systems in the proof of the next theorem. The measure μ_x of Theorem 8.13 is the measure of Lemma 8.12.

THEOREM **8.13** If f is a lower bounded (or upper bounded) Borel measurable function on ∂R, then $\overline{H}_f(x) = \underline{H}_f(x) = \int f\, d\mu_x$ for all $x \in R$.

Proof. Let $\mathscr{F}_0$ be the collection of compact (i.e., closed) subsets of ∂R. If $C \in \mathscr{F}_0$, then there is a sequence $\{f_j\}$ of continuous functions on ∂R such that $f_j \downarrow \chi_C$, the indicator function of C, with $0 \leq f_j \leq 1$ for each j. It follows from Lemma 8.8. that χ_C is a resolutive boundary function and $H_{\chi_C}(x) = \int \chi_C \, d\mu_x$. Let $\mathscr{G}$ be the collection of sets $E \subset \partial R$ such that χ_E is resolutive and $H_{\chi_E}(x) = \int \chi_E \, d\mu_x$, $x \in R$. Then $\mathscr{F}_0 \subset \mathscr{G}$. Now $\partial R \in \mathscr{G}$ since $\chi_{\partial R} = 1$ and the constant functions are resolutive. Suppose $E, F \in \mathscr{G}$ with $E \subset F$. Then $\chi_{F \sim E} = \chi_F - \chi_E$. Since χ_F and χ_E are resolutive, $\chi_{F \sim E}$ is resolutive by Lemma 8.6 and

$$H_{\chi_{F \sim E}}(x) = H_{\chi_F}(x) - H_{\chi_E}(x) = \int \chi_F \, d\mu_x - \int \chi_E \, d\mu_x = \int \chi_{F \sim E} \, d\mu_x .$$

This shows that $\mathscr{G}$ is closed under proper differences. Suppose $\{E_j\}$ is an increasing sequence in $\mathscr{G}$ and $E = \cup_j E_j$. By Lemma 8.8 χ_E is resolutive and

$$H_{\chi_E}(x) = \lim_{j \to \infty} H_{\chi_{E_j}}(x) = \lim_{j \to \infty} \int \chi_{E_j} \, d\mu_x = \int \chi_E \, d\mu_x ,$$

$x \in R$. This shows that $\mathscr{G}$ is closed under unions of increasing sequences of elements of $\mathscr{G}$ and that $\mathscr{G}$ is a d-system containing $\mathscr{F}_0$. Therefore $\mathscr{G} \supset d(\mathscr{F}_0)$ and $\mathscr{G}$ contains the Borel subsets of ∂R; that is, the indicator function of any Borel subset of ∂R is resolutive. It follows that any simple Borel measurable function χ on ∂R is resolutive with $H_\chi(x) = \int \chi \, d\mu_x$, $x \in R$. If f is a nonnegative Borel measurable function on ∂R, then there is a sequence $\{f_j\}$ of simple Borel measurable functions such that $0 \le f_j \le f$ and $f_j \uparrow f$. By Lemma 8.8.

$$\underline{H}_f(x) = \overline{H}_f(x) = \lim_{j \to \infty} H_{f_j}(x) = \lim_{j \to \infty} \int f_j \, d\mu_x = \int f \, d\mu_x , \qquad x \in R,$$

by the Lebesgue monotone convergence theorem. If f is only bounded below by α, then we can apply the result to $f - \alpha$.

THEOREM **8.14** If W is a component of the bounded open set R, the class of Borel subsets of ∂R of μ_x-measure zero is independent of $x \in W$.

Proof. Consider a Borel set $E \subset \partial R$. Then χ_E is resolutive and $H_{\chi_E}(x) = \int \chi_E \, d\mu_x = \mu_x(E)$, $x \in R$. If $\mu_{x_0}(E) = 0$ for some $x_0 \in W$, the non-negative harmonic function $H_{\chi_E} = \mu_\cdot(E)$ must vanish on the component W by the minimum principle.

Let f be a bounded Borel measurable function on ∂R and let W be a component of R. Then f is resolutive relative to R and $f|_{\partial W}$ is resolutive relative to W with Dirichlet solutions H_f^R and $H_{f|\partial W}^W$, respectively. Since $H_f^R = H_{f|\partial W}^W$ on W, the value of $H_f = H_f^R$ at a point $x \in W$ is independent of the values of f on $\partial R \sim \partial W$. Therefore $\mu_x(\partial R \sim \partial W) = 0$ for all $x \in W$; that is, each μ_x has support in the boundary of the component containing x. This argument also shows that if μ_x^W is the harmonic measure relative to x and W, then μ_x^W is just the restriction of μ_x to ∂W. If it is necessary to consider the Dirichlet solution or harmonic measure for different open sets R, we shall use H_f^R and μ_x^R to denote the dependence upon R.

The class $\mathscr{F}_x$ of sets of the form $(E \sim N) \cup (N \sim E)$ where E is a Borel subset of ∂R and N is a subset of a Borel subset of ∂R of μ_x-measure zero is a σ-algebra of subsets of ∂R. Letting $\mathscr{F} = \cap_{x \in R} \mathscr{F}_x$, $\mathscr{F}$ is a σ-algebra of subsets of ∂R which includes the Borel subsets of ∂R. Since each μ_x can be uniquely extended to $\mathscr{F}_x$, the same is true of $\mathscr{F}$; the extension of μ_x to $\mathscr{F}$ is also denoted by μ_x.

Definition. Sets in $\mathscr{F}$ are called **$\mu_.$-measurable** sets and extended real-valued functions measurable relative to $\mathscr{F}$ are called **$\mu_.$-measurable.** The measure μ_x on $\mathscr{F}$ is called the **harmonic measure** relative to R and x. A $\mu_.$-measurable function f on ∂R is **$\mu_.$-integrable** if it is integrable relative to each μ_x, $x \in R$.

LEMMA **8.15** If f is a non-negative $\mu_.$-measurable function on ∂R, then $\overline{H}_f(x) = \underline{H}_f(x) = \int f\, d\mu_x$ for all $x \in R$; if f is any $\mu_.$-measurable function on ∂R which is μ_y-integrable for some y in each component of R, then f is $\mu_.$-integrable, resolutive, and $H_f(x) = \int f\, d\mu_x$, $x \in R$.

Proof. Consider any $F \in \mathscr{F}$. For each $x \in R$, $F \in \mathscr{F}_x$ and F is of the form $(E \sim N) \cup (N \sim E)$, where E is a Borel subset of ∂R and N is a subset of a Borel set $A \subset \partial R$ with $\mu_x(A) = 0$ (E and N may depend upon x). Note that $\chi_F = \chi_E + \chi_N - \chi_{E \cap N}$. Since $0 \le \underline{H}_{\chi_N}(x) \le \overline{H}_{\chi_N}(x) \le \overline{H}_{\chi_A}(x)$ and $\overline{H}_{\chi_A}(x) = \mu_x(A) = 0$ by Theorem 8.13, the non-negative functions $\underline{H}_{\chi_N}$ and $\overline{H}_{\chi_N}$ are harmonic on the component containing x. By the minimum principle, $\underline{H}_{\chi_N} = \overline{H}_{\chi_N} = 0$ on the component containing x. Likewise $\underline{H}_{\chi_{E \cap N}} = \overline{H}_{\chi_{E \cap N}} = 0$ on the component containing x. By Lemma 8.6 $0 \le \underline{H}_{\chi_F} \le \overline{H}_{\chi_F} \le \overline{H}_{\chi_E} + \overline{H}_{\chi_N} - \underline{H}_{\chi_{E \cap N}} = H_{\chi_E}$ on the component containing x. Likewise $\overline{H}_{\chi_F} \ge \underline{H}_{\chi_F} \ge \underline{H}_{\chi_E} + \underline{H}_{\chi_N} - \overline{H}_{\chi_{E \cap N}} = H_{\chi_E}$ on the component containing x. Therefore $\underline{H}_{\chi_F} = \overline{H}_{\chi_F} = H_{\chi_E}$ on the component containing x and

$$\underline{H}_{\chi_F}(x) = \overline{H}_{\chi_F}(x) = H_{\chi_E}(x) = \int \chi_E\, d\mu_x = \int \chi_F\, d\mu_x, \qquad x \in R,$$

by Theorem 8.13. Since the integral is finite, this also shows that χ_F is resolutive. If now f is a simple $\mu_.$-measurable function, then f is resolutive by Lemma 8.6 and

$$\underline{H}_f(x) = \overline{H}_f(x) = \int f\, d\mu_x.$$

Now let f be a non-negative $\mu_.$-measurable function on ∂R. Then there is a sequence $\{f_j\}$ of $\mu_.$-measurable simple functions such that $0 \leq f_j \leq f$ and $f_j \uparrow f$. By Lemma 8.8,

$$\underline{H}_f(x) = \overline{H}_f(x) = \lim_{j\to\infty} H_{f_j}(x) = \lim_{j\to\infty} \int f_j \, d\mu_x = \int f \, d\mu_x$$

$x \in R$. If, in addition, f is μ_y-integrable for some y in each component of R, then this equation shows that $\underline{H}_f$ and $\overline{H}_f$ are finite at some point of each component and therefore harmonic on R. It follows that f is resolutive with

$$H_f(x) = \int f \, d\mu_x, \qquad x \in R.$$

Lastly, if f is any $\mu_.$-measurable function on ∂R which is μ_y-integrable for some y in each component of R, then the above argument can be applied to f^+ and f^-, the positive and negative parts of f.

COROLLARY **8.16** If W is a component of the bounded open set R, then the class of μ_x-integrable functions is independent of $x \in W$.

The following theorem is the principle result of this section.

THEOREM **8.17** (*Brelot*) Let R be a bounded open set. A boundary function f is resolutive if and only if it is $\mu_.$-integrable; in which case, $H_f(x) = \int f \, d\mu_x$ for all $x \in R$.

Proof. The sufficiency was dealt with in Lemma 8.15. Assume that f is resolutive. If we can show that f is measurable relative to $\mathscr{F}_x$ for each $x \in R$, then f is measurable relative to $\mathscr{F} = \cap_{x \in R} \mathscr{F}_x$. Let W be the component of R containing x. In order to show that f is $\mathscr{F}_x$-measurable it suffices to show that $f|_{\partial W}$ is equal a.e. (μ_x) to a Borel measurable function on ∂W since $\mu_x(\partial R \sim \partial W) = 0$ implies that every subset of $\partial R \sim \partial W$ belongs to $\mathscr{F}_x$. Note also that the finiteness of $\int f \, d\mu_x$ is independent of the values of f on $\partial R \sim \partial W$. We can therefore assume that R is connected. Let x_0 be a fixed point of R and let $\{v_j\}$ be a sequence in $\mathfrak{U}_f$ such that $\lim_{j\to\infty} v_j(x_0) = H_f(x_0)$. We can assume that the sequence $\{v_j\}$ is decreasing. Since $v_j \in \mathfrak{U}_f$, v_j is bounded below by some α_j as is the Borel measurable function $f_j(x) = \liminf_{y\to x} v_j(y)$. Noting that $v_j \in \mathfrak{U}_{f_j}$, $v_j \geq \overline{H}_{f_j} \geq \underline{H}_{f_j} \geq \alpha_j$. It follows that the functions $\overline{H}_{f_j}$ and $\underline{H}_{f_j}$ are harmonic on R. According to Theorem 8.13 f_j is resolutive and μ_x-integrable for each $x \in R$. Since $\{v_j\}$ is a decreasing sequence, so also is $\{f_j\}$ and $f^* = \lim_{j\to\infty} f_j$ is defined. Using the fact that $v_j \in \mathfrak{U}_f$ again, $f_j(x) = \liminf_{y\to x} v_j(y) \geq f(x)$ and $f^* \geq f$. Since the f_j's are Borel measurable, the same is true of f^*. Since $v_j \geq H_{f_j} \geq \overline{H}_{f^*} \geq \underline{H}_{f^*} \geq H_f$ and $\lim_{j\to\infty} v_j(x_0) = H_f(x_0)$, $\overline{H}_{f^*}(x_0) = \underline{H}_{f^*}(x_0) = H_f(x_0)$. It follows that $\overline{H}_{f^*}$ and $\underline{H}_{f^*}$

are harmonic and equal to H_f; that is, f^* is resolutive with $H_{f^*} = H_f$. Similarly, there is an increasing sequence $\{g_j\}$ of Borel measurable $\mu_{\cdot}$-integrable functions on ∂R such that $\lim_{j\to\infty} g_j = f_* \leq f$ and $H_{f*} = H_f$. Since $f_j \geq f^* \geq f \geq f_* \geq g_j$ and f_j, g_j are μ_x-integrable for each $x \in R, f^*$ and f_* are μ_x-integrable for each $x \in R$. In view of the fact that $H_{f^*}(x) - H_{f_*}(x) = \int (f^* - f_*)\, d\mu_x$ and $f^* - f_* \geq 0$ a.e. (μ_x), $f^* = f_*$ a.e. (μ_x). Therefore $f^* = f_* = f$ a.e. (μ_x) for each $x \in R$ and f is $\mathscr{F}_x$-measurable for each $x \in R$. Since f^* is $\mu_{\cdot}$-integrable, f is $\mu_{\cdot}$-integrable.

Before taking up the boundary behavior of the Perron-Wiener-Brelot solution we shall examine harmonic measures from the functional analytic point of view. In the remainder of this section R is a bounded connected open set. We have seen that the class of μ_x-integrable functions is independent of $x \in R$. This class constitutes the Banach space $L_1(\mu_x)$ (provided functions which are equal a.e. relative to μ_x are identified) consisting of μ_x-integrable functions f on ∂R with norm defined by $\|f\|_{1,\mu_x} = \int |f|\, d\mu_x$. Note that $L_1(\mu_x)$ is independent of x; it need not be true that the norm $\|\cdot\|_{1,\mu_x}$ is independent of x. For any $p > 1$, we can also consider the Banach space $L_p(\mu_x)$ of equivalence classes of μ_x-measurable boundary functions f for each

$$\|f\|_{p,\mu_x} = \left[\int |f|^p\, d\mu_x\right]^{1/p} < +\infty.$$

Since the left side of this equation is a harmonic function of x, $L_p(\mu_x)$ is also independent of x; but again the norm $\|\cdot\|_{p,\mu_x}$ will depend upon x. Consider any $p \geq 1$. There are two notions of convergence in $L_p(\mu_x)$, weak and strong convergence, which could possibly depend upon x. Consider any two points x and y in R. If $\|f\|_{p,\mu_x} = 0$, then $H_{|f|^p}(x) = 0$ and $H_{|f|^p} = 0$ on R; in particular $H_{|f|^p}(y) = \|f\|_{p,\mu_y} = 0$. This shows that $\|f\|_{p,\mu_x} = 0$ if and only if $\|f\|_{p,\mu_y} = 0$. Let $K = \{x, y\}$. By Harnack's inequality there is a constant k such that $h(x')/h(x'') \leq k$ for $x', x'' \in K$ whenever h is a positive harmonic function on R. If $\|f\|_{p,\mu_x} > 0$, then $\|f\|_{p,\mu_{\cdot}} > 0$ on R and $\|f\|_{p,\mu_x}/\|f\|_{p,\mu_y} \leq k$; that is, $\|f\|_{p,\mu_x} \leq k\|f\|_{p,\mu_y}$ with this inequality obviously holding if $\|f\|_{p,\mu_x} = 0$. This shows that the norms $\|\cdot\|_{p,\mu_x}$ and $\|\cdot\|_{p,\mu_y}$ are equivalent norms on $L_p(\mu_x) = L_p(\mu_y)$. If $\{f_j\}$ is a sequence in $L_p(\mu_x)$ for which there is an $f \in L_p(\mu_x)$ such that $\lim_{j\to\infty} \int |f_j - f|^p\, d\mu_x = 0$, then $\lim_{j\to\infty} \int |f_j - f|^p\, d\mu_y = 0$; that is, strong convergence in $L_p(\mu_x)$ is independent of x. Since the norms $\|\cdot\|_{p,\mu_x}$ and $\|\cdot\|_{p,\mu_y}$ are equivalent, the adjoint space $L_p^*(\mu_x) = L_q(\mu_x)$ ($p^{-1} + q^{-1} = 1$ if $p > 1$; $q = \infty$ if $p = 1$) is independent of x and the weak topology on $L_p(\mu_x)$ is also independent of x. It follows that if $\{f_j\}$ is a sequence in $L_p(\mu_x)$ for which there is an $f \in L_p(\mu_x)$ such that $\lim_{j\to\infty} \int f_j g\, d\mu_x = \int fg\, d\mu_x$ for all $g \in L_q(\mu_x)$, then $\lim_{j\to\infty} \int f_j g\, d\mu_y = \int fg\, d\mu_y$ for all $g \in L_q(\mu_x)$ and weak convergence in $L_p(\mu_x)$ is independent of x.

Under the same conditions on R, let $\{x_\alpha : \alpha \in A\}$ be a net in R having a limit $x \in R$. Then $\mu_{x_\alpha} \xrightarrow{w^*} \mu_x$ since $H_f(x_\alpha) = \int f\, d\mu_{x_\alpha} \to \int f\, d\mu_x = H_f(x)$, H_f being a harmonic function whenever f is a continuous function on ∂R. Lastly, we note that the measures $\{\mu_x : x \in R\}$ are mutually absolutely continuous for if $\mu_x(A) = 0$ for some $\mu_.$-measurable set $A \subset \partial R$, then $\mu_y(A) = 0$ for any $y \in R$ since the class of zero sets is independent of x.

3. Boundary Behavior of the PWB Solution

Having found that each boundary function f in a rather large class determines a harmonic function H_f via the Perron-Wiener-Brelot method, we are left with establishing the relationship between H_f and f.

Definition. A point $x \in \partial R$ is a **regular boundary point** if $\lim_{y \to x} H_f(y) = f(x)$ for every $f \in C(\partial R)$. Otherwise x is an **irregular boundary point**. R is a **regular** (or **Dirichlet**) region if every point of ∂R is a regular boundary point.

EXAMPLE 1 Let $R = B_{0,1} \sim \{0\} \subset E^2$. Then 0 is a boundary point for R but not a regular boundary point. This follows from the fact that $H_f = 0$ on R for the continuous boundary function f which takes on the value 1 at 0 and the value 0 at points of $\partial B_{0,1}$. In this case $\lim_{y \to 0} H_f(y) = 0 \neq 1 = f(0)$.

The irregular boundary point of this example is an isolated point of the boundary. It is generally true that an isolated boundary point is an irregular boundary point for bounded regions. Suppose x_0 is an isolated boundary point of R and $f \in C(\partial R)$. Then any function f' on ∂R which agrees with f on $\partial R \sim \{x_0\}$ is continuous on ∂R, no matter what its value at x_0. Since $f = f'$ except possibly on the polar set $\{x_0\}$, $H_f = H_{f'}$ on R by Lemma 8.3. Therefore $\lim_{y \to x_0} H_f(y)$ is quite independent of the value of f at x_0. It follows that any isolated point of ∂R (assuming R to be bounded) is an irregular boundary point.

We now turn to necessary and sufficient conditions in order that a boundary point be a regular boundary point.

Definition. A function w is a **barrier** at $x \in \partial R$ if w is defined on $W \cap R$ for some neighborhood W of x and possesses the following properties:

(i) w is superharmonic on $W \cap R$.
(ii) $w > 0$ on $W \cap R$.
(iii) $\lim_{\substack{y \to x \\ y \in W \cap R}} w(y) = 0$.

We shall soon prove that a boundary point is a regular boundary point if and only if there is a barrier at the point.

THEOREM **8.18** (*Bouligand*) Let R be a bounded open set. There is a barrier at $x \in \partial R$ if and only if there is a positive harmonic function h on R such that $\lim_{y \to x} h(y) = 0$ and $\liminf_{y \to x'} h(y) > 0$ for all $x' \in \partial R$, $x' \neq x$.

Proof. The sufficiency is obvious since any such function h is a barrier at x with $W = E^n$. As to the necessity, let w be a barrier at $x \in \partial R$. We define a function m on ∂R by putting $m(y) = \|y - x\|$, $y \in \partial R$. We shall show that the function $h = H_m$ has the required properties. Consider the function $\|y - x\|$ on R. This is a subharmonic function and belongs to the lower class $\mathfrak{L}_m$. Clearly, $\|y - x\| \leq H_m(y)$ on R. It follows that H_m is a strictly positive harmonic function on R and that $\liminf_{y \to x'} H_m(y) \geq \|x' - x\| > 0$ for all $x' \in \partial R$, $x' \neq x$. The only remaining thing to show is that $\lim_{y \to x} H_m(y) = 0$. Since w is a barrier at x, there is a ball $B_{x,r}$ such that w is positive on $B_{x,r} \cap R$, superharmonic on $B_{x,r} \cap R$, and $\lim_{\substack{y \to x \\ y \in B_{x,r} \cap R}} w(y) = 0$. Consider any ρ for which $0 < \rho < r$ and let $B = B_{x,\rho}$. Surface area on ∂B will be denoted by σ. We shall consider two cases depending upon whether or not $\partial B \cap R$ is empty. Assume first that $\partial B \cap R = \varnothing$. If $u \in \mathfrak{L}_m$, then $\limsup_{\substack{z \to y \\ z \in B \cap R}} u(z) \leq \rho$ for all $y \in \partial(B \cap R)$. Therefore, $u \leq \rho$ on $B \cap R$ and since this is true for every $u \in \mathfrak{L}_m$, $H_m \leq \rho$ on $B \cap R$. It follows that $\limsup_{y \to x} H_m(y) = \limsup_{\substack{y \to x \\ y \in B \cap R}} H_m(y) \leq \rho$. Now assume that $\partial B \cap R \neq \varnothing$. Choose a closed set $F \subset \partial B \cap R$ such that $\sigma((\partial B \cap R) \sim F) < \rho/M$ where $M = \sup\{\|y - x\| : y \in \partial R\}$. Note that $m \leq M$ on ∂R. Also let $k = \inf\{w(y) : y \in F\} > 0$. Define a function f on ∂B as follows:

$$f(y) = \begin{cases} M & \text{for } \; y \in (\partial B \cap R) \sim F, \\ 0 & \text{otherwise.} \end{cases}$$

Let $u \in \mathfrak{L}_m$ and consider the function

$$s = u - \rho - \frac{M}{k} w - \mathbf{PI}(f, B),$$

which is subharmonic on $B \cap R$. We now show that $\limsup_{z \to y,\, z \in B \cap R} s(z) \leq 0$ for all $y \in \partial(B \cap R)$. On the set $(\partial B \cap R) \sim F$, f is a continuous function and $\lim_{z \to y,\, z \in B \cap R} \mathbf{PI}(f, B)(z) = f(y) = M$; otherwise $-\liminf_{z \to y,\, z \in B \cap R} \mathbf{PI}(f, B)(z) \leq 0$. Since w is l.s.c. at points of F, $-\liminf_{z \to y,\, z \in B \cap R} w(z) \leq -k$ for $y \in F$; otherwise $-\liminf_{z \to y,\, z \in B \cap R} w(z) \leq 0$. Since $m \leq M$ and $u \in \mathfrak{L}_m$, $u \leq m$ and $\limsup_{z \to y,\, z \in B \cap R} u(z) \leq M$ for all $y \in \partial B \cap R$, whereas $\limsup_{z \to y,\, z \in B \cap R} u(z) \leq m(y) \leq \rho$ for $y \in \partial R \cap \bar{B}$. By consideration of points

y in F, $(\partial B \cap R) \sim F$, and $\partial R \cap \bar{B}$, it is easily seen that $\limsup_{z\to y,\, z \in B \cap R} s(z) \leq 0$ for all $y \in \partial(B \cap R)$. It follows that $s \leq 0$ on $B \cap R$; that is,

$$u \leq \rho + \frac{M}{k} w + \mathbf{PI}(f, B) \quad \text{on} \quad B \cap R.$$

But since u is an arbitrary element of $\mathfrak{L}_m$,

$$H_m \leq \rho + \frac{M}{k} w + \mathbf{PI}(f, B) \quad \text{on} \quad B \cap R.$$

Since $\mathbf{PI}(f, B)(x) =$ average of f over $\partial B \leq M(\rho/M) = \rho$ and $\mathbf{PI}(f, B)$ is continuous on B, $\limsup_{\substack{y\to x \\ y \in R \cap B}} \mathbf{PI}(f, B)(y) \leq \rho$; by hypothesis $\lim_{y\to x} w(y) = 0$. Therefore

$$0 \leq \limsup_{y\to x} H_m(y) = \limsup_{\substack{y\to x \\ y \in R \cap B}} H_m(y) \leq 2\rho$$

in the $\partial B \cap R \neq \varnothing$ case. In either case $\limsup_{y\to x} H_m(y) \leq 2\rho$. Since ρ can be made arbitrarily small, $\lim_{y\to x} H_m(y) = 0$.

It is worth noting that if there is a barrier at the point $x \in \partial R$ (R bounded) the function $H_{\|y-x\|}$ is a barrier at x. We should also note that the existence of a barrier at a boundary point is a local property of R according to the definition of a barrier; that is, the existence of a barrier at $x \in \partial R$ depends only upon the relationship between x and that part of R near x.

COROLLARY **8.19** Let R be a bounded open set. If there is a barrier at $x \in \partial R$, there is a barrier w at x defined on R such that (i) $\liminf_{y\to x'} w(y) > 0$ for all $x' \in \partial R$, $x' \neq x$ and (ii) $\inf\{w(y): y \in R \sim W_x\} > 0$ for each neighborhood W_x of x.

Proof. Take $w(y) = H_{\|y-x\|}$.

LEMMA **8.20** Let R be a bounded open set. If f is bounded above on ∂R and there is a barrier at $x \in \partial R$, then

$$\limsup_{\substack{y\to x \\ y \in R}} \overline{H}_f(y) \leq \limsup_{\substack{y\to x \\ y \in \partial R}} f(y);$$

or if f is bounded below on ∂R and there is a barrier at $x \in \partial R$, then

$$\liminf_{\substack{y\to x \\ y \in R}} \underline{H}_f(y) \geq \liminf_{\substack{y\to x \\ y \in \partial R}} f(y).$$

Proof. By Corollary 8.19 there is a barrier w at x defined on R such that $\inf\{w(y): y \in R \sim W_x\} > 0$ for each neighborhood W_x of x. If $L = \limsup_{\substack{y \to x \\ y \in \partial R}} f(y)$ and $\varepsilon > 0$, there is a neighborhood V of x such that $f(y) < L + \varepsilon$ for all $y \in \partial R \cap V$. Choose $c > 0$ such that

$$L + \varepsilon + c\Big[\inf_{y \in R \sim V} w(y)\Big] > \sup_{y \in \partial R} f(y)$$

and let $u = L + \varepsilon + cw$. Then u is superharmonic on R and u is bounded below since $w > 0$. Moreover, $\liminf_{y \to x'} u(y) \geq L + \varepsilon + c \liminf_{y \to x'} w(y)$ for $x' \in \partial R$. If $x' \in \partial R \sim V$, then $\liminf_{\substack{y \to x' \\ y \in R}} w(y) \geq \inf_{y \in R \sim V} w(y)$ and

$$\liminf_{\substack{y \to x' \\ y \in R}} u(y) \geq L + \varepsilon + c\Big[\inf_{y \in R \sim V} w(y)\Big] > \sup_{y \in \partial R} f(y) \geq f(x').$$

On the other hand, if $x' \in \partial R \cap V$, then $L + \varepsilon > f(x')$ and $\liminf_{y \to x'} u(y) \geq L + \varepsilon > f(x')$. This shows that $\liminf_{y \to x'} u(y) \geq f(x')$ for all $x' \in \partial R$ and that $u \in \mathfrak{U}_f$. Therefore $\overline{H}_f \leq u$ on R and $\limsup_{y \to x} \overline{H}_f(y) \leq \limsup_{y \to x} u(y) \leq L + \varepsilon + c \limsup_{y \to x} w(y) = L + \varepsilon$. Since ε is arbitrary, $\limsup_{\substack{y \to x \\ y \in R}} \overline{H}_f(y) \leq L = \limsup_{\substack{y \to x \\ y \in \partial R}} f(y)$.

COROLLARY **8.21** Suppose R is a bounded open set and there is a barrier at $x \in \partial R$. If f is bounded on ∂R and continuous at x, then

$$\lim_{y \to x} \overline{H}_f(y) = \lim_{y \to x} \underline{H}_f(y) = f(x).$$

Proof. By the preceding lemma $\limsup_{y \to x} \overline{H}_f(y) \leq f(x) \leq \liminf_{y \to x} \underline{H}_f(y)$, but since $\liminf_{y \to x} \underline{H}_f(y) \leq \liminf_{y \to x} \overline{H}_f(y)$, $\liminf_{y \to x} \underline{H}_f(y) \leq f(x)$. Therefore $f(x) = \liminf_{y \to x} \underline{H}_f(y) \leq \liminf_{y \to x} \overline{H}_f(y) \leq \limsup_{y \to x} \overline{H}_f(y) = f(x)$ and $\lim_{y \to x} \overline{H}_f(y)$ exists and is equal to $f(x)$.

THEOREM **8.22** Let R be a bounded open set. A point $x \in \partial R$ is a regular boundary point if and only if there is a barrier at x.

Proof. The sufficiency is immediate from the definition of a regular boundary point and the preceding corollary. Assume that x is a regular boundary point. Define $m(y) = \|y - x\|$, $y \in \partial R$. Then $H_m(y) \geq \|y - x\|$, $y \in R$, and H_m is a barrier at x if $\lim_{y \to x} H_m(y) = 0$. Since x is a regular boundary point and m is continuous on ∂R, $\lim_{y \to x} H_m(y) = m(x) = 0$.

Although we have a criterion for regularity of a boundary point in terms of a barrier, more concrete tests of regularity of a geometrical nature are desirable and available.

THEOREM **8.23** If x is a regular boundary point for the bounded open set R and R_0 is an open subset of R for which x is also a boundary point, then x is regular for R_0.

Proof. A barrier at x relative to R is also a barrier at x relative to R_0.

THEOREM **8.24** (*Poincaré*) Let R be a bounded open set. If $x \in \partial R$ and there is a ball B with $B \cap R = \varnothing$ and $x \in \partial B$, then x is a regular boundary point for R.

Proof. By taking a ball internally tangent to B at x, we can assume that $\partial B \cap \partial R = \{x\}$. If $B = B_{y,\rho}$, we can add a constant c to the fundamental harmonic function u_y having pole y so that $u_y + c$ vanishes on ∂B. The function $w = -(u_y + c)$ is a barrier at x for R. It follows that x is a regular boundary point by Theorem 8.22.

EXAMPLE **2** The boundary points of a disk, square, and annulus in E^2 are regular boundary points.

In the one example of an irregular boundary point that we have considered, namely the point 0 of the boundary of $R = B_{0,1} \sim \{0\} \subset E^2$, the region R is not simply connected. Generally speaking, irregular boundary points are more difficult to illustrate in E^2 than in E^n, $n \geq 3$.

THEOREM **8.25** If x is a boundary point for the bounded open set $R \subset E^2$ having the property that there is a line segment $I \subset \sim R$ with x as a endpoint, then x is a regular boundary point.

Proof. We consider x to be a complex number. Let $B = B_{x,\delta}$ be a ball such that $I \cap \partial B \neq \varnothing$ and $\delta < 1$. Now $\log(z - x)$ can be defined to be continuous on $B \sim I$ by taking the branch cut along the ray starting at x and in the direction of I. Let

$$w(z) = -\mathscr{Re}\left[\frac{1}{\log(z-x)}\right] = -\frac{\log|z-x|}{|\log(z-x)|^2}.$$

Now w is harmonic on $R \cap B$ since it is the real part of a function which is analytic on $R \cap B$. Moreover, $w > 0$ on $R \cap B$. It is easily seen that $\lim_{\substack{z \to x \\ z \in R \cap B}} w(z) = 0$ and it follows that there is a barrier at x.

Actually much more is true in the $n = 2$ case than is stated in Theorem 8.25. Recall that a **continuum** is a set containing more than one point which is closed and connected. For the sake of completeness the following theorem is included here but the proof is deferred until Chapter 10.

THEOREM **8.26** (*Lebesgue*) If x is a boundary point for the bounded open set $R \subset E^2$ having the property that there is a continuum $\ell \subset \sim R$ with $x \in \ell$, then x is a regular boundary point. In particular, if R is simply connected, then it is a regular region.

We now return to more general considerations which are less dependent upon the dimensionality.

THEOREM **8.27** (*Zaremba*) If x is a boundary point of the bounded open set $R \subset E^n$, $n \geq 2$, and there is a truncated closed solid cone in $\sim R$ with vertex at x, then x is a regular boundary point.

Proof. By reducing the half-angle of the cone we can assume that the truncated cone intersects ∂R only in $\{x\}$. Let C_x be a closed solid cone with vertex x such that for some $\rho > 0$ $C_x \cap \bar{B}_{x,\rho} \cap R = \emptyset$. Let $R_0 = R \cup (B_{x,\rho} \sim C_x)$. Then x is a boundary point of $R_0 \supset R$. If we can show that x is a regular boundary point for R_0, it would follow from Theorem 8.23 that x is a regular boundary point for R. To show that x is regular boundary point for R_0 it suffices to show that there is a positive superharmonic function w on $B_{x,\rho} \sim C_x$ for which $\lim_{\substack{y \to x \\ y \in B_{x,\rho} \sim C_x}} w(y) = 0$. In other words, it suffices to prove that x is a regular boundary point for $B_{x,\rho} \sim C_x$. We assume that $R = B_{x,\rho} \sim C_x$. Let $m(y) = \|y - x\|$, $y \in \partial R$. As in the proof of Theorem 8.18, $H_m(z) \geq \|z - x\| > 0$ for $z \in R$. Since H_m is a harmonic function on R, we need show only that $\lim_{z \to x} H_m(z) = 0$. According to Theorem 8.24, all boundary points of R, except possibly x, are regular boundary points. Since m is a continuous boundary function, $\lim_{z \to y} H_m(z) = \|y - x\|$, $y \in \partial R \sim \{x\}$. Let

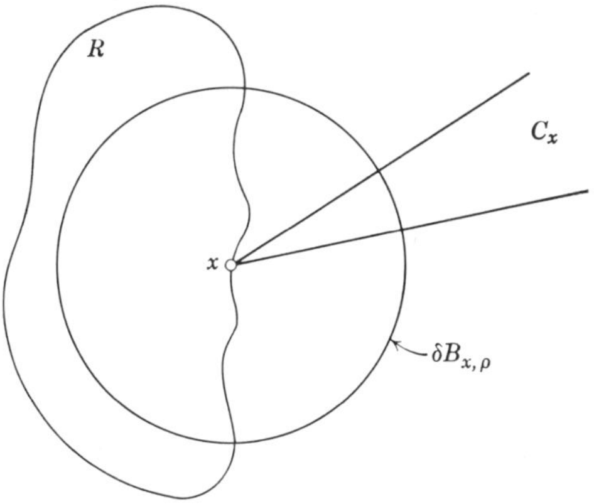

FIGURE 4

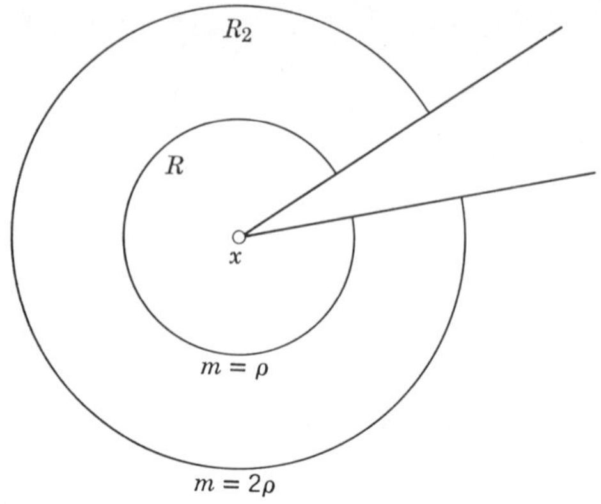

FIGURE 5

$R_2 = \{x + 2(y - x) : y \in R\}$ and let $r(y) = \|y - x\|$ for $y \in \partial R_2$. Consider the Dirichlet solution H_r corresponding to the region R_2 and the boundary function r. Note that $m = \rho$ on $\partial R \cap \partial B_{x,\rho}$ and that $r = 2\rho$ on $\partial R_2 \cap \partial B_{x,2\rho}$. We define a function u on $\bar{R}_2 \sim \{x\}$ by

$$u(y) = \begin{cases} H_r(y), & y \in R_2, \\ r(y) = \|y - x\|, & y \in \partial R_2 \sim \{x\}, \end{cases}$$

and a second function v on $\bar{R} \sim \{x\}$ by the equation $v(y) = u(x + 2(y - x))$. On $\partial R_2 \cap \partial B_{x,2\rho}$, $u = 2\rho$. Moreover, $v = 2\rho$ on $\partial R \cap \partial B_{x,\rho}$. Since u satisfies the maximum principle on R_2, $u \le \alpha' < \alpha = 2\rho$ on $\partial R \cap \partial B_{x,\rho}$. If we increase α', this inequality is not affected as long as $\alpha' < \alpha$. Therefore we can assume that $\alpha'/\alpha > \frac{1}{2}$. Then $u \le \alpha' < \alpha = v$ on $\partial R \cap \partial B_{x,\rho}$; on the remainder of $\partial R \sim \{x\}$, $u = (\frac{1}{2})v < (\alpha'/\alpha)v$. Then $u \le (\alpha'/\alpha)v$ on $\partial R \sim \{x\}$. Since $u - (\alpha'/\alpha)v$ is harmonic on R and $\limsup_{\substack{z \to y \\ z \in R}} [u(z) - (\alpha'/\alpha)v(z)] \le 0$ for all $y \in \partial R \sim \{x\}$, $u - (\alpha'/\alpha)v \le 0$ on R by Theorem 7.10. It follows that

$$u(y) \le (\alpha'/\alpha)u(x + 2(y - x)) = (\alpha'/\alpha)u(2y - x)$$

for $y \in R$. Then $\limsup_{\substack{y \to x \\ y \in R}} u(y) \le (\alpha'/\alpha) \limsup_{\substack{y \to x \\ y \in R}} u(2y - x) = (\alpha'/\alpha) \limsup_{\substack{y \to x \\ y \in R}} u(y)$. Since $\alpha'/\alpha < 1$, $\limsup_{\substack{y \to x \\ y \in R}} u(y) = \limsup_{\substack{y \to x \\ y \in R}} H_r(y) = 0$. Since this is true for all $\rho > 0$, $\lim_{\substack{y \to x \\ y \in R}} H_m(y) = 0$.

COROLLARY **8.28** If R is an open subset of E^n, $n \ge 2$, then there is an increasing sequence $\{R_j\}$ of bounded regular open sets with closures in R such that $R = \cup R_j$.

Proof. Let $\{K_j\}$ be an increasing sequence of compact sets such that $R = \cup K_j$. Consider K_1 and a finite covering of K_1 by balls $B_1^{(1)}, \ldots, B_{k_1}^{(1)}$ having closures in R. By increasing slightly the radii of some of the balls, we can assume that each boundary point of $R_1 = \bigcup_{j=1}^{k_1} B_j^{(1)}$ is the vertex of a truncated closed solid cone lying outside R_1; that is, we can assume that R_1 is a bounded regular region containing K_1. Apply the same procedure to $\bar{R}_1 \cup K_2$, etc.

Before taking up some applications, we shall look at a nontrivial example of an irregular boundary point. We shall, in fact, construct a bounded open set $R \subset E^3$ having an irregular boundary point which is not just an isolated boundary point. This region has a **Lebesgue spine** and is constructed as follows. Consider a measure on the line segment $\{(\xi, 0, 0): 0 \leq \xi \leq 1\}$ which increases linearly from $(0, 0, 0)$ to $(1, 0, 0)$. The Newtonian potential $u(x, y, z)$ of this measure at a point (x, y, z) not on the line segment is given by

$$u(x, y, z) = \int_0^1 \frac{\xi\, d\xi}{[(\xi - x)^2 + y^2 + z^2]^{1/2}}$$

$$= \int_0^1 \frac{(\xi - x)\, d\xi}{[(\xi - x)^2 + y^2 + z^2]^{1/2}} + x \int_0^1 \frac{d\xi}{[(\xi - x)^2 + y^2 + z^2]^{1/2}}.$$

Using elementary calculus, we can evaluate these integrals explicitly to obtain

$$\begin{aligned} u(x, y, z) = {} & [(1 - x)^2 + y^2 + z^2]^{1/2} - (x^2 + y^2 + z^2)^{1/2} \\ & + x \log|1 - x + [(1 - x)^2 + y^2 + z^2]^{1/2}| \\ & + x \log|x + (x^2 + y^2 + z^2)^{1/2}| - x \log(y^2 + z^2). \end{aligned}$$

The sum of the first three terms clearly tends to 1 as $(x, y, z) \to (0, 0, 0)$. Consider only those points (x, y, z) for which $x > 0$. Then it is easily seen that the fourth term on the right tends to zero as $(x, y, z) \to (0, 0, 0)$. It follows that

$$\lim_{\substack{\|(x, y, z)\| \to 0 \\ x > 0}} u(x, y, z) = 1 - \lim_{\substack{\|(x, y, z)\| \to 0 \\ x > 0}} x \log \rho^2,$$

where $\rho^2 = y^2 + z^2$. Now, if $(x, y, z) \to (0, 0, 0)$ along a path for which $x > 0$ and $x^k = \rho^2$, then $u(x, y, z) \to 1$ for any positive integer k. If, however, the approach is along a path for which $x > 0$ and $\rho = \exp(-c/2x)$, $c > 0$, then $u(x, y, z) \to 1 + c$. As the Newtonian potential of a measure with compact support, we know that $\lim_{\|(x, y, z)\| \to +\infty} u(x, y, z) = 0$ and we also know that u is harmonic off the support of the measure. Let $0 < \varepsilon < 1$ and $c_0 > 0$. Take $R = \{(x, y, z) : \varepsilon < u(x, y, z) < 1 + c_0\}$. It is easily seen that $u(x, y, z) \to +\infty$ as (x, y, z) approaches any interior point of the support of the measure defining u. Therefore R is contained in the complement of the support and u is harmonic on R. It also follows that $(0, 0, 0)$ is a boundary point of R. It remains only to show that (0, 0, 0) is an irregular boundary point. Clearly, u is the Dirichlet solution corresponding to a boundary function which is ε on the boundary of the region where $u < \varepsilon$ and which is $1 + c_0$ on the boundary of the region where $u < 1 + c_0$. Suppose $0 < c < c_0$. Then in a neighborhood of (0, 0, 0), the surface $\rho = \exp(-c/2x)$, $x > 0$, lies in the region R; along a path in this surface, $u(x, y, z) \to 1 + c < 1 + c_0$. This shows that (0, 0, 0) is not a regular boundary point for R. The fact that R is not simply-connected is not an essential point. By performing an inversion relative to a sphere, a simply connected region with essentially the same properties can be obtained.

4. Boundary Behavior of the Green Function

In the case of a ball B we know that the Green function $G_B(x, \cdot)$, $x \in B$, tends to zero upon approach to points of ∂B. We shall extend this result in this section and also show how the Green function for a bounded region can be obtained using the Perron-Wiener-Brelot method. Before stating the first result a trivial modification in notation will be made. Suppose R is an open set and f is an extended real-valued function with domain including ∂R. We use H_f for $H_{f|\partial R}$. Recall that u_x denotes the fundamental harmonic function for E^n with pole x.

LEMMA **8.29** Let R be a bounded open set. If G is the Green function for R and $x \in R$, then $G(x, \cdot) = u_x - H_{u_x}$ on R.

Proof. Since u_x is in the upper class defining H_{u_x}, $u_x - H_{u_x} \geq 0$ on R. From the definition of the Green function $u_x - H_{u_x} \geq G(x, \cdot)$ on R. Also from the definition of the Green function, $G(x, \cdot) = u_x + h_x$ where for each $x \in R$, h_x is harmonic on R. Since $G(x, \cdot) \geq 0$, $u_x \geq -h_x$ on R and it is easily seen that $-h_x$ belongs to the lower class defining H_{u_x}; that is, $-h_x \leq H_{u_x}$. Moreover, $u_x - H_{u_x} \geq u_x + h_x$ implies that $-h_x \geq H_{u_x}$. This shows that $-h_x = H_{u_x}$ and that $G(x, \cdot) = u_x - H_{u_x}$ on R.

THEOREM **8.30** If R is a bounded open set with Green function G, then $\lim_{z \to y} G(x, z) = 0$ whenever y is a regular boundary point for R and $x \in R$; if, in addition, R is regular, then $\lim_{z \to y} G(x, z) = 0$ for all $y \in \partial R$. If R is any open set having a Green function G and $x \in R$, then $\lim_{z \to y} G(x, z) = 0$ for all $y \in \partial R$ except possibly for a polar subset of ∂R.

Proof. The first assertion follows from the representation $G(x, \cdot) = u_x - H_{u_x}$ and the fact that $\lim_{z \to y} H_{u_x}(z) = u_x(y)$ whenever y is a regular boundary point for R. Now let R be any open set having a Green function G. By

Corollary 8.28 there is a sequence $\{R_j\}$ of bounded regular open sets with closures in R such that $R_j \uparrow R$. Let G_j be the Green function for R_j, $j \geq 1$. Fix $x \in R$ and let $G^*(x, \cdot)$ be equal to $G(x, \cdot)$ on R and equal to zero on $\sim R$; likewise define $G_j^*(x, \cdot)$ to be $G_j(x, \cdot)$ on R_j and zero on $\sim R_j$. Then $0 \leq G_j^*(x, \cdot) \leq G^*(x, \cdot)$. By Theorem 5.15 $G_j^*(x, \cdot) \uparrow G^*(x, \cdot)$ on R. We can assume that $x \in R_1$. Let $\bar{B}_{x,\rho} \subset R_1$ and let $B = B_{x,\rho}$. By Lemma 5.8 $G^*(x, \cdot)$ is bounded on $\sim B$ and it follows that the sequence $G_j^*(x, \cdot)$ is uniformly bounded on $\sim B$. Since each $G_j^*(x, \cdot)$ is harmonic on $R_j \sim B$ and

$$\lim_{\substack{z \to y \\ z \in R_j}} G_j^*(x, z) = 0$$

for $y \in \partial R_j$, it is easily seen that each $G_j^*(x, \cdot)$ is subharmonic on $\sim B$. Let $g = \lim_{j \to \infty} G_j^*(x, \cdot)$ on $\sim B$. Then $g = G(x, \cdot)$ on $R \sim B$ and it is easily seen that $\bar{g}(y) = \limsup_{z \to y} g(z) = \limsup_{z \to y,\, z \in R} G(x, z)$ for $y \in \partial R$. Since $g = G^*(x, \cdot)$ on $R \sim (B \cup \partial R)$ and $G^*(x, \cdot)$ is continuous on $R \sim (B \cup \partial R)$, $g = \bar{g}$ where defined except possibly on ∂R; but by Theorem 7.39 $g = \bar{g}$, except possibly on a polar subset of $\sim B$. Since $g = 0$ on ∂R and $\bar{g}(y) = \limsup_{z \to y,\, z \in R} G(x, z)$ for $y \in \partial R$, it follows that $\limsup_{z \to y \in \partial R} G(x, z) = 0$ except possibly for y in a polar subset of ∂R.

The following theorem establishes that every open subset R of E^n having a Green function G is an admissible set; that is, the condition that potentials $G\mu$ have boundary limit zero whenever μ is a measure having compact support in R is superfluous. So all results of Chapter 7 apply to any open set having a Green function.

THEOREM **8.31** Let R be an open set having a Green function G. Then there is a polar set $Z \subset \partial R$ such that $\lim_{y \to z \in \partial R \sim Z} G\mu(y) = 0$ for any measure μ having compact support in R. If R has finite Lebesgue measure when $n \geq 3$ or is bounded when $n = 2$, then $\lim_{y \to z \in \partial R \sim Z} Gf(y) = 0$ whenever f is a bounded measurable function on R. If R is a regular open set, then these statements hold with $Z = \varnothing$.

Proof. Let $\{R_j : j \geq 1\}$ be the collection of components of R and let $x_j \in R_j$, $j \geq 1$. By the preceding theorem for each j there is a polar set $Z_j \subset \partial R$ such that $\lim_{y \to z \in \partial R \sim Z_j} G(x_j, y) = 0$. If $Z = \cup Z_j$, then Z is a polar set and $\lim_{y \to z \in \partial R \sim Z} G(x_j, y) = 0$ for all j. Now let K be a compact subset of R. We shall show that $\lim_{y \to z \in \partial R \sim Z} G(\cdot, y) = 0$ uniformly on K. Note that there is an integer p such that $K \subset \bigcup_{j=1}^{p} R_j$. For each $j \leq p$ let V_j be a neighborhood of $R_j \cap K$ with $\bar{V}_j \subset R$. By Harnack's inequality for each such j there is a constant k_j such that $G(x, y) \leq k_j G(x_j, y)$ whenever $x \in R_j \cap K$ and

$y \in R_j \sim V_j$; note that this equality is trivially satisfied if $y \notin R_j$ since both sides are then zero. Therefore $G(x, y) \le k_j G(x_j, y)$ whenever $x \in R_j \cap K$ and $y \in R \sim \bigcup_{j=1}^{p} V_j$. It follows that $G(x, y) \le \max[k_j G(x_j, y) : 1 \le j \le p]$ whenever $x \in K$ and $y \in R \sim \bigcup_{j=1}^{p} V_j$ and that $\lim_{y \to z \in \partial R \sim Z} G(\cdot, y) = 0$ uniformly on K. If μ is a measure having support in K, then

$$G\mu(y) = \int_K G(x, y)\, d\mu(x)$$

and $\lim_{y \to z \in \partial R \sim Z} G\mu(y) = 0$. The proof of Lemma 6.24 applies to the second assertion after noting that $\lim_{y \to z \in \partial R \sim Z} G(\cdot, y) = 0$ uniformly on any compact subset of R.

We now return to the problem of the existence of a Green function for an open subset of E^2.

LEMMA **8.32** Let R be an open connected subset of E^2. If there is a nonconstant positive superharmonic function w on R, than for each $x \in R$ there is a ball $B_{x,\rho}$ with closure in R and a continuous superharmonic function v_x on R with the following properties:

(i) $v_x = 1$ on $B_{x,\rho}$.
(ii) $0 < v_x \le 1$ on R.
(iii) v_x is harmonic on $R \sim \bar{B}_{x,\rho}$.
(iv) There is a point $y \in R \sim \bar{B}_{x,\rho}$ such that $v_x(y) < 1$.

Proof. Consider $w(x)$. By the minimum principle there is a point $y \in R$ such that $w(y) < w(x)$. By modifying w we can assume that $w(y) < 1 < w(x)$. Now let $w^* = w \wedge 1$. Then $w^*(y) < 1 = w^*(x)$ and $w^* = 1$ in a neighborhood W of x. Choose $\rho > 0$ such that $\bar{B}_{x,\rho} \subset W \subset R$. Let $\mathscr{F}$ be the family of non-negative superharmonic functions on R which majorize 1 on $\bar{B}_{x,\rho}$. Then $w^* \in \mathscr{F}$ and it is easily seen that $\mathscr{F}$ is a saturated family over $R \sim \bar{B}_{x,\rho}$. Letting $u = \inf\{f : f \in \mathscr{F}\}$, it follows that $u = 1$ on $\bar{B}_{x,\rho}$ and that u is harmonic on $R \sim \bar{B}_{x,\rho}$ by Theorem 5.9. By Theorem 7.39 the lower regularization $\hat{u}$ of u is superharmonic on R and differs from u at most on a polar subset Z of $\partial B_{x,\rho}$. Using Theorem 4.14 and the fact that $\hat{u} = 1$ on $B_{x,\rho}$, it can be seen that $\hat{u}(z) = \liminf_{\substack{z' \to z \\ z' \in R \sim B_{x,\rho}}} \hat{u}(z')$ for all $z \in \partial B_{x,\rho}$. Choose $\delta > 0$ such that $\bar{B}_{x,\rho} \subset B_{x,\delta} \subset \bar{B}_{x,\delta} \subset R$ and let $V_x = B_{x,\delta} \sim \bar{B}_{x,\rho}$. Define a function g on ∂V_x as follows:

$$g(z) = \begin{cases} u(z), & z \in \partial B_{x,\delta}, \\ 1, & z \in \partial B_{x,\rho}. \end{cases}$$

Then $\hat{u} = g$ on ∂V_x except possibly on Z. Therefore $\liminf_{\substack{z' \to z \\ z' \in V_x}} \hat{u}(z') = g(z)$ for $z \in \partial V_x \sim Z$ and $\hat{u}$ belongs to the upper class defining H_g (relative to V_x)

by Corollary 8.4; that is, $\hat{u} \geq H_g$ on V_x. Since V_x is a regular region, $\lim_{\substack{z' \to z \\ z' \in V_x}} H_g(z') = 1$ for all $z \in \partial B_{x,\rho}$ and it follows that $\hat{u} = 1$ on $\partial B_{x,\rho}$. The function u has the required properties.

THEOREM **8.33** (*Myrberg*) If R is a nonempty open subset of E^2, then the following three conditions are equivalent:

(i) $\sim R$ is not a polar set.
(ii) There is a nonconstant positive superharmonic function defined on R.
(iii) R has a Green function.

Proof. We first show that (iii) → (i). Assume that R has a Green function. Then ∂R is not a polar set by Lemma 7.9. Since $\partial R \subset \sim R$, $\sim R$ is not a polar set. We now show that (i) → (ii). Assume that $\sim R$ is not a polar set and let $F = \sim R$, which is closed. Let $x \in R$ and let $\{B_j\}$ be a sequence of balls such that $x \in \partial B_j$ for each j and $E^2 \sim \{x\} = \cup_j B_j$. Then, for some j_0, $F \cap \bar{B}_{j_0}$ is not polar; for in the contrary case $F = \cup_j (F \cap \bar{B}_j)$ would be polar by Theorem 7.6. Let $C = F \cap \bar{B}_{j_0}$ and let B be a ball that contains B_{j_0} properly and $x \in \partial B$. Then $C \subset B$. If we can show that there is a continuous positive superharmonic function v on $\sim C$ such that $v(x) = 1$ and $v < 1$ on $B \sim C$, the restriction of v to R would have the required properties, since there would be points y of R arbitrarily close to x for which $v(y) < 1$. Consider the réduite $\mathbf{R}_C^1$ and the balayage $\hat{\mathbf{R}}_C^1$ relative to B. Since $\mathbf{R}_C^1$ and $\hat{\mathbf{R}}_C^1$ differ at most on a polar subset of C and $\mathbf{R}_C^1 = 1$ on the non-polar set C, $\hat{\mathbf{R}}_C^1$ must take on the value 1 at some point of C and $\hat{\mathbf{R}}_C^1 > 0$ on B. We also know that $\lim_{\substack{z \to y \\ z \in B}} \hat{\mathbf{R}}_C^1(z) = 0$ for all $y \in \partial B$, since $\hat{\mathbf{R}}_C^1$ is the potential of a measure (relative to the Green function G_B) with compact support $C \subset B$. It follows that $0 < \hat{\mathbf{R}}_C^1 < 1$ on $B \sim C$. Let

$$u = \begin{cases} \mathbf{R}_C^1 & \text{on} \quad B \sim C, \\ 0 & \text{on} \quad \sim B. \end{cases}$$

It is easily seen that u is a continuous subharmonic function on $\sim C$. The function $v = 1 - u$ has the desired properties. Lastly, we show that (ii) → (iii). We might as well assume that R is connected for by Theorem 5.27 R has a Green function in the contrary case. Also assume that (ii) is satisfied and consider any $x \in R$. By Lemma 5.11 it suffices to show that there is a non-negative superharmonic function on R of the form $u_x + h_x$ with h_x harmonic on R. By Lemma 8.32 there is a ball $B_{x,\rho}$ with closure in R and a continuous subharmonic function w_x on R such that $w_x = 0$ on $B_{x,\rho}$, w_x is harmonic on $R \sim \bar{B}_{x,\rho}$, and $0 < w_x < 1$ on $R \sim \bar{B}_{x,\rho}$. Choose δ such that $\delta > \rho$ and

$\bar{B}_{x,\delta} \subset R$. Define a function v_x on R as follows. Let $2\alpha = \inf\{w_x(z): z \in \partial B_{x,\delta}\} > 0$. Put

$$v_x(y) = \begin{cases} \log \dfrac{r}{\rho} & \text{if } r = \|y - x\| < \rho, \\ \left(\log \dfrac{r}{\rho}\right) \vee \left(\dfrac{w_x}{\alpha} \log \dfrac{\delta}{\rho}\right) & \text{if } \rho \le r = \|y - x\| < \delta, \\ \dfrac{w_x}{\alpha} \log \dfrac{\delta}{\rho} & \text{if } r = \|y - x\| \ge \delta. \end{cases}$$

Since w_x is continuous, v_x is continuous (in the extended sense at x). Now v_x is obviously subharmonic on $R \sim (\partial B_{x,\rho} \cup \partial B_{x,\delta})$. It is easily seen that v_x is sub-mean-valued on $\partial B_{x,\rho}$ and therefore subharmonic on $R \sim \partial B_{x,\delta}$. Since $w_x/\alpha > 1$ on $\partial B_{x,\delta}$, $v_x = (w_x/\alpha) \log \delta/\rho$ on a neighborhood of $\partial B_{x,\delta}$ and is therefore subharmonic on R. Since $w_x \le 1$ on R, $v_x \le (1/\alpha) \log \delta/\rho$ on $R \sim B_{x,\delta}$. By Corollary 8.28, there is a sequence $\{R_j\}$ of bounded regular open sets such that $R_j \uparrow R$. We can assume that $\bar{B}_{x,\delta} \subset R_1$. Let G_j be the Green function for R_j. Then $G_j \le G_{j+1}$ on R_j and we can define $G = \lim_{j\to\infty} G_j$ on R (with due attention paid to domains). Now each G_j is of the form $G_j(x, \cdot) = u_x + h_x^j$ with h_x^j harmonic on R_j. Let $h_x = \lim_{j\to\infty} h_x^j$ on R. Then $G(x, \cdot) = u_x + h_x$ where h_x is either identically $+\infty$ or harmonic on R since the sequence $\{h_x^j\}$ is increasing. If we can show that $G(x, \cdot)$ is superharmonic then h_x is harmonic and we are done. Now each $G_j(x, \cdot)$ is harmonic on $R_j \sim \{x\}$ and approaches zero upon approach to points of the boundary of R_j. If we let $G_j^*(x, \cdot)$ be equal to $G_j(x, \cdot)$ on $R_j \sim \{x\}$ and equal to zero on $R \sim R_j$, then $G_j^*(x, \cdot)$ is subharmonic on $R \sim \{x\}$. Since v_x is subharmonic on R, $v_x + G_j^*(x, \cdot)$ is subharmonic on $R \sim \{x\}$; but, in view of the fact that $v_x + G_j^*(x, \cdot) = -\log \rho + h_x^j$ on $B_{x,\rho} \sim \{x\}$, $v_x + G_j^*(x, \cdot)$ can be considered as a subharmonic function on R. Since $v_x + G_j^*(x, \cdot) \le (1/\alpha) \log \delta/\rho$ on $R \sim R_j$, $v_x + G_j^*(x, \cdot) \le c = (1/\alpha) \log \delta/\rho$ on ∂R_j by continuity. Applying the maximum principle to $v_x + G_j^*(x, \cdot)$ (on each component of R_j), we have $v_x + G_j^*(x, \cdot) \le c$ on R. Therefore $G_j(x, \cdot) \le c - v_x$ on R_j and

$$G(x, \cdot) = \lim_{j\to\infty} G_j(x, \cdot) \le c - v_x$$

on R. Since $c - v_x$ is superharmonic on R, the same is true of $G(x, \cdot)$.

We conclude this section by relating the set of regular boundary points to the Green function.

THEOREM **8.34** Let R be a bounded connected open set with Green function G. A point $x \in \partial R$ is a regular boundary point if and only if

$\lim_{z \to x} G(y, z) = 0$ for some $y \in R$. The set if irregular boundary points of R is a polar set.

Proof. If $\lim_{z \to x} G(y, z) = 0$ for some $y \in R$, then $G(y, \cdot)$ is a barrier at x and x is a regular boundary point. Conversely, if x is a regular boundary point, then $\lim_{z \to x} G(y, z) = u_y(x) - \lim_{z \to x} H_{u_y}(z) = 0$ for any $y \in R$. The second assertion follows from Theorem 8.30.

5. Applications

In this section we use the preceding development to establish some properties of the balayage $\hat{\mathbf{R}}_E^u$, considered as an operator on u or as a function of the set E. We shall also look at the Maria-Frostman domination principle which is used to compare a superharmonic function and a potential.

THEOREM **8.35** If the open set R has a Green function, Z is a polar subset of R, and $x_0 \in R \sim Z$, there is a positive superharmonic function u on R such that $u = +\infty$ on Z and $u(x_0) < +\infty$.

Proof. Assume first that $Z \subset B_{y,\rho} \subset \bar{B}_{y,\rho} \subset R$ and that $x_0 \notin \bar{B}_{y,\rho}$. Let $r > \rho$ be such that $\bar{B}_{y,r} \subset R$. As in the proof of Lemma 7.13, there is a superharmonic function w such that $w(x_0) < +\infty$, $w = +\infty$ on Z, and $w = \alpha u_y + \beta$ on $\sim B_{y,r}$. If G is the Green function for R, then $G(y, \cdot) = u_y + h_y$, where h_y is harmonic on R. If we define $w^* = w + (\alpha h_y + \beta)$, then $w^* = \alpha G(y, \cdot) + 2\beta$ on $R \sim B_{y,r}$. By Theorem 8.30 $\lim_{z \to x} G(y, z) = 0$ for all $x \in \partial R$ except possibly for a polar subset of ∂R. Also note that $\lim_{\|z\| \to +\infty} G(y, z) = 0$ if R is unbounded and $n \geq 3$ since $G(y, \cdot)$ is majorized by the Green function for E^n. It follows that $\lim_{z \to x} w^*(z) = 2\beta$ for all $x \in \partial R$ except possibly for a polar set and $\lim_{\|z\| \to +\infty} w^*(z) = 2\beta$ in case R is unbounded and $n \geq 3$. It is easily seen that w^* is bounded below and superharmonic on R. By Theorem 7.10 $w^* \geq 2\beta$ on R. It follows that $u = w^* - 2\beta = +\infty$ on Z, strictly positive on R, and finite on $R \sim B_{y,r}$. Since $w(x_0) < +\infty$, the same is true of w^* and u. Returning to the general case, let $x_0 \in R \sim Z$ and for each $y \in Z$ let B_y be a ball with center y such that $x_0 \notin \bar{B}_y \subset R$. Now $\{B_y, y \in Z\}$ is a covering of Z and a sequence of such balls correspondingto a sequence of points $\{y_j\}$ in Z suffice to cover Z. For each j let B_j be a ball such that $\bar{B}_{y_j} \subset B_j \subset \bar{B}_j \subset R$ and $x_0 \notin \bar{B}_j$. According to the first part of the proof, for each j there is a positive superharmonic function w_j on R such that $w_j(x_0) < +\infty$ and $w_j = +\infty$ on $B_{y_j} \cap Z$. Choose a sequence $\{\beta_j\}$ of positive numbers such that $\sum_j \beta_j w_j(x_0) < +\infty$. It is easily seen that the function $\sum_j \beta_j w_j$ is a positive superharmonic function which is infite at x_0 and $+\infty$ on Z.

COROLLARY **8.36** Let R be an open subset of E^n having a Green function. If $E \subset R$ and u is a non-negative superharmonic function on R, then $\hat{\mathbf{R}}_E^u = \mathbf{R}_E^u$ on $R \sim E$.

Proof. We know from Lemma 7.11 that $\mathbf{R}_E^u = u$ on E and that $\hat{\mathbf{R}}_E^u = \mathbf{R}_E^u$ on R except possibly for a polar set by Corollary 7.40. Let Z be the set of points $z \in E$ for which $\hat{\mathbf{R}}_E^u(z) \neq \mathbf{R}_E^u(z)$. Then Z is a polar subset of E. Let $x_0 \in R \sim E$. By Theorem 8.35 there is a positive superharmonic function w on R such that $w = +\infty$ on Z and $w(x_0) < +\infty$. Then $\hat{\mathbf{R}}_E^u + \varepsilon w \geq u$ on E for every $\varepsilon > 0$ and therefore $\hat{\mathbf{R}}_E^u + \varepsilon w \geq \mathbf{R}_E^u$ on R for every $\varepsilon > 0$. Since $w(x_0) < +\infty$ and ε can be any positive number, $\hat{\mathbf{R}}_E^u(x_0) \geq \mathbf{R}_E^u(x_0)$. Since we always have $\hat{\mathbf{R}}_E^u(x_0) \leq \mathbf{R}_E^u(x_0)$, there is equality at x_0.

COROLLARY **8.37** Let R be an open set having a Green function. If u is a non-negative superharmonic function on R, $E \subset R$, and Z is a polar subset of E, then $\hat{\mathbf{R}}_E^u = \hat{\mathbf{R}}_{E \sim Z}^u$.

Proof. We know from Theorem 7.12 that $\hat{\mathbf{R}}_{E\sim Z}^u \leq \hat{\mathbf{R}}_E^u$. Consider any $x \in R \sim Z$ and any positive superharmonic function w on R which majorizes u on $E \sim Z$. By Theorem 8.35 there is a positive superharmonic function v on R such that $v(x) < +\infty$ and $v = +\infty$ on Z. Then $w + \varepsilon v$ majorizes u on E for every $\varepsilon > 0$. Therefore $\hat{\mathbf{R}}_E^u(x) \leq w(x) + \varepsilon v(x)$ and letting $\varepsilon \downarrow 0$ we obtain $\hat{\mathbf{R}}_E^u(x) \leq w(x)$ for all $x \in R \sim Z$. Since Z has Lebesgue measure zero, $A(\hat{\mathbf{R}}_E^u : y, \delta) \leq A(w : y, \delta)$ whenever $\bar{B}_{y,\delta} \subset R$. Letting $\delta \downarrow 0$ we get $\hat{\mathbf{R}}_E^u \leq w$ on R, and by taking the infimum over all such w $\hat{\mathbf{R}}_E^u \leq \mathbf{R}_{E\sim Z}^u$. Taking the lower regularization of both members of this inequality results in $\hat{\mathbf{R}}_E^u \leq \hat{\mathbf{R}}_{E\sim Z}^u$.

THEOREM **8.38** Let R be an open subset of E^n having a Green function. If u is a positive superharmonic function on R, F a relatively closed subset of R, and $\{F_j\}$ an increasing sequence of relatively closed subsets of R such that $F = \cup_j F_j$, then $\lim_{j\to\infty} \hat{\mathbf{R}}_{F_j}^u = \hat{\mathbf{R}}_F^u$.

Proof. The sequence $\{\hat{\mathbf{R}}_{F_j}^u\}$ is increasing and is majorized by the superharmonic function $\hat{\mathbf{R}}_F^u$. The function $v = \lim_{j\to\infty} \hat{\mathbf{R}}_{F_j}^u$ is superharmonic and $v \leq \hat{\mathbf{R}}_F^u$. By Corollary 7.40 $\hat{\mathbf{R}}_{F_j}^u$ differs from u on F_j on at most a polar subset of F_j. Therefore v differs from u on F on at most a polar subset Z of F. Let w be a positive superharmonic function on R such that $w = +\infty$ on Z and let $\varepsilon > 0$. Then $v + \varepsilon w \geq u$ on F and therefore $v + \varepsilon w \geq \hat{\mathbf{R}}_F^u$. Since ε can be made arbitrarily small, $v \geq \hat{\mathbf{R}}_F^u$. This proves that $\lim_{j\to\infty} \hat{\mathbf{R}}_{F_j}^u = \hat{\mathbf{R}}_F^u$.

LEMMA **8.39** Let R be an open set having a Green function, let W be an open subset of R, and let f be a resolutive boundary function relative to R. The function g defined on ∂W by

$$g = \begin{cases} H_f^R & \text{on} \quad R \cap \partial W, \\ f & \text{on} \quad \partial R \cap \partial W, \end{cases}$$

is resolutive and $H_f^R = H_g^W$ on W.

Proof. Let $u \in \mathfrak{U}_f^R$. Then $u \geq H_f^R$ on R. For $x \in R \cap \partial W$, $\liminf_{z \to x, z \in W} u(z) \geq \liminf_{z \to x, z \in W} H_f^R(z) = H_f^R(x)$; while for $x \in \partial R \cap \partial W$, $\liminf_{z \to x, z \in W} u(z) \geq \liminf_{z \to x, z \in R} u(z) \geq f(x)$. Therefore $u \in \mathfrak{U}_g^W$ and $\overline{H}_f^R \geq \overline{H}_g^W$ on W. Similarly, $\underline{H}_g^W \geq \underline{H}_f^R$ on W. It follows that $0 \leq \overline{H}_g^W - \underline{H}_g^W \leq \overline{H}_f^R - \underline{H}_f^R$. The equality and harmonicity of $\overline{H}_f^R$ and $\underline{H}_f^R$ implies that of $\overline{H}_g^W$ and $\underline{H}_g^W$.

LEMMA **8.40** Let R be an open set having a Green function, let W be an open set having compact closure $\overline{W} \subset R$, and let u be a positive superharmonic function on R. The lower regularization $\hat{v}$ of the function

$$v = \begin{cases} u & \text{on} \quad R \sim W, \\ H_u^W & \text{on} \quad W, \end{cases}$$

is superharmonic on R.

Proof. Note first that H_u^W is defined, since $u \geq \overline{H}_u^W \geq \underline{H}_u^W \geq 0$ on W. Let B be a ball with $\overline{B} \subset R$. We shall first show that $v \geq H_v^B$ on B. Since $u \geq H_u^W$ on W, $v \leq u$ on R and $H_v^B \leq H_u^B \leq u = v$ on $B \sim W$. Now let

$$w = \begin{cases} v & \text{on} \quad R \sim B, \\ H_v^B & \text{on} \quad B. \end{cases}$$

Note that $w = v$ on $\overline{W} \cap \partial B$ and that $w = H_v^B \leq H_u^B \leq u = v$ on $\partial W \cap B$; that is, $w \leq v$ on $\partial(B \cap W)$. It follows that $H_w^{B \cap W} \leq H_v^{B \cap W}$ on $B \cap W$. Since $H_v^B = H_w^{B \cap W}$ and $H_v^{B \cap W} = H_u^W$ on $B \cap W$ by the preceding lemma, $H_v^B \leq H_u^W = v$ on $B \cap W$. Therefore $H_v^B \leq v$ on B. Noting that $\hat{v} \leq v$, $H_{\hat{v}}^B \leq H_v^B \leq v$ on B. In view of the fact that $H_{\hat{v}}^B$ is continuous on B, $H_{\hat{v}}^B \leq \hat{v}$ on B for any ball B with $\overline{B} \subset R$ and $\hat{v}$ is superharmonic on R.

THEOREM **8.41** Let R be an open subset of E^n having a Green function. If u is a positive superharmonic function on R and W is an open subset of R with compact closure $\overline{W} \subset R$, then $\hat{\mathbf{R}}_{R \sim W}^u = H_u^W$ on W; moreover, if $\{W_j\}$ is an increasing sequence of such sets with $R = \cup_j W_j$, then u is a potential if and only if $\lim_{j \to \infty} H_u^{W_j} = 0$ on R.

Proof. If

$$v = \begin{cases} u & \text{on} \quad R \sim W, \\ H_u^W & \text{on} \quad W, \end{cases}$$

then $\hat{v}$ is superharmonic on R. Since u is l.s.c. on ∂W, there is a sequence $\{f_j\}$ of continuous functions on ∂W such that $f_j \uparrow u$ on ∂W. Since

$$\liminf_{\substack{z \to y \\ z \in W}} v(z) = \liminf_{\substack{z \to y \\ z \in W}} H_u^W(z) \geq \liminf_{\substack{z \to y \\ z \in W}} H_{f_j}^W(z) = f_j(y)$$

for all $y \in \partial W$ that are regular boundary points for W, $\liminf_{\substack{z \to y \\ z \in W}} v(z) \geq u(y)$ for all $y \in \partial W$ except possibly for a polar subset of ∂W. Therefore $\hat{v}(y) = \liminf_{\substack{z \to y \\ z \in R}} v(z) \geq u(y)$ for all $y \in \partial W$ except possibly for a polar subset of ∂W. Since $\hat{v} = v = u$ on $R \sim \overline{W}$, $\hat{v} \geq u$ on $R \sim W$ except possibly for a polar subset of ∂W. It follows from Corollary 8.37 that $\hat{v} \geq \hat{\mathbf{R}}_{R \sim W}^u$. If w is any positive superharmonic function on R which majorizes u on $R \sim W$, then $w|_W \in \mathfrak{U}_u^W$ and $w \geq v$ on W; therefore $w \geq v$ on R. This shows that $\mathbf{R}_{R \sim W}^u \geq v$ on R and consequently that $\hat{\mathbf{R}}_{R \sim W}^u \geq \hat{v}$ on R. Hence $\hat{\mathbf{R}}_{R \sim W}^u = v = H_u^W$ on W. Now let $\{W_j\}$ be an increasing sequence of open sets having compact closures $\overline{W}_j \subset R$ such that $R = \cup W_j$. We have just shown that $H_u^{W_j} = \hat{\mathbf{R}}_{R \sim W_j}^u$ on W_j. Since $\hat{\mathbf{R}}_{R \sim W_{j+1}}^u \leq \hat{\mathbf{R}}_{R \sim W_j}^u$ by Theorem 7.12, $H_u^{W_{j+1}} \leq H_u^{W_j}$ on W_j for each j. The sequence $\{H_u^{W_j}\}$ is then a decreasing sequence of non-negative harmonic functions (with due attention given to domains) and $h = \lim_{j \to \infty} H_u^{W_j}$ defines a non-negative harmonic function on R with $u \geq h \geq 0$. The function h is obviously the greatest harmonic minorant of u.

COROLLARY **8.42** Let R be an open set having a Green function, let $F = R \sim W$, where W is an open subset of R having compact closure $\overline{W} \subset R$, and let u, v be positive superharmonic functions on R. Then $\hat{\mathbf{R}}_F^{u+v} = \hat{\mathbf{R}}_F^u + \hat{\mathbf{R}}_F^v$ on R.

Proof. Since $\hat{\mathbf{R}}_F^{u+v} = u + v$, $\hat{\mathbf{R}}_F^u = u$, and $\hat{\mathbf{R}}_F^v = v$ on F except possibly for a polar set, $\hat{\mathbf{R}}_F^{u+v} = \hat{\mathbf{R}}_F^u + \hat{\mathbf{R}}_F^v$ on F except possibly for a polar set. By the preceding theorem and Lemma 8.6 $\hat{\mathbf{R}}_F^{u+v} = H_{u+v}^W = H_u^W + H_v^W$, $\hat{\mathbf{R}}_F^u = H_u^W$, and $\hat{\mathbf{R}}_F^v = H_v^W$ on W. Therefore, $\hat{\mathbf{R}}_F^{u+v} = \hat{\mathbf{R}}_F^u + \hat{\mathbf{R}}_F^v$ on R except possibly for a polar set. The exceptional set is empty as can be seen by averaging.

We shall include in this chapter a theorem that will be needed in the last chapter. The result is known as the **domination principle** and is reconsidered in Chapter 11.

THEOREM **8.43** (*Maria-Frostman*) Let R be an open subset of E^n having a Green function G and let μ be a measure on R such that $G\mu$ is finite. If w is a positive superharmonic function on R such that $w \geq G\mu$ on the support of μ except possibly for a polar set, then $w \geq G\mu$ on R.

Proof. We can assume that R is connected. We first prove the theorem under the additional assumptions that μ has compact support K and that

$G\mu = u$ is continuous on R. Let $\{R_j\}$ be an increasing sequence of open sets with compact closures $\bar{R}_j \subset R$ such that $K \subset R_j$ and $R = \cup R_j$. Also let $W_j = R_j \sim K$. Since $u = G\mu$ is harmonic on W_j and continuous on $\overline{W}_j$, $u = H_u^{W_j}$ on W_j, where $H_u^{W_j}$ is the Dirichlet solution corresponding to W_j and $u|_{\partial W_j}$. Consider any $v \in \mathfrak{U}_u^{R_j}$, the upper class corresponding to R_j and the boundary function $u|_{\partial R_j}$. Then $\lim\inf_{\substack{y \to x \\ y \in R_j}} v(y) \geq u(x)$ for all $x \in \partial R_j$. Now consider $v + w$. For $x \in \partial R_j$, $\lim\inf_{\substack{y \to x \\ y \in W_j}} (v(y) + w(y)) \geq \lim\inf_{\substack{y \to x \\ y \in R_j}} v(y) \geq u(x)$, whereas for $x \in \partial K$, $\lim\inf_{\substack{y \to x \\ y \in W_j}} (v(y) + w(y)) \geq \lim\inf_{y \to x} w(y) \geq w(x) \geq u(x)$ except possibly for x in a polar subset of K. By Corollary 8.4 $v + w \geq H_u^{W_j}$ on W_j. Since v can be any element of $\mathfrak{U}_u^{R_j}$,

$$H_u^{R_j} + w \geq H_u^{W_j} = u \quad \text{on} \quad W_j;$$

but since $w \geq u$ on K, except possibly for a polar set, $H_u^{R_j} + w \geq u$ on R_j, except possibly for a polar set. Now the sequence $\{H_u^{R_j}\}$ decreases on R and defines a non-negative harmonic function $h = \lim_{j \to \infty} H_u^{R_j}$ on R. Clearly $h \leq u$ on R and since u is a potential, $h = 0$ on R. This proves that $w \geq u$ on R except possibly for a polar subset. Hence $w(x) = \lim_{\delta \to 0} A(w : x, \delta) \geq \lim_{\delta \to 0} A(u : x, \delta) = u(x)$ for all $x \in R$. We shall now remove the continuity condition. By Theorem 6.21 there is an increasing sequence $\{K_j\}$ of compact subsets of K such that $G\mu|_{K_j}$ is continuous on R and $\mu(K \sim K_j) \to 0$ as $j \to \infty$. Since $G\mu|_{K_j} \leq G\mu \leq w$ on K except possibly for a polar set, $G\mu|_{K_j} \leq w$ on R by the first part of the proof. Let $\mu_j = \mu|_{K_j}$ and $\nu_j = \mu|_{K \sim K_j}$. Then $G\mu = G\mu_j + G\nu_j$ and the sequence $\{G\mu_j\}$ is increasing. If x is any point in $R \sim K$, then $G(x, \cdot)$ is bounded on K and

$$G\nu_j(x) = \int G(x, y) \, d\nu_j(y) \leq \left[\sup_{y \in K} G(x, y)\right] \mu(K \sim K_j).$$

Since $\mu(K \sim K_j) \to 0$ as $j \to \infty$, $\lim_{j \to \infty} G\nu_j = 0$ on $R \sim K$. If $q = \lim_{j \to \infty} G\nu_j$, then the lower regularization q^* of q is a non-negative superharmonic function which must vanish on R, since it vanishes on $R \sim K$. Since q differs from q^* at most on a polar set, $G\mu = \lim_{j \to \infty} G\mu_j + \lim_{j \to \infty} G\nu_j \leq w$ except possibly on a polar subset of R. As before, this implies that $G\mu \leq w$ on R. We need only remove the condition that the support K of μ is compact. Let $\{K_j\}$ be a sequence of compact subsets of K such that $K_j \uparrow K$. By the second part of the proof $G\mu|_{K_j} \leq G\mu \leq w$ on K except possibly for a polar set implies that $G\mu|_{K_j} \leq w$ on R. Then

$$G\mu = \lim_{j \to \infty} \int_{K_j} G(\cdot, y) \, d\mu = \lim_{j \to \infty} G\mu|_{K_j} \leq w \quad \text{on} \quad R.$$

CHAPTER 9

Dirichlet Problem for Unbounded Regions

The principal problem with unbounded regions is the lack of uniqueness of the solution of the Dirichlet problem. To resolve this problem the point at infinity must be taken into consideration. In this chapter we enlarge E^n by adjoining Δ, the point at infinity, to E^n, and this requires redefinition of such things as harmonic functions, superharmonic functions, and polar sets. The exterior Dirichlet problem for a ball is solved in this chapter by means of a Poisson type integral. The Perron-Wiener-Brelot method if used to solve the generalized Dirichlet problem and boundary behavior of the solution is examined.

1. Exterior Dirichlet Problem

In this section we return to the Dirichlet problem for general open sets $R \subset E^n$. For this purpose we enlarge the basic space E^n as follows. Let Δ be an object not in E^n and let $E^n_\Delta = E^n \cup \{\Delta\}$. We call Δ the point at infinity. We topologize E^n_Δ by adjoining to a base for the topology for E^n the class of sets of the form $E^n_\Delta \sim C$, where C is a compact subset of E^n in the usual topology. Under this topology, E^n_Δ is a compact space and, in fact, a metrizable compact space. From now on R is a subset of E^n_Δ which may include the point at infinity.

Let R be an open subset of E^n_Δ. The following definition is consistent with our earlier definition.

Definition. The extended real-valued function u is **superharmonic** on R if it is superharmonic on $R \sim \{\Delta\}$ in the usual sense, and, if $\Delta \in R$, then

(i) u is l.s.c. at Δ,
(ii) $u(\Delta) \geq L(u : x_0, \delta)$ for any x_0 and δ such that $\sim B_{x_0, \delta} \subset R$;

the function u is **subharmonic** on R if $-u$ is superharmonic on R; if both u and $-u$ are superharmonic on R, then u is **harmonic** on R.

The constant functions on E_Δ^n are obviously harmonic on E_Δ^n. The role of the point at infinity is typified by considering the function $u_x(y) = \|x - y\|^{-n+2}$ defined on E_Δ^n, $n \geq 3$, by putting $u_x(\Delta) = 0$; although this function is superharmonic on $E_\Delta^n \sim \{\Delta\}$, it is not superharmonic on E_Δ^n, since $u_x(\Delta) < L(u_x : x_0, \delta)$ for any x_0 and δ. It is, however, subharmonic on $E_\Delta^n \sim \{x\}$. We shall soon see that the only functions that are superharmonic on E_Δ^n are the constant functions.

Before looking at properties of superharmonic functions, we shall show that there are nontrivial examples of harmonic functions defined on a neighborhood of Δ. This requires a preliminary representation of such functions.

LEMMA **9.1** If h is harmonic on $E^n \sim \bar{B}_{y,\rho}$, continuous on $E^n \sim B_{y,\rho}$, and $\lim_{x\to\Delta} h(x)$ exists and is equal to α, then for $x \in E^n \sim \bar{B}_{y,\rho}$

$$h(x) = \frac{1}{\sigma_n \rho} \int_{\partial B_{y,\rho}} \frac{\|y - x\|^2 - \rho^2}{\|z - x\|^n} h(z)\, d\sigma(z) - \frac{\alpha\rho^{n-2}}{\|x - y\|^{n-2}} + \alpha.$$

Moreover, if $\alpha = L(h : y, \rho)$, then $L(h : y, r) = \alpha$ for all $r > \rho$.

Proof. If $x \in E^n \sim \bar{B}_{y,\rho}$, let $x^\star$ denote the inverse of x relative to $\partial B_{y,\rho}$. Under this inversion the region $E^n \sim B_{y,\rho}$ maps onto the region $\bar{B}_{y,\rho} \sim \{y\}$. The Kelvin transformation relative to $\partial B_{y,\rho}$ induces a function $h^\star$ on $\bar{B}_{y,\rho} \sim \{y\}$ by means of the equation

$$\frac{\|x^\star - y\|^{n-2}}{\rho^{n-2}} h^\star(x^\star) = h(x), \qquad x^\star \in \bar{B}_{y,\rho} \sim \{y\}.$$

The function $h^\star$ is harmonic on $B_{y,\rho} \sim \{y\}$ and continuous on $\bar{B}_{y,\rho} \sim \{y\}$ by Theorem 2.24. Consider first the $n \geq 3$ case. By Theorem 5.4 there is a constant $c \geq 0$ and a function v' which is harmonic on $B_{y,\rho}$ such that $h^\star = cu_y + v'$ on $B_{y,\rho} \sim \{y\}$. Since

$$h(x) = \frac{\|x^\star - y\|^{n-2}}{\rho^{n-2}} \left[\frac{c}{\|x^\star - y\|^{n-2}} + v'(x^\star)\right],$$

$\lim_{x\to\Delta} h(x) = \alpha$, and v' is harmonic on $B_{y,\rho}$,

$$\frac{c}{\rho^{n-2}} = \alpha;$$

that is, $h^\star = \alpha\rho^{n-2}u_y + v'$ on $B_{y,\rho} \sim \{y\}$. This can be written

$$h^\star = \alpha\rho^{n-2}\left(u_y - \frac{1}{\rho^{n-2}}\right) + (v' + \alpha)$$

Since $u_y - \rho^{-n+2} = 0$ on $\partial B_{y,\rho}$, $h^\star = h = v' + \alpha$ on $\partial B_{y,\rho}$. Therefore

$$h^\star(x^\star) = \frac{\alpha\rho^{n-2}}{\|x^\star - y\|^{n-2}} - \alpha + \frac{1}{\sigma_n \rho}\int_{\partial B_{y,\rho}} \frac{\rho^2 - \|x^\star - y\|^2}{\|z - x^\star\|^n} h(z)\,d\sigma(z)$$

for $x^\star \in B_{y,\rho} \sim \{y\}$. Using the fact that $\|x - y\|\;\|x^\star - y\| = \rho^2$ and that

$$\|z - x^\star\| = \frac{\rho\|z - x\|}{\|x - y\|}, \qquad z \in \partial B_{y,\rho}$$

(see the discussion preceding Theorem 1.7),

$$\frac{1}{\sigma_n \rho}\int_{\partial B_{y,\rho}} \frac{\rho^2 - \|x^\star - y\|^2}{\|z - x^\star\|^n} h(z)\,d\sigma(z) = \frac{\|y - x\|^{n-2}}{\rho^{n-2}} \frac{1}{\sigma_n \rho}\int_{\partial B_{y,\rho}} \frac{\|x - y\|^2 - \rho^2}{\|z - x\|^n} \times h(z)\,d\sigma(z)$$

Therefore

$$h(x) = \frac{\|x^\star - y\|^{n-2}}{\rho^{n-2}} h^\star(x^\star) = \alpha - \frac{\alpha\rho^{n-2}}{\|x - y\|^{n-2}} + \frac{1}{\sigma_n \rho}\int_{\partial B_{y,\rho}} \frac{\|x - y\|^2 - \rho^2}{\|z - x\|^n} \times h(z)\,d\sigma(z)$$

for $x \in E^n \sim \bar{B}_{y,\rho}$. To prove the second assertion of the lemma we first establish a relation between averages of h and averages of $h^\star$. If $\delta \geq \rho$, then

$$\begin{aligned} L(h : y, \delta) &= \frac{1}{\sigma_n}\int_{\|\theta\|=1} h(y + \delta\theta)\,d\sigma(\theta) \\ &= \frac{\rho^{n-2}}{\sigma_n \delta^{n-2}}\int_{\|\theta\|=1} \frac{\|\delta\theta\|^{n-2}}{\rho^{n-2}} h(y + \delta\theta)\,d\sigma(\theta); \end{aligned}$$

but the integrand is just $h^\star(y + (\rho^2/\delta)\,\theta)$. Hence

$$\begin{aligned} L(h : y, \delta) &= \frac{\rho^{n-2}}{\delta^{n-2}} \frac{1}{\sigma_n}\int_{\|\theta\|=1} h^\star\left(y + \frac{\rho^2}{\delta}\,\theta\right) d\sigma(\theta) \\ &= \frac{\rho^{n-2}}{\delta^{n-2}} L\left(h^\star : y, \frac{\rho^2}{\delta}\right). \end{aligned}$$

Now $L(h^\star : y, \rho^2/\delta) = \alpha\rho^{n-2}L(u_y : y, \rho^2/\delta) + L(v' : y, \rho^2/\delta) = \alpha(\delta^{n-2}/\rho^{n-2}) + v'(y)$, since u_y is equal to δ^{n-2}/ρ^{2n-4} on $\partial B_{y,\rho^2/\delta}$. Assuming that $\alpha = L(h : y, \rho) = L(h^\star : y, \rho)$, then $v'(y) = 0$. It follows that

$$L(h : y, \delta) = \frac{\rho^{n-2}}{\delta^{n-2}} L\left(h^\star : y, \frac{\rho^2}{\delta}\right) = \alpha$$

for all $\delta > \rho$. The $n = 2$ case is easier.

Note that the equation of the preceding lemma can be put into a neater form under the additional hypothesis that

$$\alpha = L(h : y, \rho) = \frac{1}{\sigma_n \rho} \int_{\partial B_{y,\rho}} \frac{h(z)}{\rho^{n-2}} \, d\sigma(z),$$

in which case

$$h(x) = \frac{1}{\sigma_n \rho} \int_{\partial B_{y,\rho}} \left[\frac{\|y - x\|^2 - \rho^2}{\|z - x\|^n} + \frac{1}{\rho^{n-2}} - \frac{1}{\|x - y\|^{n-2}} \right] h(z) \, d\sigma(z).$$

for $x \in E^n \sim \bar{B}_{y,\rho}$. This lemma gives us a representation of functions which are harmonic in a neighborhood of Δ.

Definition. If f is a Borel measurable function defined on $\partial B_{y,\rho}$, which is integrable relative to surface area, the **Poisson integral** of f relative to $E^n_\Delta \sim B_{y,\rho}$ is defined as

$$\mathbf{PI}(f, \sim B_{y,\rho})(x) = \frac{1}{\sigma_n \rho} \int_{\partial B_{y,\rho}} \left(\frac{\|y - x\|^2 - \rho^2}{\|z - x\|^n} + \frac{1}{\rho^{n-2}} - \frac{1}{\|x - y\|^{n-2}} \right) \times f(z) \, d\sigma(z)$$

for $x \in E^n_\Delta \sim \bar{B}_{y,\rho}$. The integrand is defined to be ρ^{-n+2} when $x = \Delta$.

THEOREM **9.2** If f is a Borel measurable function defined on $\partial B_{y,\rho}$ which is integrable relative to surface area, then $h = \mathbf{PI}(f, \sim B_{y,\rho})$ is harmonic on $E^n_\Delta \sim \bar{B}_{y,\rho}$ and $\limsup_{\substack{z \to x \\ z \in \sim B_{y,\rho}}} h(z) \le \limsup_{\substack{z \to x \\ z \in \partial B_{y,\rho}}} f(z)$ for all $x \in \partial B_{y,\rho}$; if, in addition, f is continuous at $x \in \partial B_{y,\rho}$, then $\lim_{\substack{z \to x \\ z \in \sim B_{y,\rho}}} h(z) = f(x)$.

Proof. To show that h is harmonic on $E^n \sim \bar{B}_{y,\rho}$ we can compute the Laplacian of both sides of the equation defining $\mathbf{PI}(f, \sim B_{y,\rho})$ and easily justify taking the Laplacian under the integral; since the integrand is harmonic, h is harmonic on $E^n \sim \bar{B}_{y,\rho}$. It is also easily seen that

$$h(\Delta) = \lim_{x \to \Delta} h(x) = \frac{1}{\sigma_n \rho^{n-1}} \int_{\partial B_{y,\rho}} f(z) \, d\sigma(z) = L(f : y, \rho).$$

By applying the inversion relative to $\partial B_{y,\rho}$ to the equation defining h we get

$$h^\star(x^\star) = \frac{L(f : y, \rho)\rho^{n-2}}{\|x^\star - y\|^{n-2}} - L(f : y, \rho) + \frac{1}{\sigma_n \rho} \int_{\partial B_{y,\rho}} \frac{\rho^2 - \|x^\star - y\|^2}{\|z - x^\star\|^n} \times f(z) \, d\sigma(z).$$

If $r > \rho$, then

$$L\left(h^\star : y, \frac{\rho^2}{r}\right) = L(f : y, \rho)\frac{r^{n-2}}{\rho^{n-2}} - L(f : y, \rho) + L(f : y, \rho),$$

since the last term on the right is just $\mathbf{PI}(f, B_{y,\rho})$ and $\mathbf{PI}(f, B_{y,\rho})(y) = L(f : y, \rho)$. In view of the fact that

$$L(h : y, r) = (\rho^{n-2}/r^{n-2})L(h^\star : y, \rho^2/r),$$

$$L(h : y, r) = L(f : y, \rho) = h(\Delta) \quad \text{for all} \quad r > \rho.$$

This does not entirely prove that h is harmonic, for we must also show that $h(\Delta) = L(h : x_0, \delta)$ whenever $\sim B_{x_0,\delta} \subset E^n \sim \bar{B}_{y,\rho}$. Consider any such x_0 and δ. Choose $r_0 > \rho$ such that $\bar{B}_{y,r_0} \subset B_{x_0,\delta}$. Then $\bar{B}_{x_0,\delta} \sim B_{y,r_0} \subset E^n \sim \bar{B}_{y,\rho}$. Since h is harmonic on the latter region, h is harmonic on $\bar{B}_{x_0,\delta} \sim B_{y,r_0}$ and

$$\int_{\partial B_{y,r_0}} D_{\mathbf{n}} h \, d\sigma(z) = \int_{\partial B_{x_0,\delta}} D_{\mathbf{n}} h \, d\sigma(z)$$

by Green's theorem. Since $L(h : y, r) = L(h : y, r_0)$ for all $r > r_0$, these two integrals must be zero by Theorem 4.22. By the same theorem $L(h : x_0, \delta') =$ constant for all $\delta' \geq \delta$. Since $\lim_{x\to\Delta} h(x) = L(f : y, \rho)$, this constant must be $L(f : y, \rho)$ and therefore $L(h : x_0, \delta') = L(f : y, \rho) = h(\Delta)$ for all $\delta' \geq \delta$. The assertions concerning the boundary behavior of h follow by examining the Kelvin transform $h^\star$ of h.

Most of the properties of superharmonic functions carry over to the present situation.

THEOREM **9.3** If u is superharmonic on the open connected set $R \subset E^n_\Delta$, then u satisfies the minimum principle on R. If R is any open subset of E^n_Δ, u is superharmonic on R, and $\liminf_{z\to x} u(z) \geq 0$ for all $x \in \partial R$, then $u \geq 0$ on R.

Proof. The first assertion is just the first assertion of Corollary 4.3 if $\Delta \notin R$. There is nothing new to prove unless we assume that $\Delta \in R$. Let $m = \inf \{u(y) : y \in R\}$ and let $M = \{y \in R : u(y) = m\}$. Then M is relatively closed by the l.s.c. of u. If $y \in M \cap E^n$, then we know that u must be equal to m in a neighborhood of y. Suppose $\Delta \in M$. Since $\Delta \in R$, there is an x_0 and δ such that $\partial B_{x_0,\rho} \subset R$ whenever $\rho \geq \delta$. Since $\Delta \in M$, $u(\Delta) = L(u : x_0, \rho)$ for all $\rho \geq \delta$. It is easily seen that $u = m$ on $\partial B_{x_0,\rho}$ for all $\rho \geq \delta$ and therefore $u = m$ on $\sim B_{x_0,\delta}$. This shows that u must be equal to m in a neighborhood of Δ. Hence M is both open and relatively closed. If R is connected, then $M = R$ or $M = \varnothing$; in the first case u is constant on R and in the second case u does not attain its minimum on R. To prove the second assertion define a function u^* on $\bar{R}$ as follows:

$$u^*(y) = \begin{cases} u(y), & y \in R, \\ \liminf_{z \to y} u(z), & y \in \partial R. \end{cases}$$

Then u^* is l.s.c. on $\bar{R}$ and $u^* \geq 0$ on ∂R. Consider any component W of R. Then $\partial W \subset \partial R$ and $u^* \geq 0$ on ∂W. Since E_Δ^n is compact, $\bar{W}$ is compact and u^* must attain its minimum on $\bar{W}$. If u^* attains a negative minimum at an interior point of W, then we have a contradiction to the minimum principle. Since $u^* \geq 0$ on ∂W, $u^* \geq 0$ on W.

COROLLARY **9.4** If u is superharmonic on E_Δ^n, then u is a constant function.

Proof. Since E_Δ^n is compact and u is l.s.c. on E_Δ^n, it must attain its minimum on E_Δ^n and is therefore constant by the minimum principle.

2. Dirichlet Problem for Unbounded Regions

Theorem 9.5 follows easily from Theorem 4.12 and the definitions given in this chapter.

THEOREM **9.5** If u and v are superharmonic on an open set $R \subset E_\Delta^n$ and $c > 0$, then

(i) cu is superharmonic on R
(ii) $u + v$ is superharmonic on R
(iii) $u \wedge v$ is superharmonic on R.

LEMMA **9.6** Let u be superharmonic on the open set $R \subset E_\Delta^n$. If W is an open subset of R with $\bar{W} \subset R$, h is continuous on $\bar{W}$ and harmonic on W, and $u \geq h$ on ∂W, then $u \geq h$ on W.

Proof. If $\Delta \notin \bar{W}$, the assertion if just that of Theorem 4.4. In any case, $\liminf_{\substack{z \to x \\ z \in W}} (u - h)(z) \geq 0$ for all $x \in \partial W$ and the result follows from Theorem 9.3.

LEMMA **9.7** If R is an open subset of E_Δ^n with $E_\Delta^n \sim B_{x_0,\delta} \subset R$ for some ball $B_{x_0,\delta}$ and u is superharmonic on R, the function

$$u^* = \begin{cases} u & \text{on} \quad \bar{B}_{x_0,\delta} \cap R, \\ \mathbf{PI}(u, \sim B_{x_0,\delta}) & \text{on} \quad E_\Delta^n \sim \bar{B}_{x_0,\delta}, \end{cases}$$

is superharmonic on R, harmonic on $E_\Delta^n \sim \bar{B}_{x_0,\delta}$, and $u \geq u^*$ on R.

Proof. Let $\{\phi_j\}$ be an increasing sequence of continuous functions on $\partial B_{x0,\delta}$ such that $\phi_j \uparrow u$ and let

$$h_j = \begin{cases} \phi_j & \text{on} \quad \partial B_{x0,\delta}, \\ \mathbf{PI}(\phi_j, \sim B_{x0,\delta}) & \text{on} \quad R \sim \bar{B}_{x0,\delta}. \end{cases}$$

By Theorem 9.2 each h_j is continuous on $E_\Delta^n \sim B_{x0,\delta}$ and harmonic on $E_\Delta^n \sim \bar{B}_{x0,\delta}$. Since $u \geq h_j = \phi_j$ on $\partial B_{x0,\delta}$, $u \geq h_j$ on $E_\Delta^n \sim B_{x0,\delta}$ by the preceding lemma; that is, $u \geq \mathbf{PI}(\phi_j, \sim B_{x0,\delta})$ on $E_\Delta^n \sim \bar{B}_{x0,\delta}$. Since $\phi_j \uparrow u$ on $\partial B_{x0,\delta}$, $u \geq \mathbf{PI}(u, \sim B_{x0,\delta}) = u^*$ on $E_\Delta^n \sim \bar{B}_{x0,\delta}$. Obviously, $u \geq u^*$ on R. The proof that u^* is superharmonic on R is the same as the proof of Lemma 4.17.

LEMMA **9.8** If $\{h_i : i \in I\}$ is a right-directed family of harmonic functions on the open set $R \subset E_\Delta^n$, on each component of R the function $\sup_{i \in I} h_i$ is either identically $+\infty$ or harmonic.

Proof. We can assume that R is connected. We might as well assume that $\Delta \in R$, for otherwise there is nothing new to prove. By Lemma 2.21 $h = \sup_{i \in I} h_i$ is either identically $+\infty$ or harmonic on $R \sim \{\Delta\}$. If $\sim B_{x0,\delta} \subset R$, then $h_i = \mathbf{PI}(h_i, \sim B_{x0,\delta})$ on $E_\Delta^n \sim \bar{B}_{x0,\delta}$ for each $i \in I$ and $h = \mathbf{PI}(h, \sim B_{x0,\delta})$ on $E_\Delta^n \sim \bar{B}_{x0,\delta}$ by Lemma 2.23. If h is harmonic on $R \sim \{\Delta\}$, it is integrable on $\partial B_{x0,\delta}$ and therefore harmonic on $E_\Delta^n \sim B_{x0,\delta}$. If h is identically $+\infty$ on $R \sim \{\Delta\}$, then $h(\Delta) = L(h : x_0, \delta) = +\infty$ and h is identically $+\infty$ on R.

Definition. A family $\mathscr{F}$ of superharmonic functions on the open set $R \subset E_\Delta^n$ is **saturated** over R if it is saturated over $R \sim \{\Delta\}$ in the usual sense and if $\Delta \in R$, then

(i) $u, v \in \mathscr{F}$ implies $u \wedge v \in \mathscr{F}$,
(ii) for every x_0, δ such that $\sim B_{x0,\delta} \subset R$ and $u \in \mathscr{F}$, the function

$$u^* = \begin{cases} u & \text{on} \quad \bar{B}_{x0,\delta} \cap R, \\ \mathbf{PI}(u, \sim B_{x0,\delta}) & \text{on} \quad E_\Delta^n \sim \bar{B}_{x0,\delta}, \end{cases}$$

belongs to $\mathscr{F}$.

THEOREM **9.9** If $\mathscr{F}$ is a family of superharmonic functions which is saturated over the open set $R \subset E_\Delta^n$, then on each component of R the function $\inf_{u \in F} u$ is either identically $-\infty$ or harmonic.
The proof of this theorem is the same as for Theorem 5.9.

We shall also need to extend the notion of polar set since we have enlarged our space.

Definition. A set $E \subset E_\Delta^n$ is a **polar set** if to each $x \in E$ there corresponds a neighborhood W of x and a superharmonic function v on W such that $v = +\infty$ on $E \cap W$.

THEOREM **9.10** $\{\Delta\}$ is a polar subset of E_Δ^2 but not of any E_Δ^n, $n \geq 3$.

Proof. First consider E_Δ^2. The function $v(x) = \log \|x\|$ is harmonic on $E^2 \sim \bar{B}_{0,1}$ and approaches $+\infty$ as $\|x\| \to +\infty$. Defining $v(\Delta) = +\infty$, v is easily seen to be superharmonic on $E_\Delta^2 \sim \bar{B}_{0,1}$. This shows that $\{\Delta\}$ is a polar subset of E_Δ^2. Now let W be any neighborhood of Δ in E_Δ^n, $n \geq 3$ and let v be superharmonic on W. Then $\sim B_{x,\delta} \subset W$ for some δ and v is superharmonic on $E^n \sim \bar{B}_{x,\delta}$. Let $\delta_1 > \delta$. By Theorem 4.22 $L(v : x, \rho)$ is a concave function of ρ^{-n+2} for $\rho \geq \delta_1$. This implies that there are constants a and b such that $L(v : x, \rho) \leq a\rho^{-n+2} + b$ for all $\rho \geq \delta_1$. Therefore $\limsup_{\rho \to +\infty} L(v : x, \rho) \leq b < +\infty$. If $v(\Delta) = +\infty$, then $L(v : x, \rho)$ would have to approach $+\infty$ as $\rho \to +\infty$ by the l.s.c. of v at Δ. Therefore $v(\Delta) < +\infty$ for any such v and $\{\Delta\}$ is not a polar subset of E_Δ^n, $n \geq 3$.

A polar subset of E_Δ^n, $n \geq 3$, is just a polar subset of E^n as previously defined; but a polar subset of E_Δ^2 may include the point Δ. Note that Theorems 7.3 and 7.4 have no generalizations in the present context in view of Corollary 9.4.

For the remainder of this chapter we shall **assume that R is an open subset of E_Δ^n for which $\sim R$ is not polar**. In view of the preceding remarks, this means that $R \sim \{\Delta\}$ has a Green function by Theorem 8.33 and the fact that any open subset of E^n, $n \geq 3$, has a Green function. Equivalently, $R \sim \{\Delta\}$ supports a nonconstant positive superharmonic function, a minimal requirement for applying the Perron-Wiener-Brelot method.

Just as before, a function u on R is said to be **hyperharmonic (hypoharmonic)** on R if on each component of R the function u is superharmonic (subharmonic) or identically $+\infty$ $(-\infty)$. If f is an extended real-valued function on ∂R, the upper and lower classes associated with R and f are defined by

$$\mathfrak{U}_f = \{u : \liminf_{z \to x} u(z) \geq f(x) \text{ for all } x \in \partial R,\ u \text{ hyperharmonic and bounded below on } R\},$$

$$\mathfrak{L}_f = \{u : \limsup_{z \to x} u(z) \leq f(x) \text{ for all } x \in \partial R,\ u \text{ hypoharmonic and bounded above on } R\},$$

respectively; the corresponding upper and lower solutions are defined as $\bar{H}_f = \inf\{u : u \in \mathfrak{U}_f\}$ and $\underline{H}_f = \sup\{u : u \in \mathfrak{L}_f\}$. Since the assertions of Lemma 8.6 all follow from the order properties of functions, all such assertions are valid in the present context and will be taken as valid without further comment. Also note that $\underline{H}_f \leq \bar{H}_f$ on R; for if $u \in \mathfrak{U}_f$ and $v \in \mathfrak{L}_f$, then

$\liminf_{z\to x}(u-v)(z)\geq\liminf_{z\to x}u(z)-\limsup_{z\to x}v(z)\geq 0$ for all $x\in\partial R$ and $u\geq v$ by Theorem 9.3. We also define f to be a **resolutive boundary function** for R if $\overline{H}_f=\underline{H}_f$ and both are harmonic functions. Since the collection of functions in $\mathfrak{U}_f$, which are superharmonic on a given component of R, forms a saturated family over that component, on each component of R the function $\overline{H}_f$ is either identically $+\infty$, identically $-\infty$, or harmonic. The same is true of $\underline{H}_f$.

LEMMA **9.11** If R is an open subset of E_Δ^n with $\sim R$ nonpolar and $\{f_j\}$ is a sequence of resolutive boundary functions that converges uniformly to f, then f is resolutive and the sequence $\{H_{f_j}\}$ converges uniformly to H_f.

Proof. The same argument used in the proof of Lemma 8.7 can be used to show that the sequence $\{H_{f_j}\}$ converges uniformly to both $\overline{H}_f$ and $\underline{H}_f$. Therefore $\overline{H}_f=\underline{H}_f$ and we need only show that $\overline{H}_f$ is harmonic. It is easy to see that a uniformly convergent sequence of harmonic functions on R converges to a harmonic function on R.

LEMMA **9.12** Let R be an open subset of E_Δ^n having nonpolar complement. If u is a bounded subharmonic function on R such that $f(x)=\lim_{z\to x}u(z)$ exists for all $x\in\partial R$, then f is a resolutive boundary function.
The proof of this lemma is precisely the same as the proof of Lemma 8.9.

Our next step is to generalize Lemma 8.10. Before doing so, a few comments about the Kelvin transformation are in order. Recall that the inversion $x \rightsquigarrow x^\star$ relative to $\partial B_{y,\rho}$ defined by

$$x^\star = y+\frac{\rho^2}{\|x-y\|^2}(x-y)$$

is a mapping of $E^n\sim\{y\}$ onto $E^n\sim\{y\}$. If $R\subset E^n\sim\{y\}$ and f is a real-valued function defined on R, then this mapping induces a transformation of functions $f\rightsquigarrow f^\star$ called the Kelvin transformation which is defined by

$$f^\star(x^\star)=\frac{\|x-y\|^{n-2}}{\rho^{n-2}}f(x), \qquad x^\star\in R^\star.$$

The Kelvin transformation is linear, preserves harmonicity, and preserves positivity. Although it is possible to extend the inversion to map E_Δ^n onto E_Δ^n, there is no particular advantage in doing so insofar as the Kelvin transform is concerned, since only the Kelvin transform as originally defined will preserve harmonicity. Even so, we shall extend the inversion by mapping y onto Δ and Δ onto y. The Kelvin transform also preserves superharmonicity as will now be shown. Suppose $R\subset E^n\sim\{y\}$ and u is superharmonic on R.

Let $R^\star$ be the inversion of R and let $u^\star$ be the Kelvin transform of u. Note that the Kelvin transform preserves order since it is linear and positivity-preserving. Let $W^\star$ be an open subset of $R^\star$ with compact closure $\overline{W}^\star \subset R^\star$ and let $h^\star$ be continuous on $\overline{W}^\star$ and harmonic on the interior with $u^\star \geq h^\star$ on $\partial W^\star$. If we can show that $u^\star \geq h^\star$ on $W^\star$, it would follow that $u^\star$ is superharmonic according to the second definition of superharmonic function. Let W be the inversion of $W^\star$ and let h be the Kelvin transform of $h^\star$. Since $u^\star \geq h^\star$ on $\partial W^\star$, $u \geq h$ on ∂W and h is harmonic on W. Since u is known to be superharmonic on R, $u \geq h$ on W. It follows that $u^\star \geq h^\star$ on $W^\star$. This proves that the Kelvin transformation preserves superharmonicity (for functions with domain in $E^n \sim \{y\}$). One other facet of the Kelvin transform will be important in discussing barriers and regular boundary points. Suppose $R \subset E_\Delta^n$, $x \in \partial R$, u is a positive superharmonic function on R such that $\lim_{z \to x} u(z) = 0$, and there is a closed ball $\overline{B}_{y,\rho} \subset \sim \overline{R}$. If $x \neq \Delta$ and $u^\star$ is the Kelvin transform of u relative to $\partial B_{y,\rho}$, then

$$\lim_{z^\star \to x^\star} u^\star(z^\star) = \lim_{z \to x} \frac{\|z-y\|^{n-2}}{\rho^{n-2}} u(z) = 0;$$

that is, a barrier u at x relative to R is transformed into a barrier $u^\star$ at $x^\star$ relative to $R^\star$ provided $x \neq \Delta$. If $x = \Delta$ and $n \geq 3$, this result does not necessarily hold because of the indeterminancy of $(\|z-y\|^{n-2}/\rho^{n-2})\, u(z)$ as $z \to \Delta$; if, however, $x = \Delta$ and $n = 2$, then a positive superharmonic function which approaches zero at Δ has a Kelvin transform that is positive and approaches zero at y. The same is true in the $n = 2$ case if $R \subset B_{y,\rho}$, $y \in \partial R$, and u is a positive superharmonic function that approaches zero at y; that is, the Kelvin transform of u is a positive superharmonic function on $R^\star$ which approaches zero at Δ.

LEMMA **9.13** If u is superharmonic on $B_{y,\rho}$, harmonic on a neighborhood of y, and $u(y) = 0$, the function $u^\star$ defined on $E_\Delta^n \sim \overline{B}_{y,\rho}$ by

$$u^\star(x^\star) = \begin{cases} \dfrac{\rho^{n-2}}{\|x^\star - y\|^{n-2}} u(x), & x^\star \in E^n \sim \overline{B}_{y,\rho}, \\ 0, & x^\star = \Delta, \end{cases}$$

is superharmonic on $E_\Delta \sim \overline{B}_{y,\rho}$.

Proof. It is clear that $u^\star$ is l.s.c. on $E_\Delta^n \sim \overline{B}_{y,\rho}$. We also know that $u^\star$ is superharmonic on $E^n \sim \overline{B}_{y,\rho}$. It remains only to show that $L(u^\star : x, \delta) \leq 0$ whenever $B_{x,\delta} \supset \overline{B}_{y,\rho}$. Since u is harmonic on a neighborhood of y and $u(y) = 0$, $L(u : y, r) = 0$ for all sufficiently small r. Since $L(u : y, r) = (\rho^{n-2}/r^{n-2})$ $L(u^\star : y, p^2/r)$, $L(u^\star : y, r) = 0$ for all sufficiently large r. By Lemma 9.1 $u^\star =$

$\mathbf{PI}(u^\star : \sim B_{y,r})$ on $\sim \bar{B}_{y,r}$ for all sufficiently large r. In other words, $u^\star$ is harmonic on $E^n_\Delta \sim \bar{B}_{y,r}$ for sufficiently large r. Then $u^\star(\Delta) = L(u^\star : x, \delta) = 0$ for any x, provided δ is sufficiently large; but since $L(u^\star : x, \delta)$ is a concave function of δ^{-n+2} or $-\log \delta$ according as $n \geq 3$ or $n = 2$, we must have $L(u^\star : x, \delta) \leq 0$ whenever $B_{x,\delta} \supset \bar{B}_{y,\rho}$. Therefore $u^\star$ is superharmonic on $E^n_\Delta \sim \bar{B}_{y,\rho}$.

LEMMA **9.14** Let R be an open subset of E^n_Δ such that $\sim \bar{R} \neq \varnothing$. If f is a continuous real-valued function on ∂R and $\varepsilon > 0$, there are continuous superharmonic functions u and v defined on a neighborhood of $\bar{R}$ such that $\sup_{x \in \partial R} |f(x) - (u(x) - v(x))| < \varepsilon$.

Proof. If $\Delta \notin \bar{R}$, the assertion of just that of Lemma 8.10. We consider the remaining two cases $\Delta \in R$ and $\Delta \in \partial R$ separately. Suppose first that $\Delta \in R$. Since $\sim \bar{R} \neq \varnothing$, there is a ball $B_{y,\rho}$ with $\bar{B}_{y,\rho} \subset \sim \bar{R}$. An inversion relative to $\partial B_{y,\rho}$ maps the region R into a region $R^\star \subset B_{y,\rho}$ with y as an interior point of $R^\star$. The Kelvin transformation also induces a continuous function $f^\star$ on $\partial R^\star$. Applying Lemma 8.10 to the compact set $\partial R^\star \cup \{y\}$ and the continuous function that agrees with $f^\star$ on $\partial R^\star$ and is zero at y, there are continuous superharmonic functions $u_0^\star$ and $v_0^\star$ on $B_{y,\rho}$ such that

$$\sup_{x \in \partial R^\star} |f^\star(x) - (u_0^\star(x) - v_0^\star(x))| < \frac{\varepsilon}{4}$$

and

$$|u_0^\star(y) - v_0^\star(y)| < \frac{\varepsilon}{4}.$$

Since $L(u_0^\star : y, r)$ and $L(v_0^\star : y, r)$ increase to $u_0^\star$ and $v_0^\star$, respectively, as $r \downarrow 0$, $L(u_0^\star : y, r) = \mathbf{PI}(u_0^\star, B_{y,r})(y)$, and $L(v_0^\star : y, r) = \mathbf{PI}(v_0^\star, B_{y,r})(y)$, we can choose r small enough so that the continuous superharmonic functions

$$u_1^\star = \begin{cases} u_0^\star & \text{on } B_{y,\rho} \sim B_{y,r}, \\ \mathbf{PI}(u_0^\star, B_{y,r}) & \text{on } B_{y,r}, \end{cases}$$

$$v_1^\star = \begin{cases} v_0^\star & \text{on } B_{y,\rho} \sim B_{y,r}, \\ \mathbf{PI}(v_0^\star, B_{y,r}) & \text{on } B_{y,r}, \end{cases}$$

satisfy

$$\sup_{x \in \partial R^\star} |f^\star(x) - (u_1^\star(x) - v_1^\star(x))| < \frac{\varepsilon}{4}$$

and

$$|u_1^\star(y) - v_1^\star(y)| < \frac{\varepsilon}{4}.$$

The functions $u_1^\star$ and $v_1^\star$ are then harmonic on a neighborhood of y. Let $u^\star = u_1^\star - u_1^\star(y)$ and $v^\star = v_1^\star - v_1^\star(y)$. Then $u^\star$ and $v^\star$ are superharmonic on $B_{y,\rho}$, harmonic on a neighborhood of y, and vanish at y. Also

$$\begin{aligned}\sup_{x \in \partial R^\star} &|f^\star(x) - (u^\star(x) - v^\star(x))| \\ &\leq \sup_{x \in \partial R^\star} |f^\star(x) - (u_1^\star(x) - v_1^\star(x))| \\ &+ |u_1^\star(y) - v_1^\star(y)| \\ &< \frac{\varepsilon}{2}.\end{aligned}$$

By the preceding lemma the Kelvin transforms u and v of $u^\star$ and $v^\star$, respectively, are superharmonic on R [provided we define $u(\Delta) = v(\Delta) = 0$). It is clear that u and v are continuous superharmonic functions on a neighborhood of $\bar{R}$. Since

$$\begin{aligned}\sup_{x \in \partial R} |f(x) - (u(x) - v(x))| &= \sup_{x^\star \in \partial R^\star} \frac{\|x^\star - y\|^{n-2}}{\rho^{n-2}} |f^\star(x^\star) - (u^\star(x^\star) - v^\star(x^\star))| \\ &\leq \sup_{x^\star \in \partial R^\star} |f^\star(x^\star) - (u^\star(x^\star) - v^\star(x^\star))|,\end{aligned}$$

this assertion is proved in the case in which $\Delta \in R$. Assume now that $\Delta \in \partial R$ and that f is continuous on ∂R. Since we can approximate f uniformly by a function that is constant in a neighborhood of Δ, it suffices to prove the assertion for such functions. It is also clear that if we can prove the assertion for continuous functions that are zero in a neighborhood of Δ, we can also prove it for functions that are constant in a neighborhood of Δ simply by adding an appropriate constant to either u or v. We can therefore assume that f is continuous on ∂R and zero in a neighborhood of Δ. Letting $f^\star$ be the Kelvin transform of f relative to a ball $B_{y,\rho} \subset \sim \bar{R}$, $f^\star$ is then continuous on $\partial R^\star$ and zero in a neighborhood of $y \in \partial R^\star$. The remainder of the proof is the same as the preceding case.

THEOREM **9.15** Let R be an open subset of E_Δ^n such that $\sim \bar{R} \neq \emptyset$. If f is a continuous real-valued function on ∂R, then f is resolutive.

Proof. Since we can approximate f uniformly on ∂R by the boundary function of a difference $u - v$ of two functions that are continuous and superharmonic on a neighborhood of R and such boundary functions themselves are resolutive by Lemma 9.12, the assertion follows from Lemma 9.11.

3. Boundary Behavior

Recall that a point $x \in \partial R$ is a **regular boundary point** if $\lim_{z \to x} H_f(z) = f(x)$ for all $f \in C(\partial R)$ and that a barrier at a point $x \in \partial R$ is a positive superharmonic function defined on $W_x \cap R$ for some neighborhood W_x of x such that $\lim_{z \to x} w(z) = 0$.

LEMMA **9.16** Let R be an open subset of E_Δ^n such that $\sim \bar{R} \neq \varnothing$. If there is a barrier at $x \in \partial R$, there is a barrier w at x defined on R with the additional property that $\inf \{w(z) : z \in R \sim V_x\} > 0$ for all neighborhoods V_x of x.

Proof. If $\Delta \notin \bar{R}$, the assertion is just that of Corollary 8.19. We can therefore assume that $\Delta \in \bar{R}$. First consider the case $\Delta \in \partial R$. If $x = \Delta$ and $n \geq 3$, there is always a barrier at x with the additional property, namely, the fundamental harmonic function $\|z - y\|^{-n+2}$ with fixed pole $y \in \sim \bar{R}$. Let $\bar{B}_{y,\rho} \subset \sim \bar{R}$ and let $R^\star$ be the inversion of R relative to $\partial B_{y,\rho}$. Let v be a barrier at $x \in \partial R$. We can assume that $x \neq \Delta$ in the $n \geq 3$ case. Then the Kelvin transform

$$v^\star(z^\star) = \frac{\|z - y\|^{n-2}}{\rho^{n-2}} v(z)$$

is a barrier at $x^\star$. By Corollary 8.19 there is a barrier $w^\star$ at $x^\star$ defined on $R^\star$ such that $\inf \{w^\star(z^\star) : z^\star \in R^\star \sim V_{x^\star}^\star\} > 0$ for all neighborhoods $V_{x^\star}^\star$ of $x^\star$. It is easily seen that the Kelvin transform w of $w^\star$ is a barrier at x defined on R with the required additional property. Now consider the case $\Delta \in R$. In this case ∂R is a compact subset of E^n. Suppose that there is a barrier v at $x \in \partial R$. The boundary function $m(z) = \|z - x\|$ is then continuous on ∂R and resolutive by Theorem 9.15. The proof of the fact that $\lim_{z \to x} H_m(z) = 0$ is precisely the same as in the proof of Theorem 8.18. This time, however, it is not true that the function $\|x - z\|$ is a member of the lower class $\mathfrak{L}_m$, since it is not subharmonic at $\Delta \in R$. We shall show that there is a positive number α such that $H_m(z) \geq \alpha \|x - z\|$ on R in a neighborhood of x. By using the minimum principle, one can show that H_m has the required additional property. By considering the Dirichlet problem for $R^\star$ and the Kelvin transform $m^\star$ of m we can show that H_m is strictly positive on R (it being true for $H_{m^\star}$). Consider any ball $B_{x,\rho_1} \supset \partial R$ and let $k = \inf \{H_m(z) : z \in \partial B_{x,\rho_1}\} > 0$. Choosing $\alpha < 1$ such that $\alpha\rho_1 < k/4$, we have $\alpha \|z - x\| < k/4$ for $z \in \partial B_{x,\rho_1}$. Moreover, there is a function $u \in \mathfrak{L}_m$ such that $u(z) > k/2$ for $z \in \partial B_{x,\rho_1}$. The function

$$w(z) = \begin{cases} u(z) \vee \alpha \|x - z\|, & z \in R \cap B_{x,\rho_1}, \\ u(z), & z \in R \sim B_{x,\rho_1}, \end{cases}$$

is then a member of $\mathfrak{L}_m$. This shows that $H_m(z) \geq \alpha \|x - z\|$ for all $z \in R \cap B_{x,\rho_1}$.

THEOREM **9.17** Let R be an open subset of E^n such that $\sim\bar{R} \neq \varnothing$. If f is bounded above on ∂R and there is a barrier at $x \in \partial R$, then $\limsup_{\substack{z \to x \\ z \in R}} H_f(z) \leq \limsup_{\substack{z \to x \\ z \in \partial R}} f(z)$. A point $x \in \partial R$ is regular if and only if there is a barrier at x. The point Δ is always a regular boundary point if $n \geq 3$. A point $x \in \partial R \sim \{\Delta\}$ is a regular boundary point if and only if $x^\star$ is a regular boundary point for $R^\star$, the inversion of R relative to $\bar{B}_{y,\rho} \subset \sim\bar{R}$; if $\Delta \in \partial R$ and $n = 2$, then Δ is a regular boundary point if and only if y is a regular boundary point for $R^\star$.

Proof. The proof of the first assertion is exactly the same as the proof of Lemma 8.20. Obviously a boundary point x is a regular boundary point if there is a barrier at x. Regarding the converse, suppose x is a regular boundary point. If $x \neq \Delta$, then $H_{\|x-z\| \wedge 1}$ is a barrier at x. If $x = \Delta$ and $n \geq 3$, there is always a barrier at x, namely a fundamental harmonic function with fixed pole. In the $n = 2$ case it is easily seen that $(H_f)^\star = H_{f^\star}$; it follows that y is a regular boundary point for $R^\star$ if Δ is a regular boundary point for R. Since the Kelvin transform of a barrier at y is a barrier at Δ, the proof of the second assertion is completed. The last assertion follows using the characterization of regularity in terms of barriers.

LEMMA **9.18** If R is an open subset of E^n_Δ such that $\sim\bar{R} \neq \varnothing$, the set F of irregular boundary points of R is a polar set. If Z is any polar subset of ∂R, there is a positive superharmonic function v defined on a neighborhood of $\bar{R}$ such that $\lim_{\substack{z \to x \\ z \in R}} v(z) = -\infty$ for all $x \in Z$.

Proof. Let $\bar{B}_{y,\rho} \subset B_{y,\delta} \subset \bar{B}_{y,\delta} \subset \sim\bar{R} \neq \varnothing$. Then $F \subset E^n_\Delta \sim \bar{B}_{y,\rho}$. The inversion $F^\star$ of the set F relative to $\partial B_{y,\rho}$ is contained in the set of irregular boundary points for $R^\star$ by Theorem 9.17 (note that y may be an irregular boundary point for $R^\star$, whereas $y^\star = \Delta$ is not an irregular boundary point for R in the $n \geq 3$ case). Since $F^\star$ is a polar subset of $B_{y,\rho}$, there is a superharmonic function $v^\star$ on $B_{y,\rho}$ such that $v^\star = +\infty$ on $F^\star$. Consider the Kelvin transform v of $v^\star$. Note that v is only defined on $E^n \sim \bar{B}_{y,\rho}$. Since $v = +\infty$ on $F \sim \{\Delta\}$, $F \sim \{\Delta\}$ is a polar set. If $n = 2$ and $\Delta \in F$, then F is a polar set since $\{\Delta\}$ is a polar subset of E^2_Δ; if $n \geq 3$, then Δ cannot belong to F as Δ is always a regular boundary point and F is a polar set. This proves the first assertion. Now let Z be any polar subset of $E^n_\Delta \sim \bar{B}_{y,\delta}$. If $x \in Z \sim \{\Delta\}$, there is a neighborhood $W \subset E^n_\Delta \sim \bar{B}_{y,\delta}$ of x and a superharmonic function w on W such that $w = +\infty$ on $W \cap Z$. Then $w^\star = +\infty$ on $W^\star \cap Z^\star$. It follows that the image of $Z \sim \{\Delta\}$ is a polar subset of $B_{y,\rho}$. Since adjoining a single point of E^n to a polar set results in a polar set, $Z^\star$ is a polar subset of

$B_{y,\rho}$. Now let j_0 be the smallest integer greater than δ, let $Z_{j_0} = Z \cap (\bar{B}_{y,j_0} \sim B_{y,\delta})$, and let $Z_j = Z \cap (\bar{B}_{y,j} \sim B_{y,j-1})$, $j > j_0$. Then $Z \sim \{\Delta\} = \cup_{j \geq j_0} Z_j$ and each $Z_j^\star$ is a polar subset of $B_{y,\rho}$. It follows that there is a superharmonic function $v_j^\star$ on $B_{y,\rho}$ such that $v_j^\star = +\infty$ on $Z_j^\star$. By replacing $v_j^\star$ on $B_{y,\rho^2/j}$ by its Poisson integral we can assume that $v_j^\star$ is harmonic on a neighborhood of y without affecting its values on $Z_j^\star$. By adding a constant we can assume that $v_j^\star(y) = 0$. Then the Kelvin transform v_j of $v_j^\star$, provided we define $v_j(\Delta) = 0$, is superharmonic on $E_\Delta^n \sim \bar{B}_{y,\rho}$ by Lemma 9.13 and $v_j = +\infty$ on Z_j. Since v_j is l.s.c. on the compact set $E_\Delta^n \sim B_{y,\delta}$, we can add a constant to render it positive on this set without affecting its infinities. Now let $\{\alpha_j\}_{j \geq j_0}$ be a sequence of positive numbers such that $\sum_{j \geq j_0} \alpha_j v_j(\Delta) < +\infty$. The function $v = \sum_{j \geq j_0} \alpha_j v_j$ is then a positive superharmonic function on $E_\Delta^n \sim \bar{B}_{y,\delta}$ such that $v = +\infty$ on $Z \sim \{\Delta\}$. If $n \geq 3$, we are done, since Δ cannot belong to any polar set. In the $n = 2$ case, however, Δ may belong to Z; if this is the case, replace $v(z)$ with $v(z) + \log \|z - y\|/\delta$.

LEMMA **9.19** If R is an open subset of E_Δ^n such that $\sim R$ is not a polar set and Z is a polar subset of ∂R, there is a positive superharmonic function v on R such that $\lim_{z \to x} v(z) = +\infty$ for all $x \in Z$.

Proof. If $\sim \bar{R} \neq \emptyset$, then the assertion is the same as in the preceding lemma. We can therefore assume that $\sim R = \partial R$ and that ∂R is not a polar set. Let $\bar{B}_{y,\rho} \subset R$. Then Z is a polar subset of the boundary of $R \sim \bar{B}_{y,\rho}$. The points of $\partial B_{y,\rho}$ are all regular boundary points for $R \sim \bar{B}_{y,\rho}$ by the local character of regularity. Since ∂R is not a polar set and $\partial(R \sim B_{y,\rho}) = \partial R \cup \partial B_{y,\rho}$, some point of ∂R must be a regular boundary point for $R \sim \bar{B}_{y,\rho}$. We also know from the preceding lemma that there is a positive superharmonic function v on $R \sim \bar{B}_{y,\rho}$ such that $\lim_{y \to z} v(y) = +\infty$ whenever $z \in Z \subset \partial R$. Moreover, if B is a ball containing $\bar{B}_{y,\rho}$ with $\bar{B} \subset R$, then we can assume that v is also harmonic on $B \sim \bar{B}_{y,\rho}$ for if this is not the case then we can replace v by the balayage of v over $R \sim \bar{B}$ relative to $R \sim \bar{B}_{y,\rho}$ without changing the boundary behavior according to Lemma 7.11. Now choose $\rho_2 > \rho_1 > \rho$ such that $\bar{B}_{y,\rho_2} \subset R$ and v is harmonic on a neighborhood of $\bar{B}_{y,\rho_2} \sim B_{y,\rho_1}$. We now construct two sequences of functions $\{u_j\}$ and $\{w_j\}$ on B_{y,ρ_2} and $R \sim B_{y,\rho_1}$, respectively, as follows. If we define

$$u_1 = \begin{cases} \mathbf{PI}(v, B_{y,\rho_2}) & \text{on} \quad B_{y,\rho_2}, \\ v & \text{on} \quad \partial B_{y,\rho_2}, \end{cases}$$

then u_1 is continuous on $\bar{B}_{y,\rho_2}$ and harmonic on the interior. Now let

$$f_1 = \begin{cases} 0 & \text{on} \quad \partial R, \\ u_1 - v & \text{on} \quad \partial B_{y,\rho_1}, \end{cases}$$

and define

$$w_1 = \begin{cases} H_{f_1} & \text{on} \quad R \sim \bar{B}_{y,\rho_1}, \\ u_1 - v & \text{on} \quad \partial B_{y,\rho_1}, \end{cases}$$

where H_{f_1} is the Dirichlet solution corresponding to f_1 and the region $R \sim \bar{B}_{y,\rho_1}$. Since all points of $\partial B_{y,\rho_1}$ are regular boundary points for $R \sim \bar{B}_{y,\rho_1}$, w_1 is continuous on $R \sim B_{y,\rho_1}$ and harmonic on the interior. Now let

$$u_2 = \begin{cases} \mathbf{PI}(v + w_1, B_{y,\rho_2}) & \text{on} \quad B_{y,\rho_2}, \\ v + w_1 & \text{on} \quad \partial B_{y,\rho_2}. \end{cases}$$

Then u_2 is continuous on $\bar{B}_{y,\rho_2}$ and harmonic on the interior. Again let

$$f_2 = \begin{cases} 0 & \text{on} \quad \partial R, \\ u_2 - v & \text{on} \quad \partial B_{y,\rho_1}, \end{cases}$$

and define

$$w_2 = \begin{cases} H_{f_2} & \text{on} \quad R \sim \bar{B}_{y,\rho_1}, \\ u_2 - v & \text{on} \quad \partial B_{y,\rho_1}. \end{cases}$$

Then w_2 is continuous on $R \sim B_{y,\rho_1}$ and harmonic on the interior. The sequences $\{u_j\}$ and $\{w_j\}$ are so defined inductively. Now on $\partial B_{y,\rho_1}$, $w_j = u_j - v$, so that

$$w_{j+1} - w_j = u_{j+1} - u_j \quad \text{on} \quad \partial B_{y,\rho_1},$$

for $j \geq 1$, whereas on $\partial B_{y,\rho_2}$, $u_{j+1} = v + w_j$, so that

$$u_{j+1} - u_j = w_j - w_{j-1} \quad \text{on} \quad \partial B_{y,\rho_2},$$

for $j \geq 2$. Now let

$$\alpha_j = \sup_{z \in \partial B_{y,\rho_1}} |w_{j+1}(z) - w_j(z)| = \sup_{z \in \partial B_{y,\rho_1}} |u_{j+1}(z) - u_j(z)|$$

and

$$\beta_j = \sup_{z \in \partial B_{y,\rho_2}} |u_{j+1}(z) - u_j(z)| = \sup_{z \in \partial B_{y,\rho_2}} |w_j(z) - w_{j-1}(z)|.$$

Consider the Dirichlet solution H_g for the region $R \sim \bar{B}_{y,\rho_1}$ and the boundary function g, which is 1 on $\partial B_{y,\rho_1}$ and 0 on ∂R. Since all points of $\partial B_{y,\rho_1}$ and some point of ∂R are regular for $R \sim \bar{B}_{y,\rho_1}$, we have $0 < H_g < 1$ on $R \sim \bar{B}_{y,\rho_1}$. It follows that there is a constant $q < 1$ such that $H_g < q$ on $\partial B_{y,\rho_2}$. If g^* is any continuous function on the boundary of $R \sim \bar{B}_{y,\rho_1}$ such that $|g^*| \leq g$, then $|H_{g^*}| < q$ on $\partial B_{y,\rho_2}$. Applying this to each of the functions $(w_{j+1} - w_j)/\alpha_j$,

$$|w_{j+1} - w_j| \leq q\alpha_j \quad \text{on} \quad \partial B_{y,\rho_2}.$$

Therefore

$$\beta_{j+1} \leq q\alpha_j.$$

Since $u_{j+1} - u_j$ is harmonic on B_{y,ρ_2},

$$|u_{j+1} - u_j| \leq \sup_{z \in \partial B_{y,\rho_2}} |u_{j+1}(z) - u_j(z)| = \beta_j$$

on $\partial B_{y,\rho_1}$ and it follows that $\alpha_j \leq \beta_j$. Hence $\alpha_{j+1} \leq q\alpha_j$ for $j \geq 1$, and the positive series $\sum \alpha_j$ is dominated by a convergent geometric series. This shows that the series

$$w_1 + \sum_{j=1}^{\infty} (w_{j+1} - w_j)$$

converges uniformly on $R \sim B_{y,\rho_1}$. Since the terms are harmonic on $R \sim \bar{B}_{y,\rho_1}$,

$$w = \lim_{j\to\infty} w_j = w_1 + \sum_{j=1}^{\infty} (w_{j+1} - w_j)$$

is harmonic on $R \sim \bar{B}_{y,\rho_1}$. Similarly $\beta_{j+1} \leq q\beta_j$ and the positive series $\sum \beta_j$ is dominated by a convergent geometric series, thus establishing that

$$u = \lim_{j\to\infty} u_j = u_1 + \sum_{j=1}^{\infty} (u_{j+1} - u_j)$$

defines a function that is harmonic on B_{y,ρ_2}. Since $w_j = u_j - v$ on $\partial B_{y,\rho_1}$, $w = u - v$ on $\partial B_{y,\rho_1}$. Also $u_{j+1} = v + w_j$ on $\partial B_{y,\rho_2}$ implies that $w = u - v$ on $\partial B_{y,\rho_2}$; that is, $u = v + w$ on both $\partial B_{y,\rho_1}$ and $\partial B_{y,\rho_2}$, but since both are harmonic on the annulus $u = v + w$ on $B_{y,\rho_2} \sim \bar{B}_{y,\rho_1}$. It follows that the function

$$v^* = \begin{cases} u & \text{on} \quad B_{y,\rho_2}, \\ v + w & \text{on} \quad R \sim \bar{B}_{y,\rho_1}, \end{cases}$$

has the required properties.

The method just used to extend a harmonic function is due to Schwarz and is known as the **alternating method.**

COROLLARY **9.20** (*Bouligand*) If R is an open subset of E_Δ^n such that $\sim R$ is not polar and if h is a bounded harmonic function on R such that $\lim_{y\to z} h(y) = 0$ for all $z \in \partial R$, except possibly on a polar set, then $h = 0$ on R.

Proof. Let Z be the set of points $z \in \partial R$ for which it is not true that $\lim_{y\to z} h(y) = 0$. Then Z is a polar subset of ∂R and by the preceding lemma there is a positive superharmonic function v on R such that $\lim_{y\to z} v(y) = +\infty$

for $z \in Z$. Then, for every $\varepsilon > 0$, $h + \varepsilon v$ is superharmonic on R and $\liminf_{y \to z} (h + \varepsilon v)(y) \geq 0$ for all $z \in \partial R$. By Theorem 9.3 $h + \varepsilon v \geq 0$ on R, and, since ε is arbitrary, $h(y) \geq 0$ at every point y at which v is finite, but since v is finite a.e. on $R \cap E^n$, $h \geq 0$ a.e. on $R \cap E^n$, and it follows from the continuity of h that $h \geq 0$ on R. Applying the same argument to $-h$, $h = 0$ on R.

LEMMA **9.21** Let R be an open subset of E^n_Δ such that $\sim R$ is not polar and let $\bar{B}_{y,\rho} \subset R$. If f is a bounded function on ∂R, then

$\bar{H}_f = \inf \{u : u$ superharmonic on R, u bounded below on R, and $\liminf_{y \to x} u(y) \geq f(x)$ for all $x \in \partial R$ which are regular boundary points for $R \sim \bar{B}_{y,\rho}\}$.

Proof. Suppose u is in the class just described. By Lemma 9.18 the set of irregular boundary points of $R \sim \bar{B}_{y,\rho}$ is a polar subset of ∂R. Let v be the positive superharmonic function of Lemma 9.19 on R corresponding to the set of irregular boundary points of $R \sim \bar{B}_{y,\rho}$. Then for every $\varepsilon > 0$, $u + \varepsilon v \in \mathfrak{U}_f$ and therefore $u + \varepsilon v \geq \bar{H}_f$. Since v is finite a.e. on $R \cap E^n$ $u \geq H_f$ a.e. on $R \cap E^n$. Therefore $u(x) = \lim_{\delta \to 0} A(u : x, \delta) \geq \lim_{\delta \to 0} A(\bar{H}_f : x, \delta) = \bar{H}_f(x)$ for all $x \in R \cap E^n$ and $u \geq \bar{H}_f$.

THEOREM **9.22** Let R be an open subset of E^n_Δ such that $\sim R$ is not polar. Then each $f \in C(\partial R)$ is resolutive.

Proof. Since f is bounded, the same is true of $\bar{H}_f$ and $\underline{H}_f$ and both are harmonic. Let $m = \sup_{z \in \partial R} |f(z)|$ and let $\bar{B}_{y,\rho} \subset R$. The Dirichlet solution H_g corresponding to $R \sim \bar{B}_{y,\rho}$ and the boundary function

$$g = \begin{cases} m + 1 & \text{on} \quad \partial B_{y,\rho}, \\ f & \text{on} \quad \partial R, \end{cases}$$

is harmonic on $R \sim \bar{B}_{y,\rho}$ and $\lim_{y \to x} H_g(y) = f(x)$ whenever $x \in \partial R$ is a regular boundary point for $R \sim \bar{B}_{y,\rho}$. Moreover, the function

$$u = \begin{cases} m + 1 & \text{on} \quad B_{y,\rho}, \\ H_g & \text{on} \quad R \sim B_{y,\rho}, \end{cases}$$

is superharmonic on R and $\lim_{z \to x} u(z) = f(x)$ whenever $x \in \partial R$ is a regular boundary point for $R \sim \bar{B}_{y,\rho}$. By the preceding lemma $u \geq \underline{H}_f$. Similarly there is a bounded subharmonic function v on R such that $v \leq \underline{H}_f$ and $\lim_{z \to x} v(z) = f(x)$ whenever $x \in \partial R$ is a regular boundary point for $R \sim \bar{B}_{y,\rho}$. It follows that $\lim_{z \to x} (\bar{H}_f - \underline{H}_f)(z) = 0$ for all $x \in \partial R$ which are regular boundary points for $R \sim \bar{B}_{y,\rho}$, but, since the irregular boundary points of $R \sim \bar{B}_{y,\rho}$ constitute a polar set, $\bar{H}_f - \underline{H}_f = 0$ on R by Corollary 9.20.

In the following discussion, R is an open subset of E_Δ^n such that $\sim R$ is not polar. Since the development of necessary and sufficient conditions for the resolutivity of a boundary function is not significantly different from the earlier development, only a sketch is given.

By the preceding theorem each $f \in C(\partial R)$ determines a harmonic function H_f on R. For each $x \in R$ the map $x \rightsquigarrow H_f(x)$ is a linear functional on $C(\partial R)$ to which the Riesz representation theorem may be applied to obtain the integral representation

$$H_f(x) = \int f \, d\mu_x,$$

where μ_x is a Borel measure on ∂R with $\mu_x(\partial R) = 1$ since $H_1(x) = 1$. If $\{f_j\}$ is an increasing sequence of boundary functions such that $\overline{H}_{f_1} > -\infty$ on R and $f = \lim_{j \to \infty} f_j$, then the proof of Lemma 8.8 carries over in the present context and results in $\overline{H}_f = \lim_{j \to \infty} \overline{H}_{f_j}$; if the f_j's are also resolutive, then

$$\overline{H}_f(x) = \underline{H}_f(x) = \int f \, d\mu_x$$

on R, where the three quantities are all $+\infty$ if any one is $+\infty$. It follows that the indicator function of compact subsets of ∂R are resolutive and, as in Theorem 8.13 μ_x-integrable Borel measurable functions are resolutive. Theorem 8.14 is also valid here, making use only of the minimum principle. If, for each $x \in R$, $\mathscr{F}_x$ is the completed Borel σ-albegra of subsets of ∂R relative to the measure μ_x, then $\mathscr{F} = \bigcap_{x \in R} \mathscr{F}_x$ is also a σ-algebra. $\mathscr{F}$ is called the class of $\mu.$-**measurable** sets, and any $\mathscr{F}$-measurable function f on ∂R such that $\int |f| \, d\mu_x < +\infty$ for some x in each component of R is called a $\mu.$-**integrable** function. As before, μ_x (extended to $\mathscr{F}$) will be called **harmonic measure** relative to x and R.

THEOREM **9.23** Let R be an open subset of E_Δ^n such that $\sim R$ is not polar. A boundary function f is resolutive if and only if it is $\mu.$-integrable, in which case, $H_f(x) = \int f \, d\mu_x$ for all $x \in R$.

THEOREM **9.24** Let R be an open subset of E_Δ^n such that $\sim R$ is not polar and let μ_y be the harmonic measure associated with $y \in R$. If Z is a polar subset of ∂R, then $\mu_y(Z) = 0$ for all $y \in R$.

Proof. Let χ be the indicator function of the polar set $Z \subset \partial R$. By Lemma 9.19 there is a positive superharmonic function v on R such that $\lim_{z \to y} v(z) = +\infty$ for all $y \in Z$. Since εv has the same property for every $\varepsilon > 0$, $0 \le \underline{H}_\chi \le \overline{H}_\chi \le \varepsilon v$ for every $\varepsilon > 0$. Letting $\varepsilon \to 0$, we get $0 = \underline{H}_\chi = \overline{H}_\chi$ a.e. on $R \cap E^n$

and since the upper and lower solutions are continuous, χ is resolutive with $H_\chi(y) = 0 = \mu_y(Z)$ for all $y \in R$.

The set of irregular boundary points for an open set $R \subset E_\Delta^n$ having non-polar complement is discussed only informally. It was shown in Lemma 9.18 that the set of irregular boundary points for R is a polar set if $\sim \bar{R} \neq \emptyset$. Suppose only that $\sim R$ is not polar and let $\bar{B}_{y,\rho} \subset R$. The basic idea in the proof of Theorem 9.22 is that a regular boundary point for $R \sim \bar{B}_{y,\rho}$ is also a regular boundary point for R. In fact, this is actually what is proved. Therefore the set of irregular boundary points for R is a subset of the set of irregular boundary points for $R \sim \bar{B}_{y,\rho}$ and is a polar set.

4. Applications

In the remainder of this chapter R is an open subset of E^n having a Green function. We will allow Δ to be a boundary point of R but not an interior point. As a subset of E_Δ^n, $\Delta \in \sim R$ and $\sim R$ is not polar in the $n \geq 3$ case; in the $n = 2$ case $E^n \sim R$ is not polar by Theorem 8.33 and $E_\Delta^n \sim R$ is not polar. The results of this chapter are therefore applicable to R as a subset of E_Δ^n. If F is a relatively closed subset of R, $H_f^{R \sim F}$ will denote the Dirichlet solution corresponding to f and $R \sim F$, regarded as a subset E_Δ^n.

THEOREM **9.25** Let R be an open subset of E^n having a Green function. If u is a positive superharmonic function on R and F is a relatively closed subset of R, then $\hat{\mathbf{R}}_F^u(y) = u(y)$ for $y \in F$ except possibly for those $y \in \partial F \cap R$ which are irregular boundary points for $R \sim F$; moreover, $\hat{\mathbf{R}}_F^u = H_{u_0}^{R \sim F}$ on $R \sim F$, where

$$u_0 = \begin{cases} u, & \text{on} \quad \partial F \cap R, \\ 0, & \text{on} \quad \partial R \cap \partial(R \sim F). \end{cases}$$

Proof. Since $u|_{R \sim F} \in \mathfrak{U}_{u_0}^{R \sim F}$, $u \geq \bar{H}_{u_0}^{R \sim F} \geq \underline{H}_{u_0}^{R \sim F} \geq 0$ and u_0 is resolutive. Define a function v on R by

$$v = \begin{cases} u & \text{on} \quad F, \\ H_{u_0}^{R \sim F} & \text{on} \quad R \sim F. \end{cases}$$

The function u_0 is easily seen to be l.s.c. on $\partial(R \sim F)$. Let $\{f_j\}$ be a sequence of continuous functions on $\partial(R \sim F)$ such that $f_j \uparrow u_0$. Since

$$\liminf_{\substack{z \to y \\ z \in R \sim F}} v(z) = \liminf_{\substack{z \to y \\ z \in R \sim F}} H_{u_0}^{R \sim F} \geq \liminf_{\substack{z \to y \\ z \in R \sim F}} H_{f_j}^{R \sim F}(z) = f_j(y)$$

for all $y \in \partial F \cap R$ except possibly for those points that are irregular boundary points for $R \sim F$, $\liminf_{\substack{z \to y \\ z \in R \sim F}} v(z) \geq u(y)$ for all $y \in \partial F \cap R$ under the same condition. Since $v = u$ on F, the lower regularization $\hat{v}$ of v agrees with u on F except possibly at those points that are irregular boundary points for $R \sim F$. Now we can show that $\hat{v}$ is superharmonic on R in the same way as was done in Lemma 8.40. Therefore $\hat{v} \geq \hat{\mathbf{R}}_F^u$ on R by Corollary 8.37. Now any positive superharmonic function w on R that majorizes u on F belongs to the upper class $\mathfrak{U}_{u_0}^{R \sim F}$ and it follows that $w \geq H_{u_0}^{R \sim F}$ on $R \sim F$. Therefore $\mathbf{R}_F^u \geq v$ on R and consequently $\hat{\mathbf{R}}_F^u \geq \hat{v}$ on R. It follows that $\hat{\mathbf{R}}_F^u = u$ on F except possibly at those points $y \in \partial F \cap R$ that are irregular boundary points for $R \sim F$. Lastly, $\hat{\mathbf{R}}_F^u = \hat{v} = H_{u_0}^{R \sim F}$ on $R \sim F$.

The proof of the following corollary is omitted as it is precisely the same as that of Corollary 8.42.

COROLLARY **9.26** Let R be an open subset of E^n having a Green function and let F be a relatively closed subset of R. If u and v are positive superharmonic functions on R, then $\hat{\mathbf{R}}_F^{u+v} = \hat{\mathbf{R}}_F^u + \hat{\mathbf{R}}_F^v$ on R.

CHAPTER 10

Fine Topology

In this chapter we consider a topology on E^n that is finer than the metric topology and more natural from the potential theoretic point of view. We continue to use "open," "neighborhood," "continuous," etc. in reference to the metric topology. The new topology is called the fine topology and concepts relative to this topology are prefixed by "fine" or "finely." Thinness of a set at a point is also introduced and related to irregularity of boundary points.

1. An Intrinsic Topology

In view of the fact that superharmonic functions form the core of potential theory an intrinsically defined topology on E^n would be better suited to the problems of potential theory.

Definition. The **fine topology on** E^n is the smallest topology on E^n for which all superharmonic functions are continuous in the extended sense.

We first remark that if u is superharmonic on an open set $R \subset E^n$, then u is finely continuous on R. This follows from the fact that $u|_B$, where B is a ball with closure in R, can be extended so as to be superharmonic on E^n by Lemma 7.13. Second, if $u(x_0) = +\infty$, it is not necessary to adjoin any sets to the metric topology to make u finely continuous in the extended sense at x_0, since u is continuous in the extended sense at x_0 in the metric topology. It is necessary, however, to enlarge the collection of neighborhoods of a point where u is finite-valued but not continuous. The requirement that u be finely continuous means that sets of the type $\{x : u(x) > \alpha\}$ and $\{x : u(x) < \beta\}$ must be finely open. Sets of the first kind, however, are open in the metric topology. Third, by taking u to be one of the functions $u_{x,\delta/2} - u_{x,\delta}$ of Examples 3 and 4 of Chapter 7 it is easily seen that each ball in E^n is finely open and that the fine topology is larger than the metric topology; that it is properly larger follows from the fact that there are discontinuous superharmonic functions by Corollary 7.15. For a subbase of the fine topology we can take the collection of sets of the form $\{x : u(x) > \alpha\}$ and $\{x : u(x) < \beta\}$, where α, β are arbitrary real numbers and u is an arbitrary superharmonic function. Since sets of the type $\{x : u(x) > \alpha\}$ are open in the metric topology, they contain infinitely many distinct points if nonempty. Consider any nonempty set of the type $V = \{x : u(x) < \alpha\}$ and $x \in V$. Since $u(x) = \liminf_{\substack{y \to x \\ y \neq x}} u(y)$ for any superharmonic function u, each such set contains infinitely many distinct points. It follows that every nonempty finely open set contains infinitely many distinct points.

THEOREM **10.1** If Z is a polar subset of E^n, no point of E^n is a fine limit point of Z; in particular, the points of Z are finely isolated points of Z.

Proof. Assume that x is a fine limit point of Z. Then x is a limit point of Z. We can also assume that $x \notin Z$. By theorem 7.14 there is a superharmonic function u defined on a neighborhood of x such that $u(x) < \lim_{\substack{y \to x \\ y \in Z}} u(y) = r < +\infty$. It follows that there is a neighborhood U of x such that $u(y) > \frac{1}{2}(r + u(x))$ for $y \in U \cap Z$; but, since u is finely continuous at x, there is a fine neighborhood U^f of x such that $u(y) < \frac{1}{2}(r + u(x))$ for $y \in U^f$. Therefore $(U \cap Z) \cap U^f = (U \cap U^f) \cap Z = \varnothing$. Since $U \cap U^f$ is a fine neighborhood of x that does not intersect Z, x is not a fine limit point of Z and we have a contradiction.

The preceding theorem can be used to show that the fine topology is not a locally compact topology. Let F^f be the fine closure of a fine neighborhood of a point. Then F^f contains infinitely many distinct points x_j, $j \geq 1$. By the preceding theorem no subsequence of the sequence $\{x_j\}$ can have a fine limit point and F^f is therefore not compact. In fact, this argument shows that a set is finely compact if and only if it is finite.

EXAMPLE **1** Consider the sequence $\{x_j\}$ of points in E^2 where $x_j = (1/j, 0)$. Then there is a fine neighborhood of $(0, 0)$ which contains no points of the sequence.

2. Thin Sets

Consider an open set R and a point $x \in \partial R$. We have seen that if there is a cone in $\sim R$ with x as its vertex (or even a line segment in $\sim R$ having x as an end-point in the $n = 2$ case), then x is a regular boundary point for the Dirichlet problem. For irregular boundary points the complement of R must somehow be thinner than a cone at x. The notion of "thinness" of a set at a point expresses this more precisely.

Definition. A set E is **thin** at x if x is not a fine limit point of E.

THEOREM **10.2** A polar set Z is thin everywhere.

Proof. By Theorem 10.1 no point of E^n is a fine limit point of Z.

In particular, a line segment in E^3 is thin at each of its points. Also note that E is thin at x whenever x is not a limit point of E.

THEOREM **10.3** A set E is thin at a limit point x of E if and only if there is a superharmonic function u on a neighborhood of x such that $u(x) < \liminf_{\substack{y \to x \\ y \in E \sim \{x\}}} u(y)$.

Proof. We can assume that $x \notin E$. Suppose that E is thin at x. Then there is a fine neighborhood U^f of x such that $U^f \cap E = \varnothing$. Since x is a limit point of E, U^f is not a metric neighborhood of x. Since the class of sets of the form $\{y : v(y) < \alpha\}$ and $\{y : v(y) > \beta\}$ for arbitrary real numbers α, β and arbitrary superharmonic functions v constitute a subbase for the fine topology, we can assume that U^f has the form $U^f = \bigcap_{i=1}^{j} \{y : u_i(y) < \alpha_i\} \cap W$ where W is a neighborhood of x and the u_i's are superharmonic. Since $x \in U^f$, $u_i(x) < \alpha_i$ and by choosing $0 < \varepsilon < \min_{1 \le i \le j} (\alpha_i - u_i(x))$ we can assume that $U^f = \bigcap_{i=1}^{j} \{y : u_i(y) < u_i(x) + \varepsilon\} \cap W$. Now let $u = \sum_{i=1}^{j} u_i$, which is superharmonic. We need only show that $u(x) < \liminf_{\substack{y \to x \\ y \in E \sim \{x\}}} u(y)$. Since the u_i's are l.s.c. at x, there is a neighborhood $U \subset W$ of x such that $u_i(y) > u_i(x) - (\varepsilon/j)$ for $1 \le i \le j$ and $y \in U$. Consider any $y \in E \cap U$. Then $u_k(y) \ge u_k(x) + \varepsilon$ for some k and

$$u(y) = \sum_{i=1}^{j} u_i(y) > \sum_{i=1}^{j} u_i(x) - \frac{j-1}{j}\varepsilon + \varepsilon = u(x) + \frac{\varepsilon}{j};$$

that is, $u(y) > u(x) + \varepsilon/j$ for all $y \in E \cap U$. Therefore $\liminf_{\substack{y \to x \\ y \in E \sim \{x\}}} u(y) > u(x)$ as was to be proved. Conversely suppose we know there is a superharmonic function u such that $u(x) < \liminf_{\substack{y \to x \\ y \in E \sim \{x\}}} u(y) = r$. We can assume that $r < +\infty$ for we can always replace u by $u \wedge [u(x) + 1]$. Consider $V^f = \{y : u(y) < \frac{1}{2}(u(x) + r)\}$. Using the l.s.c. of u, there is a neighborhood V of x such that $u(y) > \frac{1}{2}(u(x) + r)$ for $y \in V \cap E$. Then $V^f \cap (V \cap E) = (V^f \cap V) \cap E = \varnothing$. Since $V^f \cap V$ is a fine neighborhood of x, x is not a fine limit point of E and E is thin at x.

THEOREM **10.4** A set E is thin at a limit point x of E if and only if there is a superharmonic function u such that $u(x) < \lim_{\substack{y \to x \\ y \quad E \sim \{x\}}} u(y) = +\infty$.

Proof. The sufficiency follows from the preceding theorem. To prove the necessity assume that E is thin at x. We can assume that $x \notin E$. By the preceding theorem there is a superharmonic function v defined on a neighborhood of x such that $v(x) < \liminf_{\substack{y \to x \\ y \in E}} v(y)$. Since there is nothing to prove if the right side of this inequality is infinite, we can assume that $\liminf_{\substack{y \to x \\ y \in E}} v(y) < +\infty$. Let B be a ball about x on which v is defined. By the Riesz decomposition theorem there is a measure ν on B and a harmonic function h on B such that $v = G_B \nu + h$ on B. Since h is continuous at x, we can assume that $v = G_B \nu$ on B. Let $\{\varepsilon_j\}$ be a decreasing sequence of positive numbers. Since

$$v(x) = \int_{B_{x, \varepsilon_j}} G_B(x, y)\, d\nu(y) + \int_{B \sim B_{x, \varepsilon_j}} G_B(x, y)\, d\nu(y) < +\infty$$

and the second integral on the right increases to $v(x)$,

$$\lim_{j \to \infty} \int_{B_{x, \varepsilon_j}} G_B(x, y)\, d\nu(y) = \lim_{j \to \infty} \int G_B(x, y)\, d\nu_j(y) = 0,$$

where $\nu_j(M) = \nu(M \cap B_{x, \varepsilon_j})$ for each Borel set $M \subset B$. By passing to a subsequence of the ε_j's, if necessary, we can assume that $\sum_j G_B \nu_j(x) < +\infty$. Then the function $u = \sum_j G_B \nu_j$ is superharmonic on B and finite at x. Let $\delta = \liminf_{\substack{y \to x \\ y \in E}} [v(y) - v(x)] > 0$ and let $v_j = G_B \nu_j$. Then $v = v_j + h_j$, where h_j is harmonic on B_{x, ε_j}, and it follows that

$$\begin{aligned}
\liminf_{\substack{y \to x \\ y \in E}} [v_j(y) - v_j(x)] &\geq \liminf_{\substack{y \to x \\ y \in E}} [v_j(y) + h_j(y)] + \liminf_{\substack{y \to x \\ y \in E}} [-h_j(y) - v_j(x)] \\
&= \liminf_{\substack{y \to x \\ y \in E}} v(y) - h_j(x) - v_j(x) \\
&= \liminf_{\substack{y \to x \\ y \in E}} [v(y) - v(x)] \geq \delta
\end{aligned}$$

for all j and that

$$\liminf_{\substack{y \to x \\ y \in E}} \left[\sum_{j=1}^{p} v_j(y) - \sum_{j=1}^{p} v_j(x) \right] \geq p\delta.$$

Therefore

$$\liminf_{\substack{y \to x \\ y \in E}} \sum_{j=1}^{p} v_j(y) \geq \sum_{j=1}^{p} v_j(x) + p\delta \geq p\delta$$

for every positive integer p. Since $u = \sum_j v_j \geq \sum_{j=1}^{p} v_j$,

$$\liminf_{\substack{y \to x \\ y \in E}} u(y) \geq p\delta$$

for every positive integer p and therefore $\liminf_{\substack{y \to x \\ y \in E}} u(y) = +\infty$. Now extend u to E^n by Lemma 7.13.

COROLLARY **10.5** If E is thin at a limit point x of E, then

$$\lim_{r \to 0} \frac{\sigma(\partial B_{x,r} \cap E)}{\sigma(\partial B_{x,r})} = 0.$$

Proof. By the preceding theorem there is a superharmonic function u such that $u(x) < \lim_{\substack{y \to x \\ y \in E \sim \{x\}}} u(y) = +\infty$. We can assume that u is non-negative in a neighborhood of x. For sufficiently small r

$$\begin{aligned} u(x) \geq L(u : x, r) &= \frac{1}{\sigma_n r^{n-1}} \int_{\partial B_{x,r}} u(y)\, d\sigma(y) \\ &\geq \frac{1}{\sigma_n r^{n-1}} \int_{\partial B_{x,r} \cap E} u(y)\, d\sigma(y) \\ &\geq \frac{\sigma(\partial B_{x,r} \cap E)}{\sigma(\partial B_{x,r})} \left[\inf_{y \in \partial B_{x,r} \cap E} u(y) \right]. \end{aligned}$$

Since $u(x) < +\infty$ and the quantity in brackets goes to $+\infty$ as $r \to 0$,

$$\lim_{r \to 0} \frac{\sigma(\partial B_{x,r} \cap E)}{\sigma(\partial B_{x,r})} = 0$$

We conclude this section by showing that thinness is invariant under certain transformations. A mapping $\mathbf{T} : E^n \rightsquigarrow E^n$ is **measurable** if $\mathbf{T}^{-1}(E)$ is a Borel subset of E^n for each Borel set $E \subset E^n$; if, in addition $\|\mathbf{T}x - \mathbf{T}y\| \leq \|x - y\|$ for all $x, y \in E^n$, then $\mathbf{T}$ is called a **contraction mapping.**

THEOREM **10.6** If E is thin at x and $\mathbf{T}$ is a contraction mapping such that $\|\mathbf{T}x - \mathbf{T}y\| = \|x - y\|$ for all $y \in E^n$, then $\mathbf{T}(E)$ is thin at $\mathbf{T}x$.

Proof. If x is not a limit point of E, it is easily seen that $\mathbf{T}x$ is not a limit point of $\mathbf{T}(E)$ and thinness is preserved. We can assume that x is a limit point of E and that $x \notin E$. By Theorem 10.4 there is a superharmonic function v such that $v(x) < \lim_{\substack{y \to x \\ y \in E \sim \{x\}}} v(y) = +\infty$. We can assume that v is non-negative on a ball B containing x and, in fact, we can assume that $v = G_B \nu$, where ν is a measure on B. Since $G_B \nu$ differs from the Newtonian (or logarithmic) potential U^ν by a harmonic function that is continuous at x, we can assume that

$$v(y) = U^\nu(y) = \int u(y, z)\, d\nu(z),$$

where we have put $u(y, z) = u_y(z)$. We shall extend ν to E^n by putting $\nu(E^n \sim B) = 0$. Now the mapping $\mathbf{T}$ induces a measure $\nu\mathbf{T}^{-1}$ on the Borel subsets F of E^n by the equation $(\nu\mathbf{T}^{-1})(F) = \nu(\mathbf{T}^{-1}(F))$. Consider

$$U^{\nu\mathbf{T}^{-1}}(y) = \int u(y, z)\, d(\nu\mathbf{T}^{-1})(z).$$

It follows from the general theory of transformation of integrals that

$$U^{\nu\mathbf{T}^{-1}}(y) = \int u(y, \mathbf{T}z)\, d\nu(z).$$

Therefore

$$U^{\nu\mathbf{T}^{-1}}(\mathbf{T}y) = \int u(\mathbf{T}y, \mathbf{T}z)\, d\nu(z)$$

Since $\|\mathbf{T}y - \mathbf{T}z\| \le \|y - z\|$, $u(\mathbf{T}y, \mathbf{T}z) \ge u(y, z)$ and $U^{\nu\mathbf{T}^{-1}}(\mathbf{T}y) \ge U^{\nu}(y)$. Now $U^{\nu\mathbf{T}^{-1}}$ is superharmonic with $U^{\nu\mathbf{T}^{-1}}(\mathbf{T}x) = U^{\nu}(x) < +\infty$. Since $\|\mathbf{T}y - \mathbf{T}x\| = \|y - x\|$ for all y,

$$\varliminf_{\substack{z \to \mathbf{T}x \\ z \in \mathbf{T}(E)}} U^{\nu\mathbf{T}^{-1}}(z) = \varliminf_{\substack{\mathbf{T}y \to \mathbf{T}x \\ y \in E}} U^{\nu\mathbf{T}^{-1}}(\mathbf{T}y) \ge \varliminf_{\substack{y \to x \\ y \in E}} U^{\nu}(y) = +\infty.$$

This proves that $\mathbf{T}(E)$ is thin at $\mathbf{T}x$.

3. Thinness and Regularity

Regularity of boundary points relative to the Dirichlet problem can be characterized in terms of thinness. We shall first characterize thinness in terms of the balayage of certain functions.

Definition. An extended real-valued function u defined on an open set R **peaks** at x if $u(x) > \sup_{y \in R \sim V} u(y)$ for every neighborhood V of x.

If R has a Green function, then there is no problem in finding a positive superharmonic function on R which peaks at $x \in R$, namely, the Green function $G_R(x, \cdot)$ with pole at x.

THEOREM **10.7** Let R be an open set having a Green function and let u be a positive superharmonic function on R which peaks at x. A set $E \subset R$ is thin at x if and only if $\hat{\mathbf{R}}_E^u(x) < u(x)$.

Proof. We know that E is thin at x if and only if $E \sim \{x\}$ is thin at x. We also know from Corollary 8.37 that $\hat{\mathbf{R}}_{E \sim \{x\}}^u = \hat{\mathbf{R}}_E^u$. It follows that we can assume $x \notin E$. Then $\hat{\mathbf{R}}_E^u(x) = \mathbf{R}_E^u(x)$ by Corollary 8.36. Suppose $\mathbf{R}_E^u(x) < u(x)$ and let us show that E is thin at x. Since E is thin at x if x is not a limit point of E, we can assume that x is a limit point. Then there is a positive superharmonic function v on R which majorizes u on E such that $v(x) < u(x)$. Therefore

$$\liminf_{\substack{y \to x \\ y \in E}} v(y) \geq \liminf_{\substack{y \to x \\ y \in E}} u(y) \geq u(x) > v(x)$$

and E is thin at x by Theorem 10.3. Suppose now that E is thin at x. If x is not a limit point of E, it is easily seen that $\mathbf{R}_E^u(x) < u(x)$, since u peaks at x. Assuming now that x is a limit point of E, by Theorem 10.3 there is a superharmonic function v on R such that $v(x) < \liminf_{y \to x, y \in E} v(y)$. We can assume that v is positive on a ball B containing x and we can also assume that v is bounded above by replacing v by $v \wedge [v(x) + 1]$ if necessary. As in the proof of Theorem 8.35, we can assume that $v \geq 0$ on R. Let α be such that $v(x) < \alpha < \beta = \liminf_{y \to x, y \in E} v(y)$. For $\lambda > 0$, let $w_\lambda = u(x) + \lambda(v - \alpha)$. Now there is a neighborhood U of x such that $v(y) > \alpha$ for $y \in U \cap E$. Then $w_\lambda(y) \geq u(x) \geq u(y)$ for $y \in U \cap E$. Moreover, $u(x) - u(y) \geq \gamma > 0$ for $y \in R \sim U$ since u peaks at x. Since v is bounded, we can choose λ_0 such that $\lambda_0 |v(y) - \alpha| \leq \gamma \leq u(x) - u(y)$ for $y \in R \sim U$. Then $\lambda_0(v(y) - \alpha) \geq u(y) - u(x)$ for $y \in R \sim U$ and $w_{\lambda_0}(y) = u(x) + \lambda_0[v(y) - \alpha] \geq u(y)$ for $y \in R \sim U$. Since $w_{\lambda_0}(y) \geq u(y)$ for $y \in U \cap E$, $w_{\lambda_0} \geq u$ on E and $w_{\lambda_0} \geq R_E^u$. By choice of α, $u(x) > u(x) + \lambda_0 \times [v(x) - \alpha] = w_{\lambda_0}(x) \geq \mathbf{R}_E^u(x)$.

LEMMA **10.8** If B is a ball of radius $\frac{1}{2}$, there is a positive continuous superharmonic function w on B such that for each $x \in B$ the function w has a decomposition into a sum of two positive superharmonic functions one of which peaks at x.

Proof. We shall prove only the $n = 2$ case, the $n \geq 3$ case being essentially the same. Define

$$w(y) = -\int_B \log \|y - z\| \, dz, \qquad y \in B.$$

Then w is a positive continuous superharmonic function on B. Consider any $x \in B$ and choose r such that $\bar{B}_{x,r} \subset B$. Then $w = w_1 + w_2$, where

$$w_1(y) = -\int_{B_{x,r}} \log \|y - z\| \, dz, \qquad y \in B$$

$$w_2(y) = -\int_{B \sim B_{x,r}} \log \|y - z\| \, dz \qquad y \in B$$

Using Theorem 6.12, the integral defining w can be evaluated explicitly to show that w is continuous on B. All three integrals are clearly non-negative.

Theorem 6.12 can also be used to get an explicit representation of w_1 to show that w_1 peaks at x.

THEOREM **10.9** The set of points of E where E is thin is a polar set.

Proof. Since thinness is a local property and a countable union of polar sets is again polar, it suffices to consider the case for which E is a subset of a ball B of radius $\frac{1}{2}$. Suppose E is thin at $x \in E$. By the preceding lemma there is a positive continuous superharmonic function w on B having a decomposition $w = w_1 + w_2$, where the w_i's are positive superharmonic functions and w_1 peaks at x. By Theorem 10.7 $\hat{\mathbf{R}}_E^{w_1}(x) < w_1(x)$. By Theorem 7.12

$$\hat{\mathbf{R}}_E^w(x) \leq \hat{\mathbf{R}}_E^{w_1}(x) + \hat{\mathbf{R}}_E^{w_2}(x) < w_1(x) + w_2(x) = w(x) = \mathbf{R}_E^w(x)$$

and $x \in \{y : \hat{\mathbf{R}}_E^w(y) < \mathbf{R}_E^w(y)\}$. Since $\hat{\mathbf{R}}_E^w = \mathbf{R}_E^w$, except possibly on a polar set by Corollary 7.40, the points of E where E is thin are contained in the polar set $\{y : \hat{\mathbf{R}}_E^w(y) < \mathbf{R}_E^w(y)\}$.

COROLLARY **10.10** A set $Z \subset E^n$ is polar if and only if it is thin at each of its points.

Proof. By Theorem 10.2 a polar set is thin at each of its points. If Z is thin at each of its points, it is polar by the preceding theorem.

So far no conditions have been placed upon E to study thinness of E. We now relate the irregularity of a boundary point x of an open set R having a Green function to the thinness of $\sim R$ at x.

THEOREM **10.11** Let E be a closed subset of E^n. Then E is not thin at $x \in \partial E$ if and only if there is a positive superharmonic function w defined on $\sim E$ in a neighborhood of x such that $\lim_{\substack{y \to x \\ y \in \sim E}} w(y) = 0$.

Proof. Suppose E is not thin at $x \in \partial E$. Let B be a ball with center x and radius δ. We can assume that $E \subset B$. Also let $u(y) = \delta - \|y - x\|$ for $y \in B$. Then u is a positive superharmonic function on B which peaks at x. Consider the balayage $\hat{\mathbf{R}}_E^u$ relative to B. Then $\hat{\mathbf{R}}_E^u(x) = u(x)$ by Theorem 10.7 and $\liminf_{\substack{y \to x \\ y \in B \sim E}} \hat{\mathbf{R}}_E^u(y) \geq \liminf_{\substack{y \to x \\ y \in B}} \hat{\mathbf{R}}_E^u(y) \geq \hat{\mathbf{R}}_E^u(x) = u(x)$. Consider the superharmonic function $w = u - \hat{\mathbf{R}}_E^u \geq 0$ on $B \sim E$. If w vanishes at some point of $B \sim E$, it must be identically zero on $B \sim E$ by the minimum principle; since $\hat{\mathbf{R}}_E^u$ is harmonic on $B \sim E$, u would be harmonic on $B \sim E$ in contradiction to the fact that the Laplacian of u is strictly negative. It follows that $w > 0$ on $B \sim E$. Moreover, $\limsup_{\substack{y \to x \\ y \in B \sim E}} w(y) \leq u(x) - \liminf_{\substack{y \to x \\ y \in B \sim E}} \hat{\mathbf{R}}_E^u(y) \leq 0$. Therefore $\lim_{\substack{y \to x \\ y \in B \sim E}} w(y) = 0$. To prove the converse, let B be a ball with center x and let w be a positive superharmonic function on $B \sim E$ such that $\lim_{\substack{y \to x \\ y \in B \sim E}} w(y) = 0$. We obtain a contradiction by also assuming that E is thin at x; that is, by assuming that there is a subharmonic function v on B such that $v(x) = 1$ and $v(z) \leq -1$ on $B \cap (E \sim \{x\})$. Let B_1 be a second ball with center x such that $\bar{B}_1 \subset B$. Now $v < 0$ on a neighborhood W of $E \cap \partial B_1$ and for any positive number λ, $v - \lambda w \leq 0$ on $W \cap (\partial B_1 \sim E)$. Since $w > 0$ or $\partial B_1 \sim E$, $\inf_{z \in \partial B_1 \sim W} w(z) > 0$. This along with the fact that v is bounded above on $\partial B_1 \sim W$ implies that we can choose a positive λ_0 for which $v - \lambda_0 w \leq 0$ on $\partial B_1 \sim W$. Then

$$\lim_{\substack{z \to y \\ z \in B_1 \sim E}} \sup[v(z) - \lambda_0 w(z) - \varepsilon u_x(z)] \leq 0$$

for any $\varepsilon > 0$ whenever $y \in \partial B_1 \sim W$. This inequality also holds at points $y \in B_1 \cap (\partial E - \{x\})$ for in a neighborhood of such points v is strictly negative; it also holds at x, since in this case the left side is $-\infty$. This shows that $v - \lambda_0 w - \varepsilon u_x \leq 0$ on $B_1 \sim E$ for every $\varepsilon > 0$. Therefore $v \leq \lambda_0 w$ on $B_1 \sim E$ and $\limsup_{\substack{y \to x \\ y \in B_1 \sim E}} v(y) \leq 0$. Since $\limsup_{\substack{y \to x \\ y \in E \sim \{x\}}} v(y) \leq -1$, $v(x) = \limsup_{\substack{y \to x \\ y \neq x}} v(y) \leq 0$, a contradiction.

COROLLARY **10.12** Let R be an open set having a Green function. A point $x \in \partial R$ is a regular boundary point for R if and only if $\sim R$ is not thin at x.

Proof. The preceding theorem states that $\sim R$ is not thin at $x \in \partial R$ if and only if there is a barrier at x. The existence of a barrier is necessary and sufficient for regularity.

COROLLARY **10.13** A line segment $\ell \subset E^2$ is not thin at its endpoints.

Proof. We first show that ℓ cannot be thin at an interior point x. Since thinness is a local concept, any closed subinterval of ℓ containing x as an interior point will be thin at x if ℓ is thin at x. It therefore suffices to show that a closed line segment ℓ cannot be thin at an interior point. Let B be a ball containing ℓ in its interior. Regarding x as a boundary point of $B \sim \ell$, x is a regular boundary point for $B \sim \ell$ by Theorem 8.25 and ℓ is not thin at x by the preceding corollary. Suppose now that ℓ is thin at an endpoint x. Then there is a superharmonic function w such that $w(x) < \lim_{\substack{y \to x \\ y \in \ell \sim \{x\}}} w(y)$. Since superharmonicity is invariant under a rotation about x, we can construct a line segment ℓ' having x as an interior point and a superharmonic function w' such that $w'(x) < \lim_{\substack{y \to x \\ y \in \ell' \sim \{x\}}} w'(y) = +\infty$, but we have just shown that this is not possible.

THEOREM **10.14** If $E \subset E^2$ and $\sim E$ is thin at x, then there are arbitrarily small ρ such that $\partial B_{x,\rho} \subset E$.

Proof. If x is not a limit point of $\sim E$, then the assertion is trivially true. Let x be a limit point of $\sim E$. Now let (r, θ) be the polar coordinates of a point in the plane relative to the pole x and a ray ℓ having x as an endpoint. Consider the mapping $\mathbf{T} : E^2 \rightsquigarrow \ell$ taking the point (r, θ) into the point $(r, 0)$. It is easily seen that $\mathbf{T}$ is a contraction mapping such that $\|\mathbf{T}x - \mathbf{T}y\| = \|x - y\|$ for all $y \in E^2$. Suppose there is a $\rho_0 > 0$ such that $(\sim E) \cap \partial B_{x,\rho} \neq \emptyset$ for all $\rho \leq \rho_0$. Then $\mathbf{T}(\sim E)$ contains the line segment $\ell^* = \{(\rho, 0) : 0 < \rho \leq \rho_0\}$. Since $\mathbf{T}(\sim E)$ is thin at x by Theorem 10.6, ℓ^* is thin at x, but this contradicts Corollary 10.13 and there is no such ρ_0.

We can now give a proof of Theorem 8.26. Let R be an open subset of E^2, let $x \in \partial R$, and let ℓ be a continuum in $\sim R$ containing the point x. By Corollary 10.12, x is a regular boundary point for R if and only if $\sim R$ is not thin at x. Assume that $\sim R$ is thin at x. Let y be a point of ℓ different from x and let ρ be such that $\partial B_{x,\rho} \subset R$ with $\rho < \|x - y\|$. Since $\partial B_{x,\rho} \cap \ell = \emptyset$, $\ell = (B_{x,\rho} \cap \ell) \cup ((\sim \bar{B}_{x,\rho}) \cap \ell)$, contradicting the connectedness of ℓ. It follows that x is a regular boundary point under the above conditions on R, x, and ℓ.

4. Fine Limits

The fine topology seems to be the natural topology for studying the boundary behavior of superharmonic functions. Fatou's nontangential boundary limit theorem, for example, can be proved by means of the fine topology. Since such a discussion is somewhat beyond the scope of this book, we limit ourselves to two theorems that illustrate the potentiality of the fine topology.

Definition. Let x be a fine limit point of the set E. If the extended real-valued function g on $E \sim \{x\}$ has a fine limit λ at x we shall write

$$f\text{-}\lim_{\substack{y \to x \\ y \in E}} g(y) = \lambda.$$

THEOREM **10.15** (*Cartan*) If x is a fine limit point of the set E and g is an extended real-valued function having a fine limit λ at x, then there is a fine neighborhood V^f of x such that $\lim_{\substack{y \to x \\ y \in (E \sim \{x\}) \cap V^f}} g(y) = \lambda$.

Proof. We can assume that λ is finite for if this is not the case, then we can replace g by arc tan g. Moreover, we can assume that E is contained in a ball B with center x. If $\{I_j\}$ is a sequence of open intervals such that $I_j \downarrow \{x\}$, then for each j there is a fine neighborhood V_j^f of x such that $y \in (E \sim \{x\}) \cap V_j^f$ implies $g(y) \in I_j$. Now if there are infinitely many V_j^f's for which x is an interior point, we can just take $V^f = B$. We can therefore assume that there are at most a finite number of V_j^f's with this property and, in fact, that x is a limit point of $\sim V_j^f$ for every j. Since V_j^f is a fine neighborhood of x, $\sim V_j^f$ is thin at x and there is a positive superharmonic function v_j on B such that $v_j(x) < \lim_{\substack{y \to x \\ y \in \sim V_j^f}} v_j(y) = +\infty$. Let $\{\alpha_j\}$ be a sequence of positive numbers such that $\sum_j \alpha_j v_j(x) < +\infty$. Then $v = \sum_j \alpha_j v_j$ is superharmonic on B and finite at x. Now let $F_j = B \cap (\sim V_j^f) \cap \{y : \alpha_j v_j(y) > j\}$ and $F = \cup_j F_j$. Then $x \notin F$. We now show that F is thin at x. Since this is trivially true if $x \notin \bar{F}$, we can assume that x is a limit point of F. Let p be any positive integer. Then $\alpha_j v_j(y) > p$ for $y \in F_j$, $j \geq p$. For each $j < p$ there is a neighborhood W_j of x such that $\alpha_j v_j(y) \geq p$ for $y \in F_j \cap W_j$. If $W = \bigcap_{i=1}^{p-1} W_i$, then $y \in F \cap W$ implies $v(y) = \sum_j \alpha_j v_j(y) \geq p$. This shows that $\lim_{\substack{y \to x \\ y \in F}} v(y) = +\infty$ and that F is thin at x. Taking V^f to be the fine interior of $\sim F$, then V^f is a fine neighborhood of x. If $O_j = B \cap \{y : \alpha_j v_j(y) > j\}$, then $O_j \cap (\sim V_j^f) = F_j \subset O_j \cap F$ and

$$V_j^f \supset O_j \cap V_j^f \supset O_j \cap (\sim F) \supset O_j \cap V^f.$$

if $y \in O_j \cap V^f \cap (E \sim \{x\}) \subset V_j^f \cap (E \sim \{x\})$, then $g(y) \in I_j$. This proves that $\lim_{y \to x,\, y \in V^f \cap E} g(y) = \lambda$.

THEOREM **10.16** (*Brelot*) If E is a closed set which is thin at x and v is a positive superharmonic function on $\sim E$ in a neighborhood of x, then v has a fine limit at x.

Proof. If x is not a limit point of E, then there is nothing to prove since in this case v is finely continuous at x. We can assume that x is a limit point of E. There is then a superharmonic function u such that $u(x) < \lim_{\substack{y \to x \\ y \in E \sim \{x\}}} u(y) = +\infty$. Since u is finely continuous at x, $u(x) = f\text{-}\lim_{\substack{y \to x \\ y \in \sim E}} u(y) < +\infty$. Now if $\liminf_{\substack{y \to x \\ y \in \sim E}} (v + u)(y) = +\infty$, then $\lim_{\substack{y \to x \\ y \in \sim E}} (v + u)(y)$ exists and, in particular, $f\text{-}\lim_{\substack{y \to x \\ y \in \sim E}} (v + u)(y) = +\infty$. It follows that $f\text{-}\lim_{\substack{y \to x \\ y \in \sim E}} v(y)$ exists and is equal to $+\infty$. Suppose now that $\lambda = \liminf_{\substack{y \to x \\ y \in \sim E}} (v + u)(y) < +\infty$ and that $k > \lambda$. Since $\lim_{\substack{y \to x \\ y \in E \sim \{x\}}} u(y) = +\infty$, there is a neighborhood

U of x having compact closure such that $u(y) > k$ for $y \in U \cap (E \sim \{x\})$. Since E is closed, $\liminf_{\substack{z \to y \\ z \in \sim E}} (v + u)(z) > k$ for $y \in U \cap (\partial E \sim \{x\})$. Let

$$w = \begin{cases} (v + u) \wedge k & \text{on} \quad U \cap (\sim E), \\ k & \text{on} \quad U \cap (E \sim \{x\}). \end{cases}$$

Then w is superharmonic on $U \sim \{x\}$ and, since it is bounded below on this set, by Theorem 7.7 it has a superharmonic extension to U which is also denoted by w. Noting that $\liminf_{\substack{y \to x \\ y \in \sim E}} (v + u)(y) = \lambda < k$, w agrees with $v + u$ on $\sim E$ in a neighborhood of x. Since w is finely continuous at x, $f\text{-}\lim_{\substack{y \to x \\ y \in \sim E}} w(y)$ is defined, and since $v + u$ agrees with w on $\sim E$ in a neighborhood of x, $f\text{-}\lim_{\substack{y \to x \\ y \in \sim E}} (v + u)(y)$ exists. Lastly, $f\text{-}\lim_{\substack{y \to x \\ y \in \sim E}} v(y)$ exists, since $f\text{-}\lim_{\substack{y \to x \\ y \in \sim E}} u(y)$ exists.

5. Wiener's Test

The original theorem of Wiener characterized irregularity of a boundary point of a region R in terms of the capacity of certain subsets of $\sim R$ in a neighborhood of the boundary point. In view of the equivalence of irregularity and thinness of the complement, it suffices to relate thinness of $\sim R$ to the capacity of certain subsets of $\sim R$.

Consider a set $E \subset E^n$ and a fixed point x. Since thinness of E at x is a local property, we shall assume that $E \subset B_{x,\frac{1}{2}}$ in the remainder of this section. In addition, the Green function for $B = B_{x,\frac{1}{2}}$ is denoted by G. As usual, we write $u(y, z)$ for $u_y(z)$ for notational convenience.

LEMMA **10.17** If $\lambda > 1$, then there is a constant m, depending only upon λ, such that $u(y,z)/u(x,y) \leq m$ whenever $y \in \sim\{\xi : \lambda^{j-1} < u(x, \xi) < \lambda^{j+2}\}$ and $z \in \{\xi : \lambda^j \leq u(x, \xi) \leq \lambda^{j+1}\}$ for each positive integer j.

Proof. Suppose first that $n = 2$. If $u(x, y) \leq \lambda^{j-1}$ and $\lambda^j \leq u(x, z) \leq \lambda^{j+1}$, then using simple geometrical considerations we can show that $u(y, z)/u(x, y) \to 1$ as $j \to \infty$ uniformly in such y and z and the boundedness of $u(y, z)/u(x, y)$ uniformly in j follows. If now $u(x, y) \geq \lambda^{j+2}$ and $\lambda^j \leq u(x, z) \leq \lambda^{j+1}$, then $u(y, z)/u(x, y) \to 1 + \lambda^{-1}(1 - \lambda)$ as $j \to \infty$ uniformly in such y and z and again uniform boundedness in j follows. Suppose now that $n \geq 3$. If $u(x, y) \leq \lambda^{j-1}$ and $\lambda^j \leq u(x, z) \leq \lambda^{j+1}$, simple geometrical considerations result in

$$\frac{u(y, z)}{u(x, y)} \leq \left[\frac{\lambda^{1/(n-2)}}{\lambda^{1/(n-2)} - 1}\right]^{n-2}.$$

Since the right side is independent of j and such y and z, we have uniform boundedness in j. If $u(x, y) \geq \lambda^{j+2}$ and $\lambda^j \leq u(x, z) \leq \lambda^{j+1}$, then

$$\frac{u(y, z)}{u(x, y)} \leq \left[\frac{1}{\lambda^{1/(n-2)} - 1}\right]^{n-2}.$$

which is again independent of j for such y and z.

LEMMA **10.18** If R is an open set having a Green function G_R and U is a nonempty open set having compact closure $\bar{U} \subset R$, there is a measure μ on ∂U such that $\mu(\partial U) = \mathscr{C}(U)$ and $G_R \mu = 1$ on U.

Proof. Let $\{U_j\}$ be an increasing sequence of open sets with compact closures $K_j = \bar{U}_j \subset U$ such that $U_j \uparrow U$. Let μ_j be the capacitary distribution of K_j. Then $\mathscr{C}(U) = \lim_{j \to \infty} \mathscr{C}(K_j) = \lim_{j \to \infty} \mu_j(K_j)$. Since $\bar{U}$ is compact, $\mathscr{C}(\bar{U}) < +\infty$ and the measures μ_j are uniformly bounded. It follows that there is a subsequence of the sequence $\{\mu_j\}$ which we can assume to be the sequence itself and a measure μ such that $\mu_j \xrightarrow{w^*} \mu$. Clearly μ has support in ∂U. If $x \in U$, then $x \in U_j$ for sufficiently large j and it is easily seen that $G_R \mu_j(x) \to G_R \mu(x)$ as $j \to +\infty$. Since $G_R \mu_j(x) = 1$ for sufficiently large j, $G_R \mu(x) = 1$.

LEMMA **10.19** Let R be an open set having a Green function G_R and let ν be a measure having support F. If $G_R \nu \geq \alpha$ on F except possibly for a polar subset of F, then $\nu(F) \geq \alpha \mathscr{C}(F)$.

Proof. Let K be any compact subset of F and let $G_R \mu_K$ be the capacitary potential of K. Since $G_R \mu_K \leq 1$, μ_K assigns measure zero to any polar subset of R by Theorem 7.35; that is, $G_R \nu \geq \alpha$ a.e. (μ_K). By the reciprocity theorem $\nu(F) = \int 1\, d\nu \geq \int G_R \mu_K\, d\nu = \int G_R \nu\, d\mu_K \geq \alpha \mu_K(K) = \alpha \mathscr{C}(K)$. Taking the supremum over all compact sets $K \subset F$, $\nu(F) \geq \alpha \mathscr{C}(F)$.

In the remainder of this section λ will be a fixed positive number. If p and q are positive integers, we shall also let $A^{\lambda}_{(p,q)} = \{\xi : \lambda^p < u(x, \xi) < \lambda^q\}$ and $A^{\lambda}_{[p,q]} = \{\xi : \lambda^p \leq u(x, \xi) \leq \lambda^q\}$.

LEMMA **10.20** If $\lambda > 1$ and ν is a finite measure on B such that $v = G\nu$ is finite at x, there is a constant α, depending only upon λ and ν, such that

$$\int_{B \sim A^{\lambda}_{(j-1,\, j+2)}} G(y, z)\, d\nu(z) \leq \alpha$$

for all $y \in B \cap A^{\lambda}_{[j, j+1]}$.

Proof. Since $u(y, z)$ is non-negative on $B \times B$, $G(y, z) \le u(y, z)$ and it suffices to prove the inequality with G replaced by u. By Lemma 10.17 there is a constant m, independent of j, such that

$$\int_{B \sim A^{\lambda}_{(j-1,\, j+2)}} u(z, y)\, d\nu(y) \le m \int_{B \sim A^{\lambda}_{(j-1,\, j+2)}} u(x, y)\, d\nu(y) \le m \int_B u(x, y)\, d\nu(y)$$

for all $z \in B \cap A^{\lambda}_{[j, j+1]}$. Now $G(x, \cdot) = u_x + h_x$, where h_x is the solution of the Dirichlet problem on B corresponding to the boundary function $u_x|_{\partial B}$. Since u_x is bounded on ∂B, h_x is bounded on B. For $z \in B \cap A^{\lambda}_{[j, j+1]}$,

$$\int_{B \sim A^{\lambda}_{(j-1,\, j+2)}} u(z, y)\, d\nu(y) \le m v(x) - m \int_B h_x(y)\, d\nu(y).$$

Capacities are relative to B in the following theorem.

THEOREM **10.21** (*Wiener, Brelot*) Let E be any subset of E^n, x any point, $\lambda > 1$, and $E_j = E \cap A^{\lambda}_{[j, j+1]}$, $j \ge 1$. Then E is thin at x if and only if $\sum_j \lambda^j \mathscr{C}^*(E_j) < +\infty$.

Proof. Suppose first that $\sum_j \lambda^j \mathscr{C}^*(E_j) < +\infty$. Since E is thin at x if x is not a limit point of E, we can assume that x is a limit point of E. Let $\{\varepsilon_j\}$ be a sequence of positive numbers such that $\sum_j \varepsilon_j \lambda^j < +\infty$ and for each j let U_j be a nonempty open subset of B containing E_j such that $\bar{U}_j \subset B$ and

(i) $U_j \cap \{\xi : u(x, \xi) > \lambda^{j+2}\} = \varnothing$,
(ii) $\mathscr{C}(U_j) < \mathscr{C}^*(E_j) + \varepsilon_j$.

Then $\sum_j \lambda^j \mathscr{C}(U_j) < +\infty$. By Lemma 10.18 there is a measure μ_j with support in ∂U_j such that $\mu_j(\partial U_j) = \mathscr{C}(U_j)$ and $G\mu_j = 1$ on U_j. Now

$$G\mu_j(x) = \int_{\partial U_j} G(x, z)\, d\mu_j(z) \le \int_{\partial U_j} u(x, z)\, d\mu_j(z) \le \lambda^{j+2}\mu_j(\partial U_j) = \lambda^{j+2}\mathscr{C}(U_j).$$

For each positive integer k define $v_k = \sum_{j \ge k} G\mu_j$. Then $v_k(x) \ge \lambda^2 \sum_{j \ge k} \lambda^j \mathscr{C}(U_j)$. Since $G\mu_j = 1$ on $U_j \supset E_j$, $v_k \ge 1$ on $\bigcup_{j \ge k} E_j = E \cap \{\xi : \lambda^k \le u(x, \xi) < +\infty\}$ and $\liminf_{y \to x,\, y \in E \sim \{x\}} v_k(y) \ge 1$ for every positive integer k. Since the series $\sum_j \lambda^j \mathscr{C}(U_j)$ converges, we can choose k so that $v_k(x) \le \lambda^2 \sum_{j \ge k} \lambda^j \mathscr{C}(U_j) < 1 \le \liminf_{y \to x,\, y \in E \sim \{x\}} v_k(y)$. This shows that E is thin at x. To prove the converse suppose that E is thin at x. We can assume that $x \notin E$ and also that x is a limit point of E, for in the contrary case $E_j = \varnothing$ for sufficiently large j and $\sum_j \lambda^j \mathscr{C}^*(E_j)$ reduces to a finite series. Then there is a positive superharmonic function v on B such that $v(x) < \lim_{y \to x,\, y \in E} v(y) = +\infty$. We can assume that

$v = Gv$. Now if $E_j = \varnothing$, then $\mathscr{C}^*(E_j) = 0$ and E_j contributes nothing to the series $\sum_j \lambda^j \mathscr{C}^*(E_j)$. Rather than complicate the notation by eliminating such j's, we assume that $E_j \neq \varnothing$ for all j. Let $\alpha_j = \inf_{E_j} v$ and let α be a fixed positive number. Consider the open sets $V_j = \{z : v(z) > \alpha_j - \alpha\} \supset E_j, j \geq 1$, and the Borel sets $U_j = V_j \cap A^{\lambda}_{[j,j+1]}$. Since $U_j \supset E_j$, it suffices to show that $\sum_j \lambda^j \mathscr{C}(U_j) < +\infty$. Now let $\{\varepsilon_j\}$ be a sequence of positive numbers such that $\sum_j \lambda^j \varepsilon_j < +\infty$ and let K_j be a compact subset of U_j such that $\mathscr{C}(K_j) > \mathscr{C}(U_j) - \varepsilon_j$. It suffices then to show that $\sum_j \lambda^j \mathscr{C}(K_j) < +\infty$. Note that $K_j \subset U_j \subset A^{\lambda}_{[j,j+1]} \subset A^{\lambda}_{(j-1,j+2)}$. In order to show that $\sum_j \lambda^j \mathscr{C}(K_j) < +\infty$, it suffices to show that each of the series

$$\sum_j \lambda^{6j+p} \mathscr{C}(K_{6j+p}), \qquad p = 0, 1, \ldots, 5,$$

converges. We do this only for the $p = 0$ series. As noted above,

$$\begin{aligned} K_{6j} \subset U_{6j} \subset \{\xi : \lambda^{6j-1} < u(x, \xi) < \lambda^{6j+2}\} \\ = \{\xi : (\lambda^3)^{2j} < \lambda u(x, \xi) < (\lambda^3)^{2j+1}\}. \end{aligned}$$

Consider any $k \neq j$. Then

$$K_{6k} \subset \{\xi : (\lambda^3)^{2k} < \lambda u(x, \xi) < (\lambda^3)^{2k+1}\}.$$

If $k = j + i$ with $i > 0$, then $(\lambda^3)^{2k} = (\lambda^3)^{2j+2i} > (\lambda^3)^{2j+1}$ and

$$K_{6k} \subset \sim\{\xi : (\lambda^3)^{2j} < \lambda u(x, \xi) < (\lambda^3)^{2j+1}\};$$

if $k = j - i$ with $i > 0$, then $(\lambda^3)^{2k+1} = (\lambda^3)^{2j-2i+1} < (\lambda^3)^{2j}$ and

$$K_{6k} \subset \sim\{\xi : (\lambda^3)^{2j} < \lambda u(x, \xi) < (\lambda^3)^{2j+1}\}.$$

In either case $K_{6k} \subset \sim A^{\lambda}_{(6j-1,\,6j+2)}$ and, in particular, the sets $K_{6j}, j \geq 1$, are disjoint. Letting $K = \bigcup_j K_{6j} \cup \{x\}$, it is easily seen that K is compact. Now let $w = \hat{\mathbf{R}}^v_K$. Since $v(x) \geq \hat{\mathbf{R}}^v_K(x) = w(x)$, $w(x) < +\infty$. If we now let $w = Gv^*$, then v^* has support in K and, in fact, $v^*(B \sim \cup K_{6j}) = 0$, since w is finite at x. Therefore

$$w = \int_{K_{6j}} G(\cdot, z)\, dv^*(z) + \int_{\bigcup_{k \neq j} K_{6k}} G(\cdot, z)\, dv^*(z).$$

Since $\bigcup_{k \neq j} K_{6k} \subset B \sim A^{\lambda}_{(6j-1,\,6j+2)}$, by Lemma 10.20 there is a constant β, depending only upon λ and v^*, such that

$$\int_{\bigcup_{k \neq j} K_{6k}} G(y, z)\, dv^*(z) \leq \beta$$

for all $y \in B \cap A^{\lambda}_{[6j,\,6j+1]}$. Therefore,

$$w(y) \le \beta + \int_{K_{6j}} G(y, z)\, dv^*(z)$$

for all $y \in B \cap A^{\lambda}_{[6j,\,6j+1]}$. Since $w = v$ on K, except possibly for a polar set $Z \subset K$, $v \ge \alpha_j - \alpha$ on K_j and $\alpha_j \to +\infty$ as $j \to \infty$, $\lim_{\substack{y \to x \\ y \in K \sim Z}} w(y) = +\infty$. Let q be a positive integer such that $q - \beta \ge \rho > 0$. Then there is a j_0 such that $w(y) \ge q$ on K_{6j} for all $j \ge j_0$ except possibly for a polar set. Therefore

$$\int_{K_{6j}} G(y, z)\, dv^*(z) \ge \rho$$

for all $y \in K_{6j}$, $j \ge j_0$. By Lemma 10.19 $v^*(K_{6j}) \ge \rho\mathscr{C}(K_{6j})$ for all $j \ge j_0$. Therefore

$$\sum_{j \ge j_0} \lambda^{6j} v^*(K_{6j}) \ge \rho \sum_{j \ge j_0} \lambda^{6j}\mathscr{C}(K_{6j})$$

and it suffices to show that the series $\sum_j \lambda^{6j} v^*(K_{6j})$ converges; but

$$\int u(x, z)\, dv^*(z) \ge \sum_j \int_{K_{6j}} u(x, z)\, dv^*(z) \ge \sum_j \lambda^{6j} v^*(K_{6j}),$$

Since

$$+\infty > w(x) = \int G(x, z)\, dv^*(z)$$

$$= \int u(x, z)\, dv^*(z) + \int h_x(z)\, dv^*(z)$$

and the second integral on the right is finite, the series $\sum_j \lambda^{6j} v^*(K_{6j})$ converges.

COROLLARY **10.22** Let R be an open set having a Green function. A point $x \in \partial R$ is an irregular boundary point for R if and only if for some $\lambda > 1$ the series $\sum_j \lambda^j \mathscr{C}((\sim R) \cap A^{\lambda}_{[j,\,j+1]})$ converges.

CHAPTER 11

Energy

From the very beginning of potential theory the energy of a distribution of charge or mass has had a greater role than indicated by the preceding development. Assuming a knowledge of the properties of the Green function, such concepts as polar set, réduite, and capacity could be based upon energy considerations. Since these concepts have already been developed, we limit ourselves to establishing properties of the energy of a distribution and the relationship to previously developed concepts.

1. Energy Principle

Let R be a connected open set having a Green function G. If μ and ν are two measures on R, the integral $\int G\mu\,d\nu$ is always defined, even though it may be infinite. This integral is called the **mutual energy** of μ and ν and is denoted by $(\mu, \nu)_e$. Note that

$$0 \leq (\mu, \nu)_e = \int G\mu\,d\nu = \int G\nu\,d\mu = (\nu, \mu)_e \leq +\infty.$$

and that $(\mu, \nu)_e$ is a bilinear form on such pairs μ, ν. The quantity $(\mu, \mu)_e$ is called the **energy** of μ and is denoted by $\|\mu\|_e^2$; that is, $\|\mu\|_e = +\sqrt{(\mu, \mu)_e}$. The set of measures μ of finite energy $\|\mu\|_e^2$ is denoted by $\mathscr{E}^+$.

We remark that if $G\mu$ is bounded by m and μ has compact support K, then μ has finite energy and

$$\|\mu\|_e^2 = \int G\mu\,d\mu \leq m\mu(K).$$

We shall show later, by example, that $G\mu$ can take on the value $+\infty$ even though μ has finite energy.

LEMMA **11.1** If $\mu \in \mathscr{E}^+$ and λ is a measure such that $G\lambda \leq G\mu$, then $\lambda \in \mathscr{E}^+$ and $\|\lambda\|_e \leq \|\mu\|_e$.

Proof.

$$\int G\lambda \, d\lambda \le \int G\mu \, d\lambda = \int G\lambda \, d\mu \le \int G\mu \, d\mu.$$

As indicated in the introduction, polar sets can be characterized by means of measures having finite energy.

THEOREM **11.2** A set $Z \subset R$ is polar if and only if $\mu(Z) = 0$ for all $\mu \in \mathscr{E}^+$.

Proof. Assume that Z is not polar. Then Z has positive capacity relative to R by Theorem 7.33 and, in particular, positive inner capacity. It follows that there is a compact set $K \subset Z$ which has positive capacity. Letting μ_K be the capacitary distribution on K, $G\mu_K = \hat{\mathbf{R}}_K^1 \le 1$ and $0 < \mu_K(K) = \mathscr{C}(K) < +\infty$. Therefore $\mu_K \in \mathscr{E}^+$ and $\mu_K(Z) > 0$. This proves the sufficiency. Now assume that Z is a polar set. Let $\mu \in \mathscr{E}^+$ and let w be a positive superharmonic function on R such that $w = +\infty$ on Z. Such a function exists by Theorem 8.35. Let $E = \{x : w(x) = +\infty\}$ and $E_j = E \cap \{x : G\mu(x) \le j\}$, $j \ge 1$. Since μ has finite energy, $\mu(E \sim \cup_j E_j) = 0$. Fix j. Let K be any compact subset of E_j and let $v = \mu|_K$. Since $Gv \le G\mu$, $Gv \le j$ on K. Since v has support in K and $w = +\infty$ on K, $Gv \le \varepsilon w$ on the support of v for every $\varepsilon > 0$. By Theorem 8.43 $Gv \le \varepsilon w$ on R for every $\varepsilon > 0$ and it follows that $Gv = 0$. This means that v is the zero measure and $v(K) = \mu(K) = 0$ for all $K \subset E_j$. Since $\mu(E_j) = \sup_{\substack{K \subset E_j \\ K \text{ compact}}} \mu(K)$, $\mu(E_j) = 0$ for all j. Using the fact that $\mu(E \sim \cup_j E_j) = 0$ and that $\cup_j E_j \subset E$, $\mu(E) = 0$. Clearly $Z \subset E$ and $\mu(Z) = 0$.

In the course of the next proof we use the fact that if $\{\mu_j\}$ is a sequence of measures on the compact Hausdorff space X, which converges in the w^*-topology to μ, the sequence $\{\mu_j \times \mu_j\}$ on $X \times X$ converges in the w^*-topology to $\mu \times \mu$. For continuous functions on $X \times X$ of the form $f(x)\, g(y)$, $x \in X$, $y \in Y$, where f and g are continuous on X, we have

$$\int\int f(x)\, g(y)\, d\mu_j(x)\, d\mu_j(y) = \left(\int f(x)\, d\mu_j(x)\right)\left(\int g(y)\, d\mu_j(y)\right)$$

$$\to \left(\int f(x)\, d\mu(x)\right)\left(\int g(y)\, d\mu(y)\right) = \int\int f(x)\, g(y)\, d\mu(x)\, d\mu(y)$$

as $j \to \infty$. This result extends easily to finite sums of such products. Since the class of such functions is a subalgebra of $C(X \times X)$, which separates points and contains the constant functions, the Stone-Weierstrass theorem implies that the above result holds for functions in a dense subset of $C(X \times X)$. A simple argument extends the above convergence to every function in $C(X \times X)$.

THEOREM **11.3** (*Gauss-Frostman*) Let K be a compact subset of R, let f be a continuous function on K, and for each measure ν on K let

$$\Phi_f(\nu) = \int G\nu \, d\nu - 2 \int f \, d\nu.$$

Then there is a measure $\mu \in \mathscr{E}^+$ on K which minimizes Φ_f. Moreover, $G\mu \le f$ on the support of μ and $G\mu \ge f$ on K except possibly for a polar set.

Proof. Let $\alpha = \inf_\nu \Phi_f(\nu) \le 0$, $\beta = \inf_{(x, y) \in K \times K} G(x, y) > 0$, and $\gamma = \sup_K f < +\infty$. Then

$$\Phi_f(\nu) = \int \int G(x, y) \, d\nu(x) \, d\nu(y) - 2 \int f \, d\nu \ge \beta\nu^2(K) - 2\gamma\nu(K).$$

Since $\beta m^2 - 2\gamma m$ has a minimum value of $-\gamma^2/\beta$, $\alpha \ge -\gamma^2/\beta > -\infty$. Letting $\{\mu_j\}$ be a sequence of measures on K such that $\Phi_f(\mu_j) \to \alpha$,

$$\Phi_f(\mu_j) \ge \beta\mu_j^2(K) - 2\gamma\mu_j(K) = \mu_j(K)\{\beta\mu_j(K) - 2\gamma\},$$

and it follows that the sequence $\{u_j(K)\}$ is bounded. There is then a subsequence of the sequence $\{\mu_j\}$, which we can assume to be the sequence itself and a measure μ with support in K such that $\mu_j \xrightarrow{w^*} \mu$. Let k be any positive integer. Since $\mu_j \times \mu_j \xrightarrow{w^*} \mu \times \mu$ and $k \wedge G(x, y)$ is a bounded continuous function on $K \times K$,

$$\lim_{j \to \infty} \int \int (k \wedge G(x, y)) \, d\mu_j(x) \, d\mu_j(y) = \int \int (k \wedge G(x, y)) \, d\mu(x) \, d\mu(y)$$

and

$$\begin{aligned} \int \int G(x, y) \, d\mu(x) \, d\mu(y) &= \lim_{k \to \infty} \int \int (k \wedge G(x, y)) \, d\mu(x) \, d\mu(y) \\ &= \lim_{k \to \infty} \lim_{j \to \infty} \int \int (k \wedge G(x, y)) \, d\mu_j(x) \, d\mu_j(y) \\ &\le \liminf_{j \to \infty} \int \int G(x, y) \, d\mu_j(x) \, d\mu_j(y). \end{aligned}$$

Therefore

$$\begin{aligned} \alpha \le \Phi_f(\mu) &= \int \int G(x, y) \, d\mu(x) \, d\mu(y) - 2 \int f \, d\mu \\ &\le \liminf_{j \to \infty} \int \int G(x, y) \, d\mu_j(x) \, d\mu_j(y) - 2 \lim_{j \to \infty} \int f \, d\mu_j \\ &= \lim_{j \to \infty} \Phi_f(\mu_j) = \alpha. \end{aligned}$$

This shows that μ minimizes Φ_f and that $\mu \in \mathscr{E}^+$. We now show that $G\mu \le f$ on the support of μ. Suppose x is in the support of μ and $G\mu(x) > f(x)$. Then there is a neighborhood U_x such that $G\mu > f$ on $U_x \cap K$. Letting $\nu = \mu|_{U_x}$ and $0 < \varepsilon \le 1$, $\mu - \varepsilon\nu$ is a measure on K and $\Phi_f(\mu) \le \Phi_f(\mu - \varepsilon\nu)$. Now

$$\begin{aligned}\Phi_f(\mu) &= \int G\mu \, d\mu - 2 \int f \, d\mu \\ &\le \Phi_f(\mu - \varepsilon\nu) \\ &= \int G(\mu - \varepsilon\nu) \, d(\mu - \varepsilon\nu) - 2 \int f \, d(\mu - \varepsilon\nu) \\ &= \int G\mu \, d\mu - 2\varepsilon \int G\mu \, d\nu + \varepsilon^2 \int G\nu \, d\nu - 2 \int f \, d\mu + 2\varepsilon \int f \, d\nu ;\end{aligned}$$

that is,

$$\int (f - G\mu) \, d\nu + \frac{\varepsilon}{2} \int G\nu \, d\nu \ge 0.$$

Since $G\nu \le G\mu$, $\nu \in \mathscr{E}^+$ by Lemma 11.1 and $\int G\nu \, d\nu < +\infty$. We can therefore let $\varepsilon \to 0$ to obtain

$$\int (f - G\mu) \, d\nu \ge 0.$$

Since $f - G\mu < 0$ on the support of ν, $\nu(U_x) = \mu(U_x) = 0$; that is, the support of μ is contained in $\{G\mu \le f\}$. Consider now the assertion that $G\mu \ge f$ on K, except possibly for a polar set. For each $\varepsilon > 0$ let μ_{K_ε} be the capacitary distribution corresponding to the compact set $K_\varepsilon = K \cap \{f \ge G\mu + \varepsilon\}$. If $\beta > 0$, then $\mu + \beta\mu_{K_\varepsilon}$ is a measure on K and $\Phi_f(\mu) \le \Phi_f(\mu + \beta\mu_{K_\varepsilon})$. As before,

$$\int (G\mu - f) \, d\mu_{K_\varepsilon} + \frac{\beta}{2} \int G\mu_{K_\varepsilon} \, d\mu_{K_\varepsilon} \ge 0.$$

Letting $\beta \to 0$, we obtain $\int (G\mu - f) \, d\mu_{K_\varepsilon} \ge 0$. Since

$$0 \le \int (G\mu - f) \, d\mu_{K_\varepsilon} \le -\varepsilon\mu_{K_\varepsilon}(K_\varepsilon),$$

$\mu_{K_\varepsilon}(K_\varepsilon) = 0$ and $\mathscr{C}(K_\varepsilon) = 0$. It follows that K_ε is polar and that $G\mu \ge f$ on K except possibly for a polar set.

If in the definition of Φ_f we take $f = 1$, then we get Gauss's integral

$$\Phi_1(\nu) = \int (G\nu - 2) \, d\nu.$$

We show later that the measure μ which minimizes Φ_1 is just the capacitary distribution μ_K corresponding to the compact set K.

THEOREM **11.4** (*Energy Principle*) If μ and ν are measures on R, then $(\mu, \nu)_e \le \|\mu\|_e \|\nu\|_e$.

Proof. It suffices to prove the inequality when μ and ν have supports in a compact set $K \subset R$. Let f be a bounded non-negative continuous function on R such that $f \le G\mu$ and let $\mu' \in \mathscr{E}^+$ be the measure on K which minimizes Φ_f. Then $G\mu' \le f$ on the support of μ' and $G\mu' \ge f$ on K except possibly for a polar set. Since $\mu' \in \mathscr{E}^+$ and polar sets have μ'-measure zero by Theorem 11.2, $G\mu' = f$ a.e. (μ') on the support of μ'. It follows that

$$\int f\, d\mu' = \int G\mu'\, d\mu'$$

and

$$\Phi_f(\mu') = \int G\mu'\, d\mu' - 2\int f\, d\mu' = -\int G\mu'\, d\mu' = -\|\mu'\|_e^2.$$

For any $\alpha > 0$, $\Phi_f(\alpha\nu) \ge \Phi_f(\mu') = -\|\mu'\|_e^2$; that is,

$$\alpha^2 \int G\nu\, d\nu - 2\alpha \int f\, d\nu + \|\mu'\|_e^2 \ge 0$$

or

$$\alpha^2 \|\nu\|_e^2 - 2\alpha \int f\, d\nu + \|\mu'\|_e^2 \ge 0$$

for every $\alpha > 0$. Since this inequality is trivially true if $\alpha \le 0$, it holds for all α and

$$\int f\, d\nu \le \|\nu\|_e \|\mu'\|_e.$$

Since $G\mu' \le f \le G\mu$ on the support of μ' and $G\mu'$ is harmonic off the support of μ', $G\mu'$ is finite-valued and Theorem 8.43 can be applied to obtain $G\mu' \le G\mu$. By Lemma 11.1 $\|\mu'\|_e \le \|\mu\|_e$ and

$$\int f\, d\nu \le \|\nu\|_e \|\mu\|_e.$$

Taking the supremum over all continuous function $f \le G\mu$, we get $\int G\mu\, d\nu \le \|\nu\|_e \|\mu\|_e$.

It is now possible to give an example of a measure $\mu \in \mathscr{E}^+$ for which $G\mu$ can take on the value $+\infty$.

EXAMPLE 1 Let Z be a nonempty compact polar subset of R. Then Z has capacity zero. For each positive integer j let U_j be an open set containing Z with compact closure $\bar{U}_j \subset R$ such that $\mathscr{C}(\bar{U}_j) < 2^{-2j}$. Let μ_j be the capacitary distribution of $\bar{U}_j$. Since $G\mu_j \le 1$, $\mu_j \in \mathscr{E}^+$. Moreover, $G\mu_j = 1$ on $U_j \supset Z$. Letting $\mu = \sum_j \mu_j$, μ is a measure for which $G\mu = +\infty$ on Z and $\mu \in \mathscr{E}^+$, since

$$\int G\mu\, d\mu \le \sum_{j,k} \int G\mu_j\, d\mu_k \le \sum_{j,k} \|\mu_j\|_e \|\mu_k\|_e$$

$$\le \sum_{j,k} 2^{-2j}2^{-2k} = \left(\sum_j 2^{-2j}\right)^2 < 1;$$

that is, $\|\mu\|_e < 1$.

2. Mutual Energy

If μ and ν are two measures, the difference $\mu - \nu$ may not be defined, since both μ and ν may be infinite. If $\mu, \nu \in \mathscr{E}^+$, we regard $\mu - \nu$ as a formal difference. Two such formal differences $\mu - \nu$ and $\mu' - \nu'$ are regarded as equivalent if $\mu + \nu' = \mu' + \nu$ and we write $\mu - \nu \sim \mu' - \nu'$. If ν and ν' are the zero measure 0, this reduces to the usual concept of equality of measures. Strictly speaking, $\mu - \nu$ should be identified with the ordered pair (μ, ν) in the product space $\mathscr{E}^+ \times \mathscr{E}^+$ in which two such pairs (μ, ν) and (μ', ν') are identified if $\mu + \nu' = \mu' + \nu$.

Definition. $\mathscr{E}$ denotes the set $\mathscr{E}^+ - \mathscr{E}^+ = \{\mu - \nu : \mu, \nu \in \mathscr{E}^+\}$. If $\lambda_i = \mu_i - \nu_i \in \mathscr{E}$ with $\mu_i, \nu_i \in \mathscr{E}^+$, $i = 1, 2$, the **mutual energy** $(\lambda_1, \lambda_2)_e$ is defined by

$$(\lambda_1, \lambda_2)_e = (\mu_1, \mu_2)_e - (\mu_1, \nu_2)_e - (\nu_1, \mu_2)_e + (\nu_1, \nu_2)_e;$$

the **energy** $\|\lambda\|_e^2$ of $\lambda \in \mathscr{E}$ is defined by $\|\lambda\|_e^2 = (\lambda, \lambda)_e$.

It is easily seen that the definition of the mutual energy $(\lambda_1, \lambda_2)_e$ is independent of the representations of λ_1 and λ_2. The definition of the energy $\|\lambda\|_e^2$ is also permissible, for, if $\lambda = \mu - \nu$, then

$$(\lambda, \lambda)_e = \|\mu\|_e^2 - 2(\mu, \nu)_e + \|\nu\|_e^2$$

$$\ge \|\mu\|_e^2 - 2\|\mu\|_e \|\nu\|_e + \|\nu\|_e^2$$

$$= [\|\mu\|_e - \|\nu\|_e]^2 \ge 0;$$

that is, the energy is always non-negative. $\mathscr{E}$ is clearly a linear vector space over the reals with $\alpha(\mu - \nu) + \beta(\mu' - \nu') = (\alpha\mu + \beta\nu') - (\alpha\nu + \beta\nu')$. It is also easy to see that $(\lambda_1, \lambda_2)_e$ is a bilinear form.

THEOREM 11.5 If $\mu, \nu \in \mathscr{E}$, then $|(\mu, \nu)_e| \leq \|\mu\|_e \|\nu\|_e$ and $\|\mu + \nu\|_e \leq \|\mu\|_e + \|\nu\|_e$.

Proof. Since the energy is non-negative, $(\mu - \alpha\nu, \mu - \alpha\nu)_e \geq 0$ for all α; that is,

$$\|\mu\|_e^2 - 2\alpha(\mu, \nu)_e + \alpha^2 \|\nu\|_e^2 \geq 0.$$

Therefore $[-2(\mu, \nu)_e]^2 \leq 4 \|\mu\|_e^2 \|\nu\|_e^2$. Also note that

$$\|\mu + \nu\|_e^2 = (\mu + \nu, \mu + \nu)_e = \|\mu\|_e^2 + 2(\mu, \nu)_e + \|\nu\|_e^2 \leq [\|\mu\|_e + \|\nu\|_e]^2.$$

THEOREM 11.6 The energy of $\mu \in \mathscr{E}$ is zero if and only if $\mu \sim 0$.

Proof. Suppose $\|\mu\|_e = 0$. Then $(\mu, \nu)_e = 0$ for all $\nu \in \mathscr{E}$. If $\mu = \mu_1 - \mu_2$ with $\mu_1, \mu_2 \in \mathscr{E}^+$, then this can be written $(\mu_1, \nu)_e = (\mu_2, \nu)_e$ for all $\nu \in \mathscr{E}$. For each x, δ such that $\bar{B}_{x,\delta} \subset R$ there is a signed measure $\tau_{x,\delta}$ of finite energy such that the class of functions $\{G\tau_{x,\delta}\}$ is a total subset of $\mathscr{K}_R^+$ (see Theorem 6.13). Since

$$\int G\tau_{x,\delta}\, d\mu_1 = \int G\mu_1\, d\tau_{x,\delta} = \int \mu_2\, d\tau_{x,\delta} = \int G\tau_{x,\delta}\, d\mu_2$$

and the set $\{G\tau_{x,\delta}\}$ is total in $\mathscr{K}_R^+$, $\mu_1 = \mu_2$ by Theorem 6.11 and $\mu = \mu_1 - \mu_2 \sim 0$. The converse is obvious.

From the Hilbert space point of view, $\mathscr{E}$ is a pre-Hilbert space in that it is a linear vector space over the reals with $(\mu, \nu)_e$ serving as an inner product and $\|\mu\|_e$ as a norm but lacks the completeness property as a metric space. The following example shows that $\mathscr{E}$ is not complete.

EXAMPLE 2 (*Cartan*) Let $R = E^n$, $n = 3$, and for each positive integer j let $\lambda_j = \mu_j - \nu_j$ be a signed measure where μ_j is a measure of total mass 1 uniformly distributed on a sphere with center at the origin and radius $1 - 4^{-j}$ and ν_j is similarly distributed on a concentric sphere of radius 1. Using Theorem 6.12,

$$(\lambda_j, \lambda_j)_e = \int G\mu_j\, d\mu_j - 2\int G\mu_j\, d\nu_j + \int G\nu_j\, d\nu_j$$

$$= \frac{1}{4^j - 1}.$$

It follows that the series $\sum_j \|\lambda_j\|_e$ converges. Since

$$\|\sum_{j=1}^{p+k} \lambda_j - \sum_{j=1}^{p} \lambda_j\|_e = \|\sum_{j=p+1}^{p+k} \lambda_j\|_e \to 0$$

as $p \to \infty$ independently of k, the sequence of partial sums $\sum_{j=1}^{p} \lambda_j$ forms a Cauchy sequence in the energy norm. We show, however, that there is no $\lambda \in \mathscr{E}$ such that $\lim_{p\to\infty} \|\sum_{j=1}^{p} \lambda_j - \lambda\|_e = 0$. Suppose that there were such a $\lambda \in \mathscr{E}$. We would then have $\lim_{p\to\infty} (\sum_{j=1}^{p} \lambda_j, v)_e = (\lambda, v)_e$ for all $v \in \mathscr{E}^+$, that is,

$$\lim_{p\to\infty} \int Gv \, d\left(\sum_{j=1}^{p} \lambda_j\right) = \int Gv \, d\lambda' - \int Gv \, d\lambda'',$$

where $\lambda = \lambda' - \lambda''$, $\lambda', \lambda'' \in \mathscr{E}^+$. In particular, this would hold for the potentials $G\tau_{x,\delta}$ of Theorem 6.13, $\bar{B}_{x,\delta} \subset B_{0,1}$. Let $\mathscr{K}_{0B}^+$ be the set of such potentials. Then $\mathscr{K}_{0B}^+$ forms a total subset of the class $\mathscr{K}_B^+$ of continuous functions on $B_{0,1}$ having compact support in $B_{0,1}$ by Theorem 6.13. Then for $f \in \mathscr{K}_{0B}^+$

$$\lim_{p\to\infty} \int f \, d\left(\sum_{j=1}^{p} \lambda_j\right) = \lim_{p\to\infty} \int f \, d\left(\sum_{j=1}^{p} \mu_j\right) = \int f \, d\lambda' - \int f \, d\lambda''.$$

Let f be a non-negative linear combination of elements of $\mathscr{K}_{0B}^+$ which majorizes 1 on the closure of a ball of radius $1 - 4^{-p}$ with $\|f\| \le 2$. Then

$$\sum_{j=1}^{p} \mu_j(B_{0,1}) \le \int f \, d\left(\sum_{j=1}^{p} \mu_j\right)$$

and

$$+\infty = \lim_{p\to\infty} \sum_{j=1}^{p} \mu_j(B_{0,1}) \le \lim_{p\to\infty} \int f \, d\left(\sum_{j=1}^{p} \mu_j\right) \le 2\lambda'(\bar{B}_{0,1}).$$

Therefore λ' is not a Borel measure and we have a contradiction.

An improved version of the Maria-Frostman domination principle is available which allows nonfinite potentials, provided we consider only measures of finite energy. The following theorem need not hold for measures not having finite energy, as can be seen by considering a unit measure with point support.

THEOREM **11.7** (*Cartan*) If $\mu \in \mathscr{E}^+$ and w is a superharmonic function such that $G\mu \le w$ a.e. (μ) on the support of μ, then $G\mu \le w$ on R.

Proof. The function $v = w \wedge G\mu$ is a potential Gv and $v \in \mathscr{E}^+$ since $Gv \le G\mu$. Since $\{G\mu > w\} = \{G\mu > Gv\}$ and $\mu(G\mu > w) = 0$, $\int (G\mu - Gv)\, d\mu = 0 \le \int (G\mu - Gv)\, dv$. It follows that

$$\|\mu - v\|_e^2 = (\mu, \mu)_e - 2(\mu, v)_e + (v, v)_e \le 0;$$

but in view of the fact that the energy is non-negative $\|\mu - v\|_e = 0$ and by Theorem 11.6 $\mu \sim v$. Since μ and v are measures, this means that $\mu = v$; that is $w \ge w \wedge G\mu = G\mu$.

Given a sequence $\{\mu_j\}$ in $\mathscr{E}^+$, there are three modes of convergence of the sequence to a measure μ: (i) **strong convergence in the energy norm**, that is, $\lim_{j\to\infty} \|\mu_j - \mu\|_e = 0$, (ii) **weak convergence** in the pre-Hilbert space, that is, $\lim_{j\to\infty} (\mu_j, \nu)_e = (\mu, \nu)_e$ for all $\nu \in \mathscr{E}$ (equivalently, for all $\nu \in \mathscr{E}^+$), and (iii) **convergence in the w*-topology**, that is, $\lim_{j\to\infty} \int f\, d\mu_j = \int f\, d\mu$ for all continuous functions f having compact support in R. Strong convergence implies weak convergence since for any $\nu \in \mathscr{E}$, $|(\mu_j, \nu)_e - (\mu, \nu)_e| \leq \|\mu_j - \mu\|_e \|\nu\|_e$. The limit μ in (i) and (ii) must, of course, belong to $\mathscr{E}$. The sequence $\{\mu_j\}$ in $\mathscr{E}^+$ is **energy bounded** if the norms $\|\mu_j\|_e$ are uniformly bounded. The following theorem is generally true in any pre-Hilbert space.

THEOREM **11.8** Weak convergence and strong convergence are equivalent for strong Cauchy sequences in $\mathscr{E}$.

Proof. Let $\{\mu_j\}$ be a strong Cauchy sequence in $\mathscr{E}$. Since $|\|\mu_j\|_e - \|\mu_k\|_e| \leq \|\mu_j - \mu_k\|_e$, there is a constant m such that $\|\mu_j\|_e \leq m$ for all j. We need show only that weak convergence of the sequence $\{\mu_j\}$ to $\mu \in \mathscr{E}$ implies that $\lim_{j\to\infty} \|\mu_j - \mu\|_e = 0$. Let $\{\lambda_j\}$ be any sequence in $\mathscr{E}$ which converges weakly to $\lambda \in \mathscr{E}$. Since $\|\lambda\|_e^2 = (\lambda, \lambda)_e \leq |(\lambda - \lambda_j, \lambda)_e| + |(\lambda_j, \lambda)_e| \leq |(\lambda - \lambda_j, \lambda)_e| + \|\lambda\|_e \times \|\lambda_j\|_e$, $\|\lambda\|_e \leq \limsup_{j\to\infty} \|\lambda_j\|_e$. For fixed k, the sequence $\{\mu_j - \mu_k\}$ converges weakly to $\mu - \mu_k$, and as we have just shown

$$\|\mu - \mu_k\|_e \leq \limsup_{j\to\infty} \|\mu_j - \mu_k\|_e .$$

Since the sequence $\{\mu_j\}$ is Cauchy, the right side goes to zero as $k \to \infty$.

THEOREM **11.9** If $\mu \in \mathscr{E}^+$ and $G\mu$ is the limit of a monotone sequence of potentials $\{G\mu_j\}$, $\mu_j \in \mathscr{E}^+$, then the sequence $\{\mu_j\}$ converges to μ strongly.

Proof. Suppose that $G\mu_j \leq G\mu_k$. Then

$$\begin{aligned}\|\mu_k - \mu_j\|_e^2 &= \|\mu_k\|_e^2 - 2\int G\mu_k\, d\mu_j + \|\mu_j\|_e^2 \\ &\leq \|\mu_k\|_e^2 - 2\int G\mu_j\, d\mu_j + \|\mu_j\|_e^2 \\ &= \|\mu_k\|_e^2 - \|\mu_j\|_e^2 .\end{aligned}$$

This shows that the sequence $\{\|\mu_j\|_e\}$ is monotone. If the sequence $\{G\mu_j\}$ increases, then $\|\mu_j\|_e \leq \|\mu\|_e$, but if the sequence $\{G\mu_j\}$ decreases, then $\|\mu_j\|_e \leq \|\mu_1\|_e$. In either case the sequence $\{\|\mu_j\|_e\}$ is bounded and is a Cauchy sequence. It follows from the above inequality that the sequence $\{\mu_j\}$ is Cauchy. By the preceding theorem it suffices to show that the sequence $\{\mu_j\}$ converges weakly to μ, that is,

$$\lim_{j\to\infty} \int G\mu_j \, d\nu = \int G\mu \, d\nu$$

for all $\nu \in \mathscr{E}^+$, but this follows from the Lebesgue dominated convergence theorem.

If $\lambda = \mu - \nu \in \mathscr{E}$ with $\mu, \nu \in \mathscr{E}^+$, we say that λ has compact support if μ and ν have compact supports. Strictly speaking, $G(\mu - \nu) = G\mu - G\nu$ may not be defined everywhere because of the infinities of $G\mu$ and $G\nu$. We write $G(\mu - \nu)$ only when $G\mu - G\nu$ can be defined everywhere.

THEOREM **11.10** The set of $\mu \in \mathscr{E}$ having compact support in R for which $G\mu \in \mathscr{K}_R$ is dense in $\mathscr{E}$.

Proof. Consider any $\mu \in \mathscr{E}^+$. Let $\{K_j\}$ be an increasing sequence of compact subsets of R such that $R = \cup_j K_j$ and let $\mu_j = \mu|_{K_j}$. Since $G\mu_j \leq G\mu$, $\mu_j \in \mathscr{E}^+$, μ_j has compact support and the sequence $\{\mu_j\}$ converges to μ strongly by the preceding theorem. This shows that the set $\mathscr{E}_{\mathscr{K}}^+$ of $\mu \in \mathscr{E}^+$ having compact support is dense in $\mathscr{E}^+$. Now consider any $\mu \in \mathscr{E}_{\mathscr{K}}^+$. By taking volume averages of $G\mu$ we can obtain a sequence $\{\mu_j\}$ in $\mathscr{E}_{\mathscr{K}}^+$ such that $G\mu_j \uparrow G\mu$ with $G\mu_j$ continuous on R. By the preceding theorem the sequence $\{\mu_j\}$ converges strongly to μ. This shows that the set of $\mu \in \mathscr{E}_{\mathscr{K}}^+$ with continuous $G\mu$ is dense in $\mathscr{E}^+$. Now consider any such μ and let $w = G\mu$. Let $\{R_j\}$ be an increasing sequence of regular regions with $\bar{R}_j \subset R$ such that $R = \cup_j R_j$ and let $H_w^{R_j}$ be the harmonic function on R_j corresponding to the boundary function $w|_{\partial R_j}$. Now let

$$w_j = \begin{cases} H_w^{R_j} & \text{on } R_j, \\ w & \text{on } R \sim R_j. \end{cases}$$

By Theorem 8.41 $w_j \downarrow 0$ on R as $j \to \infty$. Letting $w_j = G\mu_j$, $G\mu_j \leq G\mu$ implies that $\mu_j \in \mathscr{E}^+$ and, in fact, that $\mu_j \in \mathscr{E}_{\mathscr{K}}^+$. Since $G\mu_j \downarrow 0$, the sequence $\{\mu_j\}$ converges strongly to the zero measure. It follows that the sequence $\{\mu - \mu_j\}$ in $\mathscr{E}$ converges strongly to μ. Note that $G\mu - G\mu_j = w - w_j \in \mathscr{K}_R$ since w and w_j agree outside R_j. Therefore any $\mu \in \mathscr{E}^+$ can be approximated in the energy norm by a $\nu \in \mathscr{E}$ for which $G\nu \in \mathscr{K}_R$. If $\mu = \mu_1 - \mu_2$, where $\mu_i \in \mathscr{E}^+$, $i = 1, 2$, then μ_i can be approximated by ν_i in $\mathscr{E}$ having compact support and $G\nu_i \in \mathscr{K}_R$, $i = 1, 2$.

THEOREM **11.11** Let $\{\mu_j\}$ be a sequence in $\mathscr{E}^+$. Strong convergence of the sequence implies weak convergence and energy boundedness. If the sequence $\{\mu_j\}$ is energy bounded, weak convergence and convergence in the w^*-topology are equivalent.

Proof. If the sequence $\{\mu_j\}$ converges strongly to μ, then energy boundedness follows from the inequality $|\|\mu_j\|_e - \|\mu\|_e| \leq \|\mu_j - \mu\|_e$. Suppose the sequence $\{\mu_j\}$ is energy bounded and converges weakly to $\mu \in \mathscr{E}$. Let m be such that $\|\mu_j\|_e \leq m$ for all j. Consider the signed measures $\tau_{x,\delta}$ of Theorem 6.13. Since $G\tau_{x,\delta}$ is the Green potential of a signed measure of finite energy whenever $\bar{B}_{x,\delta} \subset R$,

$$\lim_{j\to\infty} \int G\tau_{x,\delta}\, d\mu_j = \int G\tau_{x,\delta}\, d\mu.$$

Therefore $\lim_{j\to\infty} \int f\, d\mu_j = \int f\, d\mu$ for f in a total subset $\mathscr{K}_0^+$ of $\mathscr{K}_R$. Consider any compact set $C \subset R$. We shall show that the sequence $\{\mu_j(C)\}$ is bounded. To do this let f^* be a non-negative linear combination of elements of $\mathscr{K}_0^+$ which majorizes 1 on C. Since $\int f^*\, d\mu_j \to \int f^*\, d\mu$ as $j \to \infty$ and $\mu_j(C) \leq \int f^*\, d\mu_j$, the sequence $\{\mu_j(C)\}$ is bounded. Now let $g \in \mathscr{K}_R^+$. If $\varepsilon > 0$ choose $f \in \mathscr{K}_0^+$ such that $\|g - f\| < \varepsilon$. Since f and g have compact supports, there is a compact set $C \subset R$ containing the supports of f and g and a constant α such that $\mu_j(C) \leq \alpha$ for all j. Since

$$\begin{aligned}\left|\int g\, d\mu_j - \int g\, d\mu\right| &\leq \int |g - f|\, d\mu_j + \left|\int f\, d\mu_j - \int f\, d\mu\right| + \int |f - g|\, d\mu \\ &\leq \varepsilon\alpha + \left|\int f\, d\mu_j - \int f\, d\mu\right| + \varepsilon\mu(C)\end{aligned}$$

and $f \in \mathscr{K}_0^+$, $\lim_{j\to\infty} \int g\, d\mu_j = \int g\, d\mu$; that is, the sequence $\{\mu_j\}$ converges to μ in the w^*-topology. Assume $\|\mu_j\|_e \leq m$ for all j and the sequence $\{\mu_j\}$ converges to μ in the w^*-topology. This assumption implies that $G\mu \leq \liminf_{j\to\infty} G\mu_j$ by Lemma 7.36. By Fatou's lemma for each positive integer p

$$\int G\mu\, d\mu_p \leq \int \left(\liminf_{j\to\infty} G\mu_j\right) d\mu_p \leq \liminf_{j\to\infty} \int G\mu_j\, d\mu_p \leq m^2.$$

Therefore

$$\int G\mu\, d\mu \leq \int \left(\liminf_{p\to\infty} G\mu_p\right) d\mu \leq \liminf_{p\to\infty} \int G\mu\, d\mu_p \leq m^2$$

and $\mu \in \mathscr{E}^+$. To show that $\lim_{j\to\infty} (\mu_j, \nu)_e = (\mu, \nu)_e$ for all $\nu \in \mathscr{E}$, it suffices to establish this for all ν in a dense subset of $\mathscr{E}$, but by theorem 11.10 the set of $\nu \in \mathscr{E}$ having compact support in R and $G\nu \in \mathscr{K}_R$ is dense in $\mathscr{E}$. For such ν

$$\lim_{j\to\infty} (\mu_j, \nu)_e = \lim_{j\to\infty} \int G\nu\, d\mu_j = \int G\nu\, d\mu = (\mu, \nu)_e$$

by the w^*-convergence.

THEOREM **11.12** (*Cartan*) $\mathscr{E}^+$ is complete.

Proof. Let $\{\mu_j\}$ be a sequence in $\mathscr{E}^+$ which is Cauchy in the energy norm. For any $\nu \in \mathscr{E}$ the inequality

$$|(\mu_j, \nu)_e - (\mu_k, \nu)_e| \le \|\mu_j - \mu_k\|_e \|\nu\|_e$$

implies that the sequence $\{(\mu_j, \nu)_e\}$ is Cauchy. Taking $\nu = \tau_{x,\delta}$, where $\bar{B}_{x,\delta} \subset R$, this means that $L(f) = \lim_{j\to\infty} \int f\, d\mu_j$ exists for each f in a total subset $\mathscr{K}_0^+$ of $\mathscr{K}_R^+$ and therefore for all f in $\mathscr{L}_0^+$, the set of linear combinations with positive coefficients of elements of $\mathscr{K}_0^+$. The linear functional $L(f)$ can be extended to $\mathscr{K}_R^+$ as follows. Consider any $g \in \mathscr{K}_R^+$ having compact support C and a neighborhood W of C having compact closure $\overline{W} \subset R$. As in the proof of Theorem 11.11, there is a constant α such that $\mu_j(\overline{W}) \le \alpha$ for all j. From the definition of a total set, there is a sequence $\{g_k\}$ in $\mathscr{L}_0^+$, with the g_k's having support in W, such that $\|g_k - g\| \to 0$ as $k \to \infty$. Since

$$\begin{aligned} |L(g_p) - L(g_q)| &= \lim_{j\to\infty} \left| \int g_p\, d\mu_j - \int g_q\, d\mu_j \right| \\ &\le \alpha \|g_p - g_q\|, \end{aligned}$$

we can define $L(g) = \lim_{j\to\infty} L(g_j)$. It follows that there is a measure μ such that

$$\lim_{j\to\infty} \int f\, d\mu_j = \int f\, d\mu \quad \text{for all} \quad f \in \mathscr{K}_R^+;$$

that is, the sequence $\{\mu_j\}$ converges to μ in the w^*-topology. Since the sequence is energy bounded, it converges weakly to μ by Theorem 11.11. Strong convergence then follows from Theorem 11.8.

3. Balayage of Measures

The pre-Hilbert space structure of $\mathscr{E}$ can be used to develop the balayage of potentials. This is done by projecting a measure $\mu \in \mathscr{E}^+$ on a convex subset of $\mathscr{E}^+$. Recall that a subset $\mathfrak{F}$ of $\mathscr{E}$ is **convex** if $\rho x + (1-\rho)y \in \mathfrak{F}$ whenever $x, y \in \mathfrak{F}$ and $0 < \rho < 1$.

LEMMA **11.13** If $\mathfrak{F}$ is a nonempty closed convex subset of $\mathscr{E}^+$ and $\mu \in \mathscr{E}^+$, then there is a unique measure $\mu_{\mathfrak{F}} \in \mathfrak{F}$ such that $\|\mu - \mu_{\mathfrak{F}}\|_e < \|\mu - \nu\|_e$ for all $\nu \in \mathfrak{F}$, $\nu \ne \mu_{\mathfrak{F}}$.

Proof. Consider any two elements $\alpha, \beta \in \mathfrak{F}$. Then $\gamma = \frac{1}{2}(\alpha + \beta) \in \mathfrak{F}$, since $\mathfrak{F}$ is convex. Now

$$\begin{aligned} \|\mu - \gamma\|_e^2 = \|(\mu - \alpha) + (\alpha - \gamma)\|_e^2 &= \|\mu - \alpha\|_e^2 + \|\alpha - \gamma\|_e^2 + 2(\mu - \alpha, \alpha - \gamma)_e \\ &= \|\mu - \alpha\|_e^2 + \tfrac{1}{4}\|\alpha - \beta\|_e^2 + (\mu - \alpha, \alpha - \beta)_e. \end{aligned}$$

Similarly,

$$\|\mu - \gamma\|_e^2 = \|\mu - \beta\|_e^2 = \tfrac{1}{4}\|\alpha - \beta\|_e^2 + (\mu - \beta, \beta - \alpha)_e$$

Therefore

$$\begin{aligned} 2\|\mu - \gamma\|_e^2 &= \|\mu - \alpha\|_e^2 + \|\mu - \beta\|_e^2 + \tfrac{1}{2}\|\alpha - \beta\|_e^2 + (\beta - \alpha, \alpha - \beta)_e \\ &= \|\mu - \alpha\|_e^2 + \|\mu - \beta\|_e^2 - \tfrac{1}{2}\|\alpha - \beta\|_e^2. \end{aligned}$$

Now let $m = \inf_{\nu \in \mathfrak{F}} \|\mu - \nu\|_e$. If $\varepsilon > 0$ and α, β are such that $\|\mu - \alpha\|_e^2 < m^2 + \varepsilon$ and $\|\mu - \beta\|_e^2 < m^2 + \varepsilon$, then

$$2m^2 \leq 2m^2 + 2\varepsilon - \tfrac{1}{2}\|\alpha - \beta\|_e^2,$$

since $\|\mu - \gamma\|_e \geq m$. If $\|\mu - \alpha\|_e = \|\mu - \beta\|_e = m$, then $\alpha = \beta$; that is, there is at most one α which minimizes $\|\mu - \nu\|_e$ for $\nu \in \mathfrak{F}$. Moreover, if $\{\nu_j\}$ is a sequence in $\mathfrak{F}$ such that $\|\mu - \nu_j\|_e \downarrow m$, then this argument shows that $\{\nu_j\}$ is a Cauchy sequence in $\mathfrak{F}$. Since $\mathscr{E}^+$ is complete and $\mathfrak{F}$ is closed, there is an $\alpha \in \mathfrak{F}$ such that

$$\|\mu - \alpha\|_e = m = \inf_{\nu \in \mathfrak{F}} \|\mu - \nu\|_e.$$

Definition. If $\mathfrak{F}$ is a nonempty closed convex subset of $\mathscr{E}^+$ and $\mu \in \mathscr{E}^+$, the measure $\mu_{\mathfrak{F}}$ of the preceding lemma is called the **projection** of μ on $\mathfrak{F}$.

LEMMA **11.14** Let $\mathfrak{F}$ be a nonempty closed convex subset of $\mathscr{E}^+$ and $\mu \in \mathscr{E}^+$. Then μ_0 is the projection of μ on $\mathfrak{F}$ if and only if $(\mu - \mu_0, \nu - \mu_0)_e \leq 0$ for all $\nu \in \mathfrak{F}$.

Proof. Let α be any element of $\mathfrak{F}$. First note that $\|\mu - \alpha\|_e^2 \leq \|\mu - \nu\|_e^2$ for all $\nu \in \mathfrak{F}$ if and only if $2(\mu - \alpha, \nu - \alpha)_e \leq \|\alpha - \nu\|_e^2$ for all $\nu \in \mathfrak{F}$. We show that the latter holds if and only if $2(\mu - \alpha, \nu - \alpha)_e \leq 0$ for all $\nu \in \mathfrak{F}$. The sufficiency is trivial. Suppose $2(\mu - \alpha, \nu - \alpha)_e \leq \|\alpha - \nu\|_e^2$ for all $\nu \in \mathfrak{F}$. If $0 < \rho < 1$ and $\nu \in \mathfrak{F}$, then $\nu^* = (1 - \rho)\alpha + \rho\nu = \alpha + \rho(\nu - \alpha) \in \mathfrak{F}$ and

$$2(\mu - \alpha, \rho(\nu - \alpha))_e \leq \|\rho(\nu - \alpha)\|_e^2$$

or

$$2(\mu - \alpha, \nu - \alpha)_e \leq \rho\|\nu - \alpha\|_e^2.$$

Letting $\rho \to 0$, we obtain $2(\mu - \alpha, \nu - \alpha)_e \leq 0$.

Recall that a convex set $\mathfrak{F}$ is a **cone with vertex** α if $\alpha + \beta \in \mathfrak{F}$ implies $\alpha + \rho\beta \in \mathfrak{F}$ for all $\rho \geq 0$.

LEMMA **11.15** Let $\mathfrak{F}$ be a closed cone in $\mathscr{E}^+$ with vertex 0 and let $\mu \in \mathscr{E}^+$. Then μ_0 is the projection of μ on $\mathfrak{F}$ if and only if

(i) $(\mu - \mu_0, \nu)_e \leq 0$ for all $\nu \in \mathfrak{F}$,
(ii) $(\mu - \mu_0, \mu_0)_e = 0$.

Proof. Suppose that μ_0 is the projection of μ on $\mathfrak{F}$. Then

$$(\mu - \mu_0, \nu - \mu_0)_e \leq 0$$

for all $\nu \in \mathfrak{F}$ by the preceding lemma. For any $\nu \in \mathfrak{F}$, $\mu_0 + \nu \in \mathfrak{F}$ and $(\mu - \mu_0, \nu)_e \leq 0$. This proves (i). In particular, $(\mu - \mu_0, \mu_0)_e \leq 0$. Putting $\nu = 0$ in the inequality $(\mu - \mu_0, \nu - \mu_0)_e \leq 0$ results in $(\mu - \mu_0, \mu_0)_e \geq 0$ and (ii) is proved. Conversely, suppose μ_0 satisfies (i) and (ii). Then $(\mu - \mu_0, \nu - \mu_0)_e = (\mu - \mu_0, \nu)_e - (\mu - \mu_0, \mu_0)_e \leq 0$ and μ_0 is the projection of μ on $\mathfrak{F}$ by the preceding lemma.

If K is a compact subset of R, $\mathfrak{F}_K$ denotes the set of measures $\nu \in \mathscr{E}^+$ with support in K. $\mathfrak{F}_K$ is easily seen to be a cone in $\mathscr{E}^+$ with vertex 0. To show that $\mathfrak{F}_K$ is closed we make use of the fact that a convex subset of a pre-Hilbert space is strongly closed if and only if it is weakly closed. This theorem can be found in any book on functional analysis. Consider a sequence $\{\mu_j\}$ in $\mathfrak{F}_K$ which converges strongly to $\mu \in \mathscr{E}^+$. This implies energy boundedness and weak convergence of the sequence. In the presence of energy boundedness, weak convergence and w^*-convergence are equivalent by Theorem 11.11. Since the μ_j's have support in K, the same is true of μ by virtue of the w^*-convergence; that is, $\mathfrak{F}_K$ is closed. According to Lemma 11.13, it makes sense to speak of the projection $\mu_{\mathfrak{F}_K}$ of $\mu \in \mathscr{E}^+$ on $\mathfrak{F}_K$.

THEOREM **11.16** If K is a compact subset of R, $\mathfrak{F}_K$, the closed cone of measures in $\mathscr{E}^+$ having support in K, and $\mu \in \mathscr{E}^+$, then μ_0 is the projection of μ on $\mathfrak{F}_K$ if and only if

(i) $G\mu_0 \leq G\mu$ on R,
(ii) $G\mu_0 = G\mu$ on K except possibly for a polar set.

Proof. Suppose μ_0 is the projection of μ on $\mathfrak{F}_K$. By the preceding lemma, μ_0 satisfies

(i′) $\int (G\mu - G\mu_0)\, d\nu \leq 0$ for all $\nu \in \mathfrak{F}_K$,
(ii′) $\int (G\mu - G\mu_0)\, d\mu_0 = 0$.

Consider the set $\{G\mu > G\mu_0\}$ and any $\nu \in \mathfrak{F}_K$. Since the restriction of ν to this set also belongs to $\mathfrak{F}_K$,

$$\int_{\{G\mu > G\mu 0\}} (G\mu - G\mu_0)\, d\nu \leq 0.$$

This means that $\nu(G\mu > G\mu_0) = 0$ for all $\nu \in \mathfrak{F}_K$. In particular, $G\mu \leq G\mu_0$ a.e. (μ_0). In conjunction with (ii′), this means that $G\mu = G\mu_0$ a.e. (μ_0) on the support of μ_0. By Theorem 11.7. $G\mu_0 \leq G\mu$ on R. This shows that μ_0 satisfies (i). Since $G\mu \leq G\mu_0$ a.e. (ν) for all $\nu \in \mathfrak{F}_K$ and $G\mu_0 \leq G\mu$, the set of points $K \cap \{G\mu \neq G\mu_0\}$ has measure zero for all $\nu \in \mathfrak{F}_K$ and therefore for all $\nu \in \mathscr{E}^+$ since the restriction of such ν to K belongs to $\mathfrak{F}_K$. An application of Theorem 11.2 verifies that μ_0 satisfies (ii). Conversely, suppose μ_0 satisfies (i) and (ii). Then (ii) implies that $G\mu_0 = G\mu$ a.e. (ν) for all $\nu \in \mathfrak{F}_K$. Properties (i′) and (ii′) follow immediately.

COROLLARY **11.17** If $\mu \in \mathscr{E}^+$ and K is a compact subset of R, then $G\mu_{\mathfrak{F}_K} = \hat{\mathbf{R}}_K^{G\mu}$.

Proof. Since $G\mu \geq \hat{\mathbf{R}}_K^{G\mu}$, $\hat{\mathbf{R}}_K^{G\mu}$ is the potential of a measure $\mu_0 \in \mathscr{E}^+$. By Corollary 7.40 $G\mu_0 = \hat{\mathbf{R}}_K^{G\mu} = G\mu$ on K, except possibly for a polar set. It follows from the preceding theorem that $\mu_0 = \mu_{\mathfrak{F}_K}$; that is, $\hat{\mathbf{R}}_K^{G\mu} = G\mu_{\mathfrak{F}_K}$.

If $\mu \in \mathscr{E}^+$ and K is a compact subset of R, the projection $\mu_{\mathfrak{F}_K}$ of μ on $\mathfrak{F}_K$ is known as the balayage of μ. For measures μ of finite energy, the operation of taking the balayage of the superharmonic function $G\mu$ is equivalent to taking the potential of the balayage of the measure.

The Gauss-Frostman theorem, Theorem 11.3, can be easily generalized to potentials $G\mu$, $\mu \in \mathscr{E}^+$, by defining

$$\Phi_{G\mu}(\nu) = \int G\nu \, d\nu - 2 \int G\mu \, d\nu.$$

THEOREM **11.18** Let $\mathfrak{F}$ be a nonempty, closed, convex set in $\mathscr{E}^+$ and let $\mu \in \mathscr{E}^+$. Then $\mu_{\mathfrak{F}}$ is the unique measure that minimizes $\Phi_{G\mu}$ over $\mathfrak{F}$.

Proof. $\|\mu - \nu\|_e^2 = \Phi_{G\mu}(\nu) + \|\mu\|_e^2$.

4. Capacity and Energy

We now relate energy and the capacity of a compact subset $K \subset R$. If K is a nonpolar compact subset of R, let $\mathfrak{A}_K$ be the set of measures $\nu \in \mathscr{E}^+$ having support in K for which $\nu(K) = 1$. $\mathfrak{A}_K$ is nonempty by Theorem 11.2. The proof that $\mathfrak{F}_K$ is a closed convex set in $\mathscr{E}^+$ is easily modified to show that $\mathfrak{A}_K$ is also a closed convex set in $\mathscr{E}^+$. Letting 0 denote the zero measure on R, $0_{\mathfrak{A}_K}$ denotes the projection of 0 on $\mathfrak{A}_K$. As usual, μ_K denotes the capacitary distribution on K.

THEOREM **11.19** If K is a nonpolar compact subset of R, then $0_{\mathfrak{A}_K}$ is the unique measure in $\mathfrak{A}_K$ having minimum energy,

$$0_{\mathfrak{A}_K} = \frac{\mu_K}{\mathscr{C}(K)} \quad \text{and} \quad \|0_{\mathfrak{A}_K}\|_e^2 = \frac{1}{\mathscr{C}(K)}.$$

Proof. Since $0_{\mathfrak{A}_K}$ is the unique measure in $\mathfrak{A}_K$ minimizing $\|0 - \nu\|_e = \|\nu\|_e$ for $\nu \in \mathfrak{A}_K$, $0_{\mathfrak{A}_K}$ has minimum energy among elements of $\mathfrak{A}_K$. It remains only to relate $0_{\mathfrak{A}_K}$ to μ_K. Let μ be a measure having compact support such that $G\mu = 1$ on K and $G\mu \leq 1$ on R. Such a measure can be obtained by letting $G\mu = \hat{\mathbf{R}}_U^1$, where U is a neighborhood of K having compact closure $\overline{U} \subset R$. Then $\mu_{\mathfrak{F}_K}$ is just μ_K by Corollary 11.17. According to Theorem 11.18, μ_K is the unique measure in $\mathfrak{F}_K$, minimizing

$$\Phi_{G\mu}(\nu) = \int (G\nu - 2G\mu)\, d\nu$$

over $\mathfrak{F}_K$. Since $G\mu = 1$ on K, which contains the supports of the measures $\nu \in \mathfrak{F}_K$, μ_K is the unique measure in $\mathfrak{F}_K$ which minimizes Gauss's integral:

$$\phi_1(\nu) = \int (G\nu - 2)\, d\nu = \Phi_0(\nu) - 2\nu(K)$$

over $\mathfrak{F}_K$. Consider those $\nu \in \mathfrak{F}_K$ for which $\nu(K) = \rho$. Now $\Phi_0(\nu) = \int G\nu\, d\nu$ is minimized by $\nu = \rho 0_{\mathfrak{A}_K}$. The minimum of $\Phi_0(\nu)$, subject to the condition $\nu(K) = \rho$, is then $\rho^2 \|0_{\mathfrak{A}_K}\|_e^2$. In order to minimize $\Phi_{G\mu}(\nu)$ over $\mathfrak{F}_K$, it is then a matter of choosing ρ_0 to minimize $\rho^2 \|0_{\mathfrak{A}_K}\|_e^2 - 2\rho$. It follows that $\rho_0 = 1/\|0_{\mathfrak{A}_K}\|_e^2$ and that $\mu_K = 0_{\mathfrak{A}_K}/\|0_{\mathfrak{A}_K}\|_e^2$. Since $\mathscr{C}(K) = \mu_K(K)$ and $0_{\mathfrak{A}_K}(K) = 1$, $\mathscr{C}(K) = 1/\|0_{\mathfrak{A}_K}\|_e^2$ and $\mu_K = \mathscr{C}(K) 0_{\mathfrak{A}_K}$.

CHAPTER 12

Martin Boundary

In this chapter we re-examine the Poisson integral representation of a positive harmonic function on a ball by means of a measure on the boundary. In order to do this, we shall embed the connected open set $R \subset E^n$ in a compact space R_M^* called the Martin compactification of R. The set $\Delta_M = R_M^* \sim R$ is then called the Martin boundary of R. It is shown that each positive harmonic function on R has an integral representation relative to a measure on Δ_M.

The basic idea of the construction of the Martin boundary is as follows: let $\{R_j\}$ be an increasing sequence of open sets with compact closures $\bar{R}_j \subset R$ such that $R = \cup R_j$. Consider a positive harmonic function h on R and the balayage $\hat{\mathbf{R}}_{R_j}^h$. We know that $h = \hat{\mathbf{R}}_{R_j}^h$ on R_j and that there is a measure μ_j, with support in ∂R_j, such that

$$\hat{\mathbf{R}}_{R_j}^h(x) = \int G(y, x)\, d\mu_j(y), \qquad x \in R.$$

Consider a fixed $x \in R$. Since, roughly speaking, the function $G(x, y)$, $y \in \partial R_j$, tends to zero as $j \to \infty$, the measures must increase to keep the above integral equal to $\hat{\mathbf{R}}_{R_j}^h(x)$; that is, the sequence $\{\mu_j\}$ is unbounded and one cannot hope to obtain a measure μ as some kind of limit of the μ_j's. Martin, however, circumvented this difficulty as follows. Let x_0 be a fixed point of R_1. Then

$$\begin{aligned}\hat{\mathbf{R}}_{R_j}^h(x) &= \int \frac{G(y, x)}{G(y, x_0)}\, G(y, x_0)\, d\mu_j(y)\\ &= \int M(y, x)\, d\mu_j^*(y),\end{aligned}$$

where $M(y, x) = G(y, x)/G(y, x_0)$ and for each Borel set $E \subset R$

$$\mu_j^*(E) = \int_E G(y, x_0)\, d\mu_j(y).$$

Note that the sequence $\{\mu_j^*\}$ is bounded since $\mu_j^*(R) = \hat{\mathbf{R}}_{R_j}^h(x_0) = h(x_0) < +\infty$. Now, at least, one can define a measure μ as a w^*-limit of the sequence $\{\mu_j^*\}$ or a subsequence thereof. Since the support of μ_j^* expands as $j \to \infty$, we must provide R with a boundary by embedding R in a larger space R_M^*, preferably compact. If the functions $M(\cdot, x)$ can be extended so as to be continuous on R_M^*, then one is in a position to let $j \to \infty$ in the above representation to obtain

$$h(x) = \int M(y, x)\, d\mu(y).$$

In working out the details, we shall first examine the problem of constructing R_M^* so that the functions $M(\cdot, x)$ can be continuously extended from R to R_M^*.

1. Martin Boundary

Throughout this chapter, R will be a connected open subset of E^n which has a Green function G. Connectedness is imposed in order to have $G > 0$ on $R \times R$. Moreover, $\Sigma(R)$ will denote the collection of extended real-valued functions on R which are continuous in the extended sense.

Definition. A compact topological space R^* is a **compactification** of R if (i) R is a dense subset of R^* and (ii) the relative topology on R coincides with the original topology. The set $\Delta = R^* \sim R$ is called the **ideal boundary** of R relative to R^*.

Definition. Let $\mathscr{Q} \subset \Sigma(R)$. A compactification $R_{\mathscr{Q}}^*$ of R is called a $\mathscr{Q}$-**compactification** if

(i) each $f \in \mathscr{Q}$ has a continuous extension f^* to $R_{\mathscr{Q}}^*$
(ii) the functions $f^*, f \in \mathscr{Q}$, separate the points of $\Delta_{\mathscr{Q}} = R_{\mathscr{Q}}^* \sim R$.

THEOREM **12.1** (*Constantinescu and Cornea*) If $\mathscr{Q} \subset \Sigma(R)$, then R has a $\mathscr{Q}$-compactification which is uniquely determined up to a homeomorphism.

Proof. Let $\mathscr{K}$ be the collection of continuous functions on R with compact support and let $I = [-\infty, +\infty]$. I is a compact space if to the usual topology on $(-\infty, +\infty)$ we adjoin intervals of the type $[-\infty, a)$ and $(a, +\infty]$. If for each $f \in \mathscr{Q} \cup \mathscr{K}$ we let $I_f = I$, then $I^{\mathscr{Q} \cup \mathscr{K}} = \prod_{f \in \mathscr{Q} \cup \mathscr{K}} I_f$ is a compact space if provided with the usual product topology. As is well-known, the projection map $\pi_f : I^{\mathscr{Q} \cup \mathscr{K}} \rightsquigarrow I_f$ is continuous for each $f \in \mathscr{Q} \cup \mathscr{K}$. Consider the mapping $\psi : R \rightsquigarrow I^{\mathscr{Q} \cup \mathscr{K}}$ defined by $\psi(x) = \{f(x)\}_{f \in \mathscr{Q} \cup \mathscr{K}}, x \in R$. It is easily seen that ψ

is continuous on R. Clearly, $\pi_f \circ \psi = f$ for $f \in \mathscr{Q} \cup \mathscr{K}$. Consider any two points $x, y \in R$ with $x \neq y$. Then there is a function $f \in \mathscr{K} \subset \mathscr{Q} \cup \mathscr{K}$ such that $f(x) \neq f(y)$ and it follows that $\psi(x) = \{f(x)\}_{f \in \mathscr{Q} \cup \mathscr{K}} \neq \{f(y)\}_{f \in \mathscr{Q} \cup \mathscr{K}} = \psi(y)$. Therefore, ψ is a one-to-one map. Now let $R_{\mathscr{Q}}^* = \overline{\psi(R)}$, the closure being relative to the topology on $I^{\mathscr{Q} \cup \mathscr{K}}$, and endow $R_{\mathscr{Q}}^*$ with the topology induced by that on $I^{\mathscr{Q} \cup \mathscr{K}}$. Then $\psi(R)$ is a dense subset of $R_{\mathscr{Q}}^*$. We now show that R is homeomorphic to $\psi(R) \subset R_{\mathscr{Q}}^*$. Since we know that ψ is a continuous and one-to-one map onto $\psi(R)$, we need only show that ψ is an open mapping. Let U be an open subset of R with compact closure in R and let $x \in U$. Then there is an $f_0 \in \mathscr{K}$ such that $f_0(x) \neq 0$ and having compact support in U. Define $V^* = \{y^* \in R_{\mathscr{Q}}^* : \pi_{f_0}(y^*) \neq 0\}$. Note that $\psi(x) \in V^*$ since $\pi_{f_0}(\psi(x)) = f_0(x) \neq 0$. Note also that V^* is open in $R_{\mathscr{Q}}^*$ since π_{f_0} is continuous. Since $\bar{U}$ is a compact subset of R and ψ is continuous, $\psi(\bar{U})$ is a compact subset of $R_{\mathscr{Q}}^*$ and $V^* \sim \psi(\bar{U})$ is open. Suppose $(V^* \sim \psi(\bar{U})) \cap \psi(R) \neq \varnothing$. Then there is a $y \in R \sim \bar{U}$ such that $\{f(y)\}_{f \in \mathscr{Q} \cup \mathscr{K}} \in V^*$; in particular, $f_0(y) \neq 0$. But since f_0 has support in U and $y \in R \sim \bar{U}$, $f_0(y) = 0$, a contradiction. Therefore

$$(V^* \sim \psi(\bar{U})) \cap \psi(R) = \varnothing.$$

But since $V^* \sim \psi(\bar{U})$ is open in $R_{\mathscr{Q}}^*$ and $\psi(R)$ is dense in $R_{\mathscr{Q}}^*$, $V^* \sim \psi(\bar{U}) = \varnothing$ or $V^* \subset \psi(\bar{U})$. It follows that $\psi^{-1}(V^*) \subset \bar{U}$ and since $\psi^{-1}(V^*)$ is open, $\psi^{-1}(V^*) \subset U$. Therefore $V^* \subset \psi(U)$ and ψ is an open mapping onto $\psi(R)$. This proves that ψ is a homeomorphism from R onto $\psi(R)$. If corresponding points of R and $\psi(R)$ are identified, then $R_{\mathscr{Q}}^*$ is a compactification of R and R is a dense open subset of $R_{\mathscr{Q}}^*$. Since the point $x \in R$ is identified with the point $\psi(x) \in \psi(R)$, the function π_f can be considered as a continuous extension of f since $\pi_f(\psi(x)) = f(x)$. Now consider two different points x^*, $y^* \in R_{\mathscr{Q}}^* \sim R$. Since $R_{\mathscr{Q}}^* = \bar{R}$, there are nets $\{x_\alpha^* : \alpha \in \mathfrak{A}\}$ and $\{y_\beta^* : \beta \in \mathfrak{B}\}$ in R such that $\lim_{\mathfrak{A}} x_\alpha^* = x^*$ and $\lim_{\mathfrak{B}} y_\beta^* = y^*$. Suppose $f \in \mathscr{K}$ has compact support K. Then $R_{\mathscr{Q}}^* \sim K$ is a neighborhood of both x^* and y^* on which π_f vanishes. It follows that $\lim_{\mathfrak{A}} \pi_f(x_\alpha^*) = \pi_f(x^*) = \pi_f(y^*) = \lim_{\mathfrak{B}} \pi_f(y_\beta^*) = 0$ for all $f \in \mathscr{K}$. Since $x^* \neq y^*$, there is an $f \in \mathscr{Q}$ such that $\pi_f(x^*) \neq \pi_f(y^*)$. This completes the proof that $R_{\mathscr{Q}}^*$ is a $\mathscr{Q}$-compactification of R. Now let $\tilde{R}^*$ be any other $\mathscr{Q}$-compactification of R. Let $\tilde{\psi}$ be the map from $\tilde{R}^*$ to $I^{\mathscr{Q} \cup \mathscr{K}}$ defined by $\tilde{\psi}(x^*) = \{\tilde{f}^*(x^*)\}_{f \in \mathscr{Q} \cup \mathscr{K}}$ where $\tilde{f}^*$ is the continuous extension of f to $\tilde{R}^*$. Since $\tilde{\psi}$ is continuous and $\tilde{R}^*$ is compact, $\tilde{\psi}(\tilde{R}^*)$ is a compact subset of $I^{\mathscr{Q} \cup \mathscr{K}}$. From $R \subset \tilde{R}^*$, we have $\overline{\tilde{\psi}(R)} \subset \tilde{\psi}(\tilde{R}^*)$. Since R is dense in $\tilde{R}^*$, we actually have $\overline{\tilde{\psi}(R)} = \tilde{\psi}(\tilde{R}^*)$. In view of the fact that the functions $\tilde{f}^*$, $f \in \mathscr{Q} \cup \mathscr{K}$, separate the points of $\tilde{R}^*$, $\tilde{\psi}$ is a one-to-one continuous map of $\tilde{R}^*$ onto $\overline{\tilde{\psi}(R)}$. Since $\tilde{R}^*$ is compact, $\tilde{\psi}$ is actually a homeomorphism and this shows that $\tilde{R}^*$ is homeomorphic to $\overline{\tilde{\psi}(R)}$. But note that $\psi(R) = \tilde{\psi}(R)$. Therefore, $\tilde{R}^*$ is homeomorphic to $\overline{\tilde{\psi}(R)} = \overline{\psi(R)} = R_{\mathscr{Q}}^*$.

Recall that R is a connected open subset of E^n having a Green function G. In this chapter we consider R as a subset of E^n_Δ and ∂R denotes its boundary relative to E^n_Δ. The point at infinity Δ may belong to ∂R but not to R.

The development of the Martin boundary will require establishing bounds for ratios of the form $\hat{\mathbf{R}}^w_V/\hat{\mathbf{R}}^1_V$ in terms of bounds on w on ∂V. More specifically, let F be a nonpolar compact subset of R and let $V = R \sim F$. If w is any finite-valued positive superharmonic function on R, we let w_V denote the function on $\partial V = \partial F \cup \partial R$ by

$$w_V = \begin{cases} w & \text{on} \quad \partial F, \\ 0 & \text{on} \quad \partial R. \end{cases}$$

If $w|_{\partial F}$ is a bounded continuous function on ∂F, then w_V is a bounded continuous function on ∂V and as such is resolutive for the open set V with the Dirichlet solution $H^V_{w_V}$. By Theorem 9.25 $H^V_{w_V} = \hat{\mathbf{R}}^w_F$ on V. Since F is nonpolar, ∂F is nonpolar and some point $x \in \partial F$ must be a regular boundary point for V; since $\lim_{\substack{y\to x\\ y\in V}} \hat{\mathbf{R}}^1_F(y) = \lim_{\substack{y\to x\\ y\in V}} H^V_{1_V}(y) = 1$, $\hat{\mathbf{R}}^1_F > 0$ on V. If $a \le w \le b$ on ∂F, then $a1_V \le w_V \le b1_V$ on ∂V and $aH^V_{1_V} = H^V_{a1_V} \le H^V_{w_V} \le H^V_{b1_V} = bH^V_{1_V}$ on V. It follows that $a\hat{\mathbf{R}}^1_F \le \hat{\mathbf{R}}^w_F \le b\hat{\mathbf{R}}^1_F$ on V and that $\hat{\mathbf{R}}^w_F/\hat{\mathbf{R}}^1_F$ has the same bounds on V that w has on ∂F. Moreover, if $a > 0$, then $\hat{\mathbf{R}}^w_F > 0$ on V and $\hat{\mathbf{R}}^1_F/\hat{\mathbf{R}}^w_F \le a^{-1}$ on V.

LEMMA **12.2** If K is a compact subset of R, W is a neighborhood of K with $\overline{W} \subset R$, and $x_0 \in K$, then $\sup\{G(y, x)/G(y, x_0) : x \in K, y \in R \sim W\} < +\infty$.

Proof. Let U be a neighborhood of K with compact closure $F = \overline{U} \subset W$. Then ∂F is not polar. Now G is continuous on $\partial F \times K$ by Theorem 5.26. Since G is strictly positive, there are positive constants a and b such that $a \le G(y, x) \le b$ for any $y \in \partial F$ and $x \in K$. Let x be a fixed point of K and let $V = R \sim F$. We need the fact that $\hat{\mathbf{R}}^{G(\cdot, x)}_F = G(\cdot, x)$ on V. For $y \in \partial F$, $\lim_{\substack{z\to y\\ z\in V}} G(z, x) = G(y, x)$; for $y \in \partial R$, $\lim_{\substack{z\to y\\ z\in V}} G(z, x) = 0$ except possibly on a polar set by Theorem 8.30. We also have $\lim_{\substack{z\to y\\ z\in V}} H^V_{G_V(\cdot, x)}(z) = G(y, x)$ for $y \in \partial F$ except possibly for a polar set and $\lim_{\substack{z\to y\\ z\in V}} H^V_{G_V(\cdot, x)}(z) = 0$ for $y \in \partial R$ except possibly for a polar set according to the discussion following Theorem 9.24. Therefore

$$\lim_{\substack{z\to y\\ z\in V}} [G(z, x) - H^V_{G_V(\cdot, x)}(z)] = 0$$

for $y \in \partial V$, except possibly for a polar set. Since $G(\cdot, x)$ is bounded outside any ball about x by Lemma 5.8, the function in brackets is bounded on V;

as it is harmonic on V, it is identically zero on U by Corollary 9.20; that is, $G(\cdot, x) = H^V_{G_V(\cdot, x)} = \hat{\mathbf{R}}_F^{G(\cdot, x)}$ on V. By the remarks preceding this lemma, for $x \in K$

$$0 < \frac{G(\cdot, x)}{G(\cdot, x_0)} = \frac{\hat{\mathbf{R}}_F^{G(\cdot, x)}}{\hat{\mathbf{R}}_F^{G(\cdot, x_0)}} = \frac{\hat{\mathbf{R}}_F^{G(\cdot, x)}}{\hat{\mathbf{R}}_F^1} \frac{\hat{\mathbf{R}}_F^1}{\hat{\mathbf{R}}_F^{G(\cdot, x_0)}} \leq ba^{-1} \quad \text{on} \quad V.$$

Throughout this chapter, x_0 is a fixed reference point of R and α_0 is a positive number such that the set $F_0 = \{y : G(y, x_0) \geq \alpha_0\}$ is compact. Now let Φ be a real-valued function on $(-\infty, +\infty)$ with the following properties: (i) Φ has a continuous second derivative, (ii) $\Phi(t) = t$ for $t \leq \alpha_0$, (iii) $\Phi(t) = \alpha_0 + \frac{1}{2}$ for $t \geq \alpha_0 + 1$ and (iv) $\Phi''(t) \leq 0$ for all t. We now show that $u(y) = \Phi(G(y, x_0))$ is superharmonic on R. Now F_0 has a nonempty interior which contains x_0. Since $\Phi(G(\cdot, x_0))$ is constant in a neighborhood of x_0, u is superharmonic thereon. If $x \neq x_0$, then

$$\Delta u = \Phi'(G(\cdot, x_0))\Delta G(\cdot, x_0) + \Phi''(G(\cdot, x_0))\left(\sum_{i=1}^n [G_{x_i}(\cdot, x_0)]^2\right)$$

in a neighborhood of x. Since $G(\cdot, x_0)$ is harmonic on $R \sim \{x_0\}$ and $\Phi'' \leq 0$, $\Delta u \leq 0$ on $R \sim \{x_0\}$. This proves that $\Phi(G(\cdot, x_0))$ is superharmonic on R. Since Φ has a continuous second derivative and $\Phi(G(\cdot, x_0))$ is constant in a neighborhood of x_0, $\Phi(G(\cdot, x_0))$ has continuous second partials on R. Note that $\Phi(G(y, x_0)) = G(y, x_0)$ whenever $y \notin F_0$.

To construct the Martin compactification, we let

$$M = \left\{f \in \Sigma(R) : f = \frac{G(\cdot, x)}{\Phi(G(\cdot, x_0))} \quad \text{for some} \quad x \in R\right\}.$$

Definition. R_M^*, the M-compactification of R, is called the **Martin compactification** of R. The set $\Delta_M = R_M^* \sim R$ is called the **Martin boundary** of R. For $y \in R$, we put

$$k(y, \cdot) = \frac{G(y, \cdot)}{\Phi(G(y, x_0))}.$$

By definition of the Martin compactification the function $k(\cdot, x)$ has a continuous extension to R_M^* for each $x \in R$. Then $\lim_{\substack{z \to y \\ z \in R}} k(z, x)$, $x \in R$, is defined for each $y \in \Delta_M$.

LEMMA **12.3** The function K defined on $R_M^* \times R$ by

$$K(y, x) = \begin{cases} k(y, x), & y \in R, \\ \lim\limits_{\substack{z \to y \\ z \in R}} k(z, x), & y \in \Delta_M = R_M^* \sim R, \end{cases}$$

is strictly positive and continuous on $R_M^* \times R$ (in the extended sense at points on the diagonal). For each $y \in \Delta_M$, $K(y, \cdot)$ is harmonic on R. If C is any compact subset of R and F any closed subset of $R_M^* \sim C$, then

$$\sup\{K(y, x): x \in C, y \in F\} < +\infty.$$

Proof. The last assertion is proved first. Let W be any open subset of R with compact closure $\overline{W} \subset R$ such that $C \subset W$. It suffices to prove that

$$\sup\{K(y, x) : x \in C, y \in R_M^* \sim W\} < +\infty;$$

that is, we can assume that $F = R_M^* \sim W$. Let V be an open set with compact closure $\overline{V} \subset R$ such that $F_0 = \{y \in R: G(y, x_0) \geq \alpha_0\} \subset V$. For $y \in R \sim V$, $K(y, x) = G(y, x)/\Phi(G(y, x_0)) = G(y, x)/G(y, x_0)$. It follows from Lemma 12.2 that

$$\begin{aligned}
&\sup\{K(y,x) : x \in C, y \in (F \sim V) \cap R\} \\
&= \sup\left\{\frac{G(y, x)}{G(y, x_0)} : x \in C, y \in (F \sim V) \cap R\right\} \\
&< +\infty.
\end{aligned}$$

Since $K(y, x)$ is jointly continuous in y and x on $(F \cap \overline{V}) \times C$,

$$\sup\{K(y, x): x \in C, y \in (F \cap \overline{V}) \cap R\} < +\infty$$

Therefore

$$\sup\{K(y, x): x \in C, y \in F \cap R\} < +\infty.$$

If $x \in C$ and $y \in \Delta_M$, then $K(y, x) = \lim_{\substack{z \to y \\ z \in R \sim W}} K(z, x)$ and the boundedness of K on $(F \cap R) \times C$ implies the boundedness of K on $F \times C$. We now show that $K(y, \cdot)$ is harmonic on R for each $y \in \Delta_M$. Let V and W be open subsets of R with compact closures $\overline{V}$, $\overline{W} \subset R$ such that $\overline{W} \subset V$. Let $\{y_\alpha : \alpha \in \mathfrak{A}\}$ be a net in $R \sim V$ such that $\lim_{\mathfrak{A}} y_\alpha = y$. Now the functions $K(y_\alpha, \cdot)$ are uniformly bounded on W by the preceding part of the proof. Moreover, the functions $K(y_\alpha, x) = G(y_\alpha, x)/\Phi(G(y_\alpha, x_0))$ are harmonic functions of x on W. Since $K(y, \cdot) = \lim_{\mathfrak{A}} K(y_\alpha, \cdot)$ on W, $K(y, \cdot)$ is harmonic on W by Theorem 2.18. This proves that $K(y, \cdot)$, $y \in \Delta_M$, is harmonic on R. Since $K(y, x) = G(y, x)/\Phi(G(y, x_0))$ on $R \times R$, it is clear that $K(y, x)$ is jointly continuous in y and x on $R \times R$ (in the extended sense at points $(x, x) \in R \times R$). We need only prove joint continuity at points (y', x'), where $y' \in \Delta_M$ and $x' \in R$. Again let V and W be open subsets of R with compact closures $\overline{V}$, $\overline{W} \subset R$ such that $x' \in W \subset \overline{W} \subset V$. Consider $K(y, \cdot)$ as a function on W for $y \in R_M^* \sim V$. Now $\{K(y, \cdot); y \in R_M^* \sim V\}$ is a family of uniformly bounded harmonic functions on W and is therefore an equi-continuous family on W by Theorem 2.18; that is, given $\varepsilon > 0$ there is a $\delta > 0$ such that $|K(y, x) - K(y, x')| < \varepsilon/2$ whenever $\|x - x'\| < \delta$ and $y \in R_M^* \sim V$. Since

$$|K(y, x) - K(y', x')| \le |K(y, x) - K(y, x')| + |K(y, x') - K(y', x')|$$
$$< \varepsilon/2 + |K(y, x') - K(y', x')|$$

whenever $\|x - x'\| < \delta$ and $y \in R_M^* \sim V$ and since $K(\cdot, x')$ has a continuous extension to R_M^*, $\lim_{(y, x)\to(y', x')} K(y, x) = K(y', x')$. It remains only to show that $K(y, \cdot) > 0$ on R for each $y \in R_M^*$. If $y \in R \sim F_0$, then

$$K(y, x_0) = \frac{G(y, x_0)}{\Phi(G(y, x_0))} = \frac{G(y, x_0)}{G(y, x_0)} = 1.$$

For any $y \in \Delta_M$, $K(y, x_0) = \lim_{\substack{z\to y\\ z \in R\sim F_0}} K(z, x_0) = 1$ and, since $K(y, \cdot)$ is harmonic on R, $K(y, \cdot) > 0$ on R. If $y \in R$, then $K(y, x) = G(y, x)/\Phi(G(y, x_0)) > 0$ on R. It follows that $K(y, \cdot) > 0$ on R for all $y \in R_M^*$.

The function $K(y, \cdot)$ is called the **Martin function with pole** y. The purpose of the following lemmas is to develop a potential theory for the kernel $K(y, x)$.

LEMMA **12.4** If $f \in M$ and f^* is the continuous extension of f to R_M^*, then there are measures λ_1 and λ_2 having compact support in R such that

$$f^*(y) = \frac{\int K(y, z)\, d\lambda_1(z)}{\int K(y, z)\, d\lambda_2(z)}, \qquad y \in R_M^*.$$

Proof. For some $x \in R$, $f(y) = G(y, x)/\Phi(G(y, x_0))$. By choice of Φ, $\Phi(G(y, x_0))$ is superharmonic on R and, in fact, the potential of a measure λ_2 having compact support, since $\Phi(G(y, x_0)) = G(y, x_0)$ for $y \in R \sim F_0$. Letting λ_1 denote the unit measure with support $\{x\}$,

$$f(y) = \frac{\int G(y, z)\, d\lambda_1(z)}{\int G(y, z)\, d\lambda_2(z)} = \frac{\int [G(y, z)/\Phi(G(y, x_0))]\, d\lambda_1(z)}{\int [G(y, z)/\Phi(G(y, x_0))]\, d\lambda_2(z).}.$$

Suppose the compact set $C \subset R$ contains the supports of λ_1 and λ_2 and W is a neighborhood of C having compact closure $\overline{W} \subset R$. Since $K(y, x)$ is continuous on $(R_M^* \sim W) \times C$, the right side of the above equation has a continuous extension to $R_M^* \sim W$ by Lemma 12.3 and the Lebesgue dominated convergence theorem. Therefore

$$f^*(y) = \frac{\int K(y, z)\, d\lambda_1(z)}{\int K(y, z)\, d\lambda_2(z)}, \qquad y \in R_M^*.$$

COROLLARY **12.5** If $y_1, y_2 \in R_M^*$ and $y_1 \neq y_2$, then $K(y_1, \cdot)$ and $K(y_2, \cdot)$ are not proportional.

Proof. We shall consider three cases. Suppose first that $y_1, y_2 \in R$, $y_1 \neq y_2$. Then $K(y_1, \cdot) = G(y_1, \cdot)/\Phi(G(y_1, x_0))$ and $K(y_2, \cdot)$
$$= G(y_2, \cdot)/\Phi(G(y_2, \cdot))$$
are not proportional because of the infinities at y_1 and y_2, respectively. If $y_1 \in R$ and $y_2 \in \Delta_M$, then $K(y_1, \cdot)$ has a pole at y_1 whereas $K(y_2, \cdot)$, being harmonic on R, is finite at y_1. Lastly, suppose $y_1, y_2 \in \Delta_M$ with $y_1 \neq y_2$. Assume $K(y_1, \cdot) = \beta K(y_2, \cdot)$ on R, where $\beta > 0$. If $f \in M$ and f^* is the continuous extension of f to R_M^*, then

$$f^*(y_1) = \frac{\int K(y_1, z)\, d\lambda_1(z)}{\int K(y_1, z)\, d\lambda_2(z)} = \frac{\int \beta K(y_2, z)\, d\lambda_1(z)}{\int \beta K(y_2, z)\, d\lambda_2(z)} = f^*(y_2),$$

where λ_1 and λ_2 are the measures of the preceding lemma; but since the class of such functions f^* separates the points of Δ_M we have a contradiction.

Definition. If μ is a Borel measure on R_M^*, the *K*-**potential** of μ is defined by

$$K\mu(x) = \int K(y, x)\, d\mu(y), \qquad x \in R,$$

provided $K\mu$ is not identically $+\infty$ on R.

The fact that $K(y, x)$ is not a symmetric function is apparent from its definition and should be kept in mind. Note also that $K\mu$ is a superharmonic function, if defined, whenever μ is a measure on R_M^*. This follows from the fact that if we put $\mu_1 = \mu|_R$ and $\mu_2 = \mu|_{\Delta_M}$ then for $x \in R$

$$0 \leq K\mu(x) = \int G(y, x) \frac{1}{\Phi(G(y, x_0))}\, d\mu_1(x) + \int K(y, x)\, d\mu_2(y);$$

the second integral on the right is finite by the boundedness of K on $\Delta_M \times \{x\}$ and therefore harmonic on R as in Lemma 6.8, whereas the first integral on the right is known to be superharmonic since it is not identically $+\infty$.

LEMMA **12.6** If f is a function on R having compact support C and continuous second partials, there is a signed measure λ on R with support C such that $f(y) = \int K(y, x)\, d\lambda(x)$; moreover, $K\mu$ is λ-integrable whenever defined.

Proof. The function $g(y) = f(y)\Phi(G(y, x_0))$ has support C and continuous second partials. Consider Poisson's equation $\tilde{\Delta} u = -\kappa_n(-\Delta g/\kappa_n)$, where $\tilde{\Delta}$ is the generalized Laplacian and κ_n is a constant. An obvious solution is just g, but by Theorem 6.25 $G(-\Delta g/\kappa_n)$, the Green potential of a measure having density $-\Delta g/\kappa_n$, is a solution. Since Δg vanishes on $\sim C$,

$$\lim_{z \to x} G(-\Delta g/\kappa_n)(z) = 0$$

for all $x \in \partial R$, except possibly for a polar set by Theorem 8.31. By Theorem 6.22 $G(-\Delta g/\kappa_n)$ is continuous on R. By Theorem 6.26 $g - G(-\Delta g/\kappa_n)$ is harmonic on R. Note that $G(-\Delta g/\kappa_n)$ is bounded on C and harmonic on $\sim C$. It follows from Theorem 8.43, that $G(-\Delta g/\kappa_n)$ is bounded on R. Since g is also bounded on R, $g = G(-\Delta g/\kappa_n)$ by Corollary 9.20. For each Borel set $E \subset R$ let

$$\lambda(E) = \int_E \frac{-\Delta g(x)}{\kappa_n}\, dx.$$

Then

$$f(y)\Phi(G(y, x_0)) = \int G(y, x)\, d\lambda(x), \qquad y \in R,$$

or

$$f(y) = \int K(y, x)\, d\lambda(x), \qquad y \in R.$$

Lastly, suppose $K\mu$ is defined. Then $K\mu$ is superharmonic. Since λ has a bounded density which vanishes outside C and $K\mu$ is Lebesgue integrable on compact sets, $K\mu$ is λ-integrable.

LEMMA **12.7** If μ and ν are measures on R_M^* for which $K\mu$ and $K\nu$ are defined and $K\mu = K\nu$, then $\mu = \nu$ on R.

Proof. Let f be any function with compact support and continuous second partials. By the preceding lemma there is a signed measure λ with compact support such that $K\mu$ and $K\nu$ are λ-integrable and

$$f(y) = \int K(y, x)\, d\lambda(x).$$

Applying Fubini's theorem (to the positive and negative parts of λ),

$$\begin{aligned}\int f(y)\, d\mu(y) &= \int\left(\int K(y, x)\, d\lambda(x)\right) d\mu(y)\\ &= \int\left(\int K(y, x)\, d\mu(y)\right) d\lambda(x)\\ &= \int K\mu(x)\, d\lambda(x).\end{aligned}$$

Since the same can be done with $\int f(y)\,dv(y)$ and $K\mu = Kv$,

$$\int f(y)\,d\mu(y) = \int f(y)\,dv(y)$$

for all functions f having compact support and continuous second partials. Therefore $\mu = v$ on R.

For purposes of the next theorem we make the convention that $a/(1 + a) = 1$ if $a = +\infty$.

THEOREM **12.8** R_M^* is metrizable; if $\{x_j\}_{j \geq 1}$ is a countable dense subset of R, then

$$\rho(y_1, y_2) = \sum_{j=1}^{\infty} \frac{1}{2^j} \left| \frac{K(y_1, x_j)}{1 + K(y_1, x_j)} - \frac{K(y_2, x_j)}{1 + K(y_2, x_j)} \right|$$

is a metric on R_M^* compatible with the topology on R_M^*.

Proof. If $\rho(y_1, y_2) = 0$, then $K(y_1, x_j) = K(y_2, x_j)$ for all $j \geq 1$. Since the sequence $\{x_j\}$ is dense in R and $K(y_i, \cdot)$ is continuous on R, $K(y_1, \cdot) = K(y_2, \cdot)$ on R. It follows from Corollary 12.5 that $y_1 = y_2$. The other requirements of a metric are easy to verify. We now show that the metric ρ is compatible with the topology on R_M^*. Let y_0 be a point of R_M^* and let W_0 be a neighborhood of y_0. We must show that there is an $\varepsilon > 0$ such that $\{y: \rho(y_0, y) < \varepsilon\} \subset W_0$. Assume that no such ε exists. If $\{\varepsilon_i\}$ is a decreasing sequence of positive numbers with $\lim_{i \to \infty} \varepsilon_i = 0$, there is a sequence $\{y_i\}$ in $R_M^* \sim W_0$ such that

$$\rho(y_0, y_i) = \sum_{j=1}^{\infty} \frac{1}{2^j} \left| \frac{K(y_0, x_j)}{1 + K(y_0, x_j)} - \frac{K(y_i, x_j)}{1 + K(y_i, x_j)} \right| < \varepsilon_i .$$

Since $R_M^* \sim W_0$ is compact, we can assume that the sequence $\{y_i\}$ has a limit $y_1 \in R_M^* \sim W_0$. By Fatou's lemma

$$\begin{aligned} \rho(y_0, y_1) &= \sum_{j=1}^{\infty} \frac{1}{2^j} \liminf_{i \to \infty} \left| \frac{K(y_0, x_j)}{1 + K(y_0, x_j)} - \frac{K(y_i, x_j)}{1 + K(y_i, x_j)} \right| \\ &\leq \liminf_{i \to \infty} \sum_{j=1}^{\infty} \frac{1}{2^j} \left| \frac{K(y_0, x_j)}{1 + K(y_0, x_j)} - \frac{K(y_i, x_j)}{1 + K(y_i, x_j)} \right| \\ &= \liminf_{i \to \infty} \rho(y_0, y_i) = 0. \end{aligned}$$

It follows that $y_0 = y_1$, but this is a contradiction, since $y_0 \in W_0$ and $y_1 \in R_M^* \sim W_0$. It remains only to show that if $\varepsilon > 0$ and $y_0 \in R_M^*$ then there is a neighborhood W_0 of y_0 such that $W_0 \subset \{y: \rho(y_0, y) < \varepsilon\}$. Choose j_0 such that

$$\sum_{j=j_0+1}^{\infty} \frac{1}{2^j} < \frac{\varepsilon}{4}$$

Then

$$\sum_{j=j_0+1}^{\infty} \frac{1}{2^j} \left| \frac{K(y_0, x_j)}{1 + K(y_0, x_j)} - \frac{K(y, x_j)}{1 + K(y, x_j)} \right| < \frac{\varepsilon}{2}$$

for all $y \in R_M^*$, since each absolute value is bounded by 2. Now the function $K(y, x)/)1 + K(y, x))$ is continuous on $R_M^* \times R$ in the ordinary sense. There is therefore a neighborhood W_0 of y_0 such that

$$\left| \frac{K(y_0, x_j)}{1 + K(y_0, x_j)} - \frac{K(y, x_j)}{1 + K(y, x_j)} \right| < \frac{\varepsilon}{2}$$

for all $y \in W_0$ and $j = 1, 2, \ldots, j_0$. It follows that

$$\sum_{i=1}^{j_0} \frac{1}{2^j} \left| \frac{K(y_0, x_j)}{1 + K(y_0, x_j)} - \frac{K(y, x_j)}{1 + K(y, x_j)} \right| \le \frac{\varepsilon}{2} \sum_{j=1}^{j_0} \frac{1}{2^j} \le \frac{\varepsilon}{2}$$

for all $y \in W_0$. Hence $W_0 \subset \{y : \rho(y_0, y) < \varepsilon\}$.

2. The Martin Representation

In this section we derive Martin's representation of a positive harmonic function as an integral of K relative to a measure on Δ_M. The representation itself is easy to obtain but the measure need not be unique. The problem of selecting some canonical measure among those representing the function so as to achieve uniqueness is much more difficult.

LEMMA **12.9** If u is a positive superharmonic function on R and F is a relatively closed subset of R, then there is a measure μ on $\bar{F}$ (the closure in R_M^*) such that $\hat{\mathbf{R}}_F^u = \int K(z, \cdot)\, d\mu(z)$ on R.

Proof. Let $\{F_j\}$ be an increasing sequence of compact subsets of R such that $F = \cup F_j$. For each j, there is a measure ν_j with support in F_j such that

$$\hat{\mathbf{R}}_{F_j}^u = \int G(z, \cdot)\, d\nu_j(z) = \int \frac{G(z, \cdot)}{\Phi(G(z, x_0))} \Phi(G(z, x_0))\, d\nu_j(z).$$

For each Borel set $E \subset R_M^*$, let

$$\mu_j(E) = \int_{E \cap R} \Phi(G(z, x_0))\, d\nu_j(z).$$

Then

$$\hat{\mathbf{R}}_{F_j}^u = \int K(z, \cdot)\, d\mu_j(z).$$

Let x_1 be any point of R such that $u(x_1) < +\infty$. Since K is strictly positive and continuous (in the extended sense) on $R_M^* \times R$ by Lemma 12.3,

$$\beta_1 = \inf_{z \in R_M^*} K(z, x_1) > 0$$

and

$$u(x_1) \geq \hat{\mathbf{R}}_{F_j}^u(x_1) = \int K(z, x_1)\, d\mu_j(z) \geq \beta_1 \mu_j(F_j) = \beta_1 \mu_j(\overline{\mathrm{F}}).$$

It follows that there is a measure μ on R_M^* with support in $\overline{F}$ and a subsequence of the sequence $\{\mu_j\}$ (which we can assume to be the sequence itself) such that $\mu_j \xrightarrow{w^*} \mu$. Note that $\{\hat{\mathbf{R}}_{F_j}^u\}$ is an increasing sequence of superharmonic functions and that $\lim_{j\to\infty} \hat{\mathbf{R}}_{F_j}^u = \hat{\mathbf{R}}_F^u$ by Theorem 8.38. Therefore

$$\hat{\mathbf{R}}_{F_j}^u = \int K(z, \cdot)\, d\mu_j(z) \uparrow \hat{\mathbf{R}}_F^u$$

as j increases to $+\infty$. Since $K(\cdot, x)$ is continuous only in the extended sense, we cannot simply let $j \to \infty$. To circumvent this difficulty let x be an arbitrary point of R and let $\overline{B}_{x,\delta} \subset R$. Let ν_δ be a unit measure uniformly distributed on $\partial B_{x,\delta}$. Then

$$\int \hat{\mathbf{R}}_{F_j}^u(y)\, d\nu_\delta(y) = \int \left(\int K(z, y)\, d\mu_j(z) \right) d\nu_\delta(y)$$

$$= \int \left(\int K(z, y)\, d\nu_\delta(y) \right) d\mu_j(z).$$

Consider the function v defined on R_M^* by

$$v(z) = \int K(z, y)\, d\nu_\delta(y), \qquad z \in R_M^*.$$

Using the continuity and boundedness properties of K, we can easily show that v is continuous on $R_M^* \sim \partial B_{x,\delta}$. For $z \in R$

$$v(z) = \int K(z, y)\, d\nu_\delta(y) = \int \frac{G(z, y)}{\Phi(G(z, x_0))}\, d\nu_\delta(y)$$

$$= \frac{L(G(z, \cdot) : x, \delta)}{\Phi(G(z, x_0))},$$

It follows from Theorem 5.26 that $L(G(z,\cdot):x,\delta)$ is continuous at points z of $\partial B_{x,\delta}$ (in the ordinary sense, since the average of a superharmonic function is always finite). Since v is continuous on R_M^*,

$$\int \hat{\mathbf{R}}_F^u(y)\,dv_\delta(y) = \int \left(\int K(z,y)\,dv_\delta(y)\right) d\mu(z);$$

that is, $L(\hat{\mathbf{R}}_F^u : x,\delta) = L(K\mu : x,\delta)$. Since $\hat{\mathbf{R}}_F^u$ and $K\mu$ are superharmonic functions ($K\mu$ is defined since $u \geq \hat{\mathbf{R}}_F^u = \lim_{j\to\infty} \hat{\mathbf{R}}_{F_j}^u = \lim_{j\to\infty} K\mu_j \geq K\mu$ as in Lemma 7.36), we can let $\delta \downarrow 0$ to obtain $\hat{\mathbf{R}}_F^u(x) = K\mu(x)$ for all $x \in R$.

As a special case, we may take $F = R$ and obtain an integral representation

$$u = \int K(z,\cdot)\,d\mu(z)$$

for any positive superharmonic function u on R. In particular, if h is a positive harmonic function on R, there is a measure μ on R_M^* such that

$$h = \int K(z,\cdot)\,d\mu(z) \quad \text{on} \quad R.$$

In this case each of the measures μ_j of the preceding proof have support in ∂F_j and it is clear that μ has support in Δ_M. Hence each positive harmonic function on R can be represented as an integral relative to a measure on the Martin boundary. We now take up the problem of uniqueness of this measure.

As has been previously pointed out, if ν is a measure on R_M^*, the equation $u(x) = \int K(z,x)\,d\nu(z)$ defines a non-negative superharmonic function on R. Suppose C is a compact subset of R. Then $\hat{\mathbf{R}}_C^u$ is a potential on R and there is a measure ν_C on C such that $\hat{\mathbf{R}}_C^u = G\nu_C$; this equation can also be written $\hat{\mathbf{R}}_C^u = K\nu_C^K$, where for each Borel set $E \subset R_M^*$

$$\nu_C^K(E) = \int_E \Phi(G(z,x_0))\,d\nu_C(z).$$

Both ν_C and ν_C^K have support in C and both have support in ∂C if u is harmonic on the interior of C. In the propositions that follow ν_C and ν_C^K denote measures on C associated in this way with the measure ν on R_M^*.

The basic idea used in selecting a "canonical" measure μ for the representation

$$h = \int K(z,\cdot)\,d\mu(z)$$

of the positive harmonic function h is that $\hat{\mathbf{R}}_{C_j}^h \uparrow h$ on R if $\{C_j\}$ is an appropriate sequence of compact sets and that the measure $\mu_{C_j}^K$ in the representation $\hat{\mathbf{R}}_{C_j}^h = K\mu_{C_j}^K$ is known to be unique. This leads to showing that the sequence $\{\mu_{C_j}^K\}$ has a w^*-limit which is a measure on a certain subset Δ_1 of Δ_M; this

w^*-limit turns out to be unique among those measures on Δ_1 which can be used to represent h.

We also associate with each compact set $C \subset R$ an operator $\mathbf{T}_C$ mapping $C(R_M^*)$ into $C(\Delta_M)$ which is defined as follows. For each $y \in \Delta_M$ let δ_y be a unit measure with support $\{y\}$ and let δ_{yC}^K be the measure having support in ∂C as defined above. For $f \in C(R_M^*)$ let

$$\mathbf{T}_C f(y) = \int f \, d\delta_{yC}^K, \qquad y \in \Delta_M.$$

LEMMA **12.10** $\mathbf{T}_C$ is a bounded linear operator from $C(R_M^*)$ to $C(\Delta_M)$; moreover, $\mathbf{T}_C 1 = 1$ and $\|\mathbf{T}_C\| = 1$ if $F_0 \subset \text{int } C$.

Proof. Consider the K-potential $K\delta_y(x) = \int K(z, x)\, d\delta_y(z) = K(y, x)$ for $y \in \Delta_M$. The function $K\delta_y(x) = K(y, x)$ is uniformly continuous on $\Delta_M \times C$. If $y_0 \in \Delta_M$ and $\varepsilon > 0$, there is a neighborhood W_0 of y_0 such that

$$|K\delta_y - K\delta_{y_0}| < \varepsilon$$

on C for $y \in \Delta_M \cap W_0$. Now let y_1 be a fixed point of Δ_M, let

$$\alpha = \inf_{x \in C} K(y_1, x) > 0,$$

and let $v = \alpha^{-1} \delta_{y_1}$. Then

$$Kv(x) = \int K(z, x)\alpha^{-1}\, d\delta_{y_1}(z) = \alpha^{-1} K(y_1, x) \geq 1$$

for $x \in C$. For $y \in \Delta_M \cap W_0$, $K\delta_y < K\delta_{y_0} + \varepsilon Kv$ and $K\delta_{y_0} < K\delta_y + \varepsilon Kv$ on C. Since $\hat{\mathbf{R}}_C^{u+v} \leq \hat{\mathbf{R}}_C^u + \hat{\mathbf{R}}_C^v$ for any positive superharmonic functions u and v and since $\hat{\mathbf{R}}_C^u \leq \hat{\mathbf{R}}_C^v$ whenever $u \leq v$ on C,

$$K\delta_{yC}^K \leq K\delta_{y_0C}^K + \varepsilon K v_C^K,$$
$$K\delta_{y_0C}^K \leq K\delta_{yC}^K + \varepsilon K v_C^K,$$

for $y \in \Delta_M \cap W_0$. Suppose now that f has compact support and continuous second partials on R. According to Lemma 12.6, there is a signed measure λ such that $f = \int K(\cdot, x)\, d\lambda(x)$ on R. For such f

$$\begin{aligned}
&\left|\int f\, d\delta_{yC}^K - \int f\, d\delta_{y_0C}^K\right| \\
&\quad = \left|\int\left(\int K(z, x)\, d\lambda(x)\right) d\delta_{yC}^K(z) - \int\left(\int K(z, x)\, d\lambda(x)\right) d\delta_{y_0C}^K(z)\right| \\
&\quad = \left|\int K\delta_{yC}^K(x)\, d\lambda(x) - \int K\delta_{y_0C}^K(x)\, d\lambda(x)\right| \\
&\quad \leq \varepsilon \int K v_C^K\, d|\lambda|
\end{aligned}$$

for $y \in \Delta_M \cap W_0$. Since the latter integral is finite and independent of y, $\mathbf{T}_C f$ is continuous on Δ_M whenever f has compact support and continuous second partials. Note that $\mathbf{T}_C$ is a positive operator and therefore

$$|\mathbf{T}_C f(y)| \leq \|f\| \mathbf{T}_C 1(y) = \|f\| \delta_{yC}^K(R_M^*).$$

Recall that $F_0 = \{z \in R : G(z, x_0) \geq \alpha_0\}$. Now for $y \in \Delta_M$

$$K(y, x_0) \geq \hat{\mathbf{R}}_C^{K\delta_y}(x_0) = \int K(z, x_0)\, d\delta_{yC}^K(z) \geq [\inf_{z \in \partial C} K(z, x_0)] \delta_{yC}^K(R_M^*).$$

Therefore

$$\delta_{yC}^K(R_M^*) \leq \frac{\sup_{y \in \Delta_M} K(y, x_0)}{\inf_{z \in \partial C} K(z, x_0)}.$$

This shows that $\mathbf{T}_C$ is a bounded operator. Now consider any $f \in C(R_M^*)$. Now note that $\mathbf{T}_C f$ depends only upon the values of f on ∂C. We can therefore assume that f has compact support. Let $\{f_j\}$ be a sequence of functions on R_M^* having compact support in R and continuous second partials on R such that $\|f_j - f\| \to 0$ as $j \to \infty$. Since $\|\mathbf{T}_C f_j - \mathbf{T}_C f\| \leq \text{const} \times \|f_j - f\|$, the sequence $\{\mathbf{T}_C f_j\}$ of continuous functions on Δ_M converges uniformly to $\mathbf{T}_C f$ and therefore $\mathbf{T}_C f$ is continuous on Δ_M. If we assume that $F_0 \subset \operatorname{int} C$, then $K(z, x_0) = 1$ for $z \notin F_0$ and for $y \in \Delta_M$

$$1 = K(y, x_0) = \hat{\mathbf{R}}_C^{K\delta_y}(x_0) = \int K(z, x_0)\, d\delta_{yC}^K(z) = \int d\delta_{yC}^K(z) = \mathbf{T}_C 1(y).$$

LEMMA **12.11** If $\{\mu_x : x \in \Delta_M\}$ is a family of measures on Δ_M such that $\mu.(E)$ is measurable on Δ_M for each Borel set $E \subset \Delta_M$, μ is a measure on Δ_M, and ν is a measure on Δ_M such that $\int f\, d\mu = \int (\int f\, d\mu_x)\, d\nu(x)$ for all $f \in C(R_M^*)$, then $\int f\, d\mu_C^K = \int (\int f\, d\mu_{xC}^K)\, d\nu(x)$ for all $f \in C(R_M^*)$.

Proof. If $\mu' = \int \mu_{xC}^K\, d\nu(x)$, then

$$\begin{aligned} K\mu' &= \lim_{j \to \infty} \int (K(y, \cdot) \wedge j)\, d\mu'(y) \\ &= \lim_{j \to \infty} \int \left(\int (K(y, \cdot) \wedge j)\, d\mu_{xC}^K(y) \right) d\nu(x) \\ &= \int \left(\int K(y, \cdot)\, d\mu_{xC}^K(y) \right) d\nu(x) \\ &= \int K\mu_{xC}^K\, d\nu(x). \end{aligned}$$

Note that $K\mu_C^K = K\mu$ and $K\mu_{xC}^K = K\mu_x$ on C, except possibly for the points of ∂C which are irregular boundary points for $R \sim C$ by Theorem 9.25. If $y \in C$ is not an irregular boundary point for $R \sim C$, then

$$\begin{aligned}
K\mu'(y) &= \int K\mu_{xC}^K(y)\, d\nu(x) = \int K\mu_x(y)\, d\nu(x) \\
&= \int \left(\int K(z, y)\, d\mu_x(z) \right) d\nu(x) \\
&= \lim_{j \to \infty} \int \left(\int (K(z, y) \wedge j)\, d\mu_x(z) \right) d\nu(x) \\
&= \lim_{j \to \infty} \int (K(z, y) \wedge j)\, d\mu(z) \\
&= K\mu(y) = K\mu_C^K(y).
\end{aligned}$$

Therefore $K\mu' = K\mu_C^K$ on C except possibly for a polar subset of ∂C. By Theorem 8.43 $K\mu' = K\mu_C^K$ on R. By Lemma 12.7 $\mu' = \mu_C^K$.

Definition. A positive K-potential $K\mu$ is called **extremal** if $K\mu = K\mu_1 + K\mu_2$ for measures μ_1 and μ_2 implies that $K\mu_1$ is proportional to $K\mu_2$. Let

$$\Delta_1 = \{y \in \Delta_M : K(y, \cdot) = K\delta_y \text{ is extremal}\}$$

and $\Delta_0 = \Delta_M \sim \Delta_1$. The points of Δ_1 are called **minimal boundary points** of the Martin boundary.

LEMMA **12.12** If $K\mu$ is extremal, then the support of μ consists of a single point $y \in R \cup \Delta_1$.

Proof. Let F be the support of μ and consider any $y \in F$. For each positive integer j let

$$U_j = \left\{ z \in R_M^* : \rho(y, z) < \frac{1}{j} \right\}.$$

Also let $\mu_j = \mu|_{U_j}$ and $\nu_j = \mu|_{R_M^* \sim U_j}$. Then $\mu = \mu_j + \nu_j$ and $K\mu = K\mu_j + K\nu_j$. Since $K\mu$ is extremal, $K\mu_j$ is proportional to $K\mu$ and there is a measure μ_j' on U_j such that $K\mu = K\mu_j'$. Consider any point $x \in R \sim \{y\}$ such that $K\mu(x) < +\infty$. For sufficiently large j, $x \notin U_j$ and

$$\inf_{z \in U_j} K(z, x) \geq \tfrac{1}{2} K(y, x) = \alpha.$$

Then for all large j

$$+\infty > K\mu(x) = \int_{U_j} K(z, x)\, d\mu_j'(z) \geq \alpha \mu_j'(U_j) = \alpha \mu_j'(R_M^*)$$

Hence there is a measure μ' having support $\{y\}$ and a subsequence of the sequence $\{\mu_j'\}$, which we can assume to be the sequence itself, such that $\mu_j' \xrightarrow{w^*} \mu'$. Clearly $\mu' = \alpha'\delta_y$ for some α' and $K\mu(x) = K\mu_j'(x) \to \alpha' K\delta_y(x)$. Therefore $K\mu = \alpha' K\delta_y = \alpha' K(y, \cdot)$ on $R \sim \{y\}$. If there were two different points $y_1, y_2 \in F$, then we would have $\alpha_1' K(y_1, \cdot) = \alpha_2' K(y_2, \cdot)$ for constants α_1', α_2' on $R \sim \{y_1, y_2\}$ and therefore on R by continuity. However, this would mean that $K(y_1, \cdot)$ and $K(y_2, \cdot)$ are proportional, a contradiction to Corollary 12.5.

LEMMA **12.13** If $y \in \Delta_1$ and $\{R_j\}$ is an increasing sequence of open sets with compact closures $C_j = \bar{R}_j \subset R$ such that $R = \cup R_j$, then $\delta_{yC_j}^K \xrightarrow{w^*} \delta_y$; conversely, if $y \in \Delta_M$ and $\delta_{yC_j}^K \xrightarrow{w^*} \delta_y$ for some such sequence $\{R_j\}$, then $y \in \Delta_1$.

Proof. Suppose $y \in \Delta_1$. We can assume that $F_0 \subset R_1$. By Lemma 12.10 $\delta_{yC_j}^K(R_M^*) = \mathbf{T}_{C_j} 1(y) = 1$ for all j. It follows that there is a measure μ on Δ_M and a subsequence $\{\delta_{yC_j}^*\}$ of the sequence $\{\delta_{yC_j}^K\}$ such that $\delta_{yC_j}^* \xrightarrow{w^*} \mu$. Clearly $K\delta_{yC_j}^* \to K\mu$ as $j \to \infty$; but, since $K\delta_{yC_j}^K = K\delta_y$ on R_j for all j, $K\mu = K\delta_y$ and $K\delta_{yC_j}^* \to K\delta_y$ on R as $j \to \infty$. Since $y \in \Delta_1$, $K\delta_y$ is extremal and therefore $K\mu$ is extremal. It follows from Corollary 12.5 and Lemma 12.12 that $\mu = \delta_y$. We know that $\delta_{yC_j}^* \xrightarrow{w^*} \delta_y$ and it remains only to show that $\delta_{yC_j}^K \xrightarrow{w^*} \delta_y$. By the above argument each subsequence of the sequence $\{\delta_{yC_j}^K\}$ has a subsequence that converges in the w^*-topology to δ_y. Assume that the sequence $\{\delta_{yC_j}^K\}$ does not converge in the w^*-topology to δ_y. Then there is a function $f \in C(R_M^*)$ such that

$$\lim_{j \to \infty} \int f \, d\delta_{yC_j}^K \neq \int f \, d\delta_y.$$

It follows that there is a subsequence $\{\delta_{yC_j}^{**}\}$ such that

$$\lim_{j \to \infty} \int f \, d\delta_{yC_j}^{**} = A \neq \int f \, d\delta_y,$$

but since some subsequence of the sequence $\{\delta_{yC_j}^{**}\}$ converges in the w^*-topology to δ_y, we have $A = \int f \, d\delta_y$, a contradiction. This proves that $\delta_{yC_j}^K \xrightarrow{w^*} \delta_y$. Now assume that $\delta_{yC_j}^K \xrightarrow{w^*} \delta_y$ for some such sequence $\{R_j\}$. We wish to show that $y \in \Delta_1$; that is, $K\delta_y$ is extremal. Suppose $K\delta_y = K\mu + K\nu$ for measures μ and ν on R_M^*. It follows from Corollary 9.26 that

$$\hat{\mathbf{R}}_{C_j}^{K\delta_y} = \hat{\mathbf{R}}_{C_j}^{K\mu + K\nu} = \hat{\mathbf{R}}_{C_j}^{K\mu} + \hat{\mathbf{R}}_{C_j}^{K\nu}$$

and therefore $K\delta_{yC_j}^K = K\mu_{C_j}^K + K\nu_{C_j}^K$. By Lemma 12.7 $\delta_{yC_j}^K = \mu_{C_j}^K + \nu_{C_j}^K$. By passing to a subsequence, if necessary, we can assume that there are measures

μ_0 and ν_0 such that $\mu_{C_j}^K \xrightarrow{w^*} \mu_0$ and $\nu_{C_j} \xrightarrow{w^*} \nu_0$. Since $\delta_{yC_j}^K \xrightarrow{w^*} \delta_y$ by hypothesis, $\delta_y = \mu_0 + \nu_0$. If follows that μ_0 and ν_0 are proportional to δ_y. Note that $K\mu = \lim_{j\to\infty} K\mu_{C_j}^K$, since $K\mu_{C_j}^K$ agrees with $K\mu$ on $R_j \subset C_j$. Therefore

$$K\mu = \lim_{j\to\infty} K\mu_{C_j}^K = \lim_{j\to\infty} \int K(z, \cdot)\, d\mu_{C_j}^K(z) = \int K(z, \cdot)\, d\mu_0(z).$$

Since μ_0 is proportional to δ_y, $K\mu$ is proportional to $K\delta_y$; that is, $K\delta_y$ is extremal.

THEOREM **12.14** Δ_1 and Δ_0 are Borel subsets of R_M^*.

Proof. Let $\mathscr{F}$ be a countable dense subset of $C(R_M^*)$ and let $\{C_j\}$ be a sequence as in the preceding lemma. We show that

$$\Delta_1 = \{y \in \Delta_M : \mathbf{T}_{C_j} f(y) \to f(y) \text{ as } j \to \infty, f \in \mathscr{F}\}.$$

Let Λ be the set on the right. If follows from the preceding lemma that $\Delta_1 \subset \Lambda$. Suppose $y \in \Lambda$. Then $\mathbf{T}_{C_j} f(y) \to f(y)$ as $j \to \infty$ for all $f \in \mathscr{F}$. We can assume that $F_0 \subset \text{int } C$ and therefore that $\|\mathbf{T}_{C_j}\| = 1$. Consider any $g \in C(R_M^*)$ and $\varepsilon > 0$. Then there is an $f \in \mathscr{F}$ such that $\|g - f\| < \varepsilon/4$. Now

$$\begin{aligned}|\mathbf{T}_{C_j} g(y) - g(y)| &\le |\mathbf{T}_{C_j}(g - f)(y)| + |g(y) - f(y)| + |\mathbf{T}_{C_j} f(y) - f(y)| \\ &\le 2\|g - f\| + |\mathbf{T}_{C_j} f(y) - f(y)|.\end{aligned}$$

Since $y \in \Lambda$ and $f \in \mathscr{F}$, there is a j_0 such that the last term on the right is less than $\varepsilon/4$ for all $j \ge j_0$. Therefore $|\mathbf{T}_{C_j} g(y) - g(y)| < \varepsilon$ for all $j \ge j_0$. Hence $\mathbf{T}_{C_j} g(y) \to g(y)$ as $j \to \infty$ for all $g \in C(R_M^*)$; that is, $\delta_{yC_j}^K \xrightarrow{w^*} \delta_y$ and $y \in \Delta_1$ by the preceding lemma.

LEMMA **12.15** If μ is a measure on Δ_1 and $\{R_j\}$ is an increasing sequence of open sets with compact closures $C_j = \bar{R}_j \subset R$ such that $R = \cup R_j$, then $\mu_{C_j}^K \xrightarrow{w^*} \mu$.

Proof. We can assume that $F_0 \subset R_1$ and that $\|\mathbf{T}_{C_j}\| = 1$. If $f \in C(R_M^*)$, then $\{\mathbf{T}_{C_j} f\}$ is a uniformly bounded sequence of functions on Δ_M which converges to f on Δ_1. Therefore

$$\int f\, d\mu = \lim_{j\to\infty} \int \mathbf{T}_{C_j} f\, d\mu = \lim_{j\to\infty} \int \left(\int f(z)\, d\delta_{yC_j}^K(z) \right) d\mu(y) = \lim_{j\to\infty} \int f\, d\mu_{C_j}^K$$

by Lemma 12.11.

In order to prove the next theorem we have need of some facts from measure theory. Let ν be a measure on Δ_M. A sequence $\{f_j\}$ of measurable functions on Δ_M is said to converge in ν-measure to f if for every $\varepsilon > 0$,

$$\lim_{j\to\infty} \nu(y \in \Delta_M : |f_j(y) - f(y)| \geq \varepsilon) = 0.$$

If the sequences $\{f_j\}$ and $\{g_j\}$ converge in ν-measure to f and g, respectively, the sequence $\{f_j + g_j\}$ converges in ν-measure to $f + g$. Moreover, if the sequence $\{f_j\}$ converges in ν-measure to f, some subsequence converges a.e. (ν) to f. If the sequence $\{f_j\}$ converges in ν-measure to some function f and the sequence $\{f_j + g_j\}$ converges in ν-measure to the zero function, the sequence $\{g_j\}$ converges in ν-measure to f also. Lastly, if the sequence $\{f_j\}$ converges in ν-measure to f and there is an integrable function g such that $|f_j| \leq g$, then $\lim_{j\to\infty} \int f_j \, d\nu = \int f \, \delta\nu$.

THEOREM **12.16** For each measure μ on Δ_M there is a unique measure ν on Δ_1 such that $K\mu = K\nu$.

Proof. The uniqueness follows from the preceding lemma since the limit of a w^*-convergent sequence of measures in unique. Let $\{C_j\}$ be a sequence of compact sets as in Lemma 12.13. As before, we can assume that $F_0 \subset R_1$. Then

$$\begin{aligned}\mu_{C_j}^K(R_M^*) &= \int 1 \, d\mu_{C_j}^K = \int \left(\int 1 \, d\delta_{yC_j}^K(z)\right) d\mu(y) = \int \mathbf{T}_{C_j} 1(y) \, d\mu(y) \\ &= \mu(R_M^*).\end{aligned}$$

From the sequence $\{\mu_{C_j}^K\}$ we can select a subsequence, which we can assume to be the sequence itself, such that $\mu_{C_j}^K \xrightarrow{w^*} \nu$ for some measure ν on Δ_M. Since $\hat{\mathbf{R}}_{C_j}^{K\mu} \uparrow K\mu$ on R,

$$K\mu = \lim_{j\to\infty} K\mu_{C_j}^K = K\nu.$$

We need only show that $\nu(\Delta_0) = 0$. To do this let $\mathscr{F}$ be the set of functions $f \in C(R_M^*)$ such that the sequence $\{\mathbf{T}_{C_j} f\}$ converges in ν-measure to f; that is, $f \in \mathscr{F}$ if for every $\varepsilon > 0$

$$\lim_{j\to\infty} \nu(y \in \Delta_M : |\mathbf{T}_{C_j} f(y) - f(y)| \geq \varepsilon) = 0.$$

We first show that $\mathscr{F} = C(R_M^*)$. $\mathscr{F}$ is clearly a linear subspace of $C(R_M^*)$. Since $\mathbf{T}_{C_j} 1 = 1$ for all j, $\mathscr{F}$ contains the constant functions. $\mathscr{F}$ is also a lattice) Suppose $g_1, g_2 \in \mathscr{F}$ and consider $f = g_1 \wedge g_2$. Since $f \leq g_i$, $\mathbf{T}_{C_j} f \leq \mathbf{T}_{C_j} g_i$ and $\mathbf{T}_{C_j} f \leq \mathbf{T}_{C_j} g_1 \wedge \mathbf{T}_{C_j} g_2$. Since the sequence $\{\mathbf{T}_{C_j} g_i\}$ is uniformly bounded by $\|g_i\|$ and converges in measure to g_i,

$$\lim_{j\to\infty} \int (\mathbf{T}_{C_j} g_1 \wedge \mathbf{T}_{C_j} g_2) \, d\nu = \int (g_1 \wedge g_2) \, d\nu = \int f \, d\nu.$$

Moreover,

$$\int \mathbf{T}_{C_j} f \, d\nu = \int \left(\int f(z) \, d\delta^K_{yC_j}(z) \right) d\nu(y) = \int f \, d\nu^K_{C_j}.$$

Since $K\mu = K\nu$, $K\mu^K_{C_j} = K\nu^K_{C_j}$ and $\mu^K_{C_j} = \nu^K_{C_j}$. Therefore

$$\begin{aligned} &\lim_{j \to \infty} \int |\mathbf{T}_{C_j} g_1 \wedge \mathbf{T}_{C_j} g_2 - \mathbf{T}_{C_j} f| \, d\nu \\ &\qquad = \lim_{j \to \infty} \int (\mathbf{T}_{C_j} g_1 \wedge \mathbf{T}_{C_j} g_2) \, d\nu - \lim_{j \to \infty} \int f \, d\mu^K_{C_j} \\ &\qquad = \int f \, d\nu - \int f \, d\nu = 0; \end{aligned}$$

that is, the sequence $\{\mathbf{T}_{C_j} g_1 \wedge \mathbf{T}_{C_j} g_2 - \mathbf{T}_{C_j} f\}$ converges in the mean, and therefore in ν-measure, to the zero function. Since the sequence $\{\mathbf{T}_{C_j} g_1 \wedge \mathbf{T}_{C_j} g_2\}$ converges in ν-measure to $f = g_1 \wedge g_2$, the sequence $\{\mathbf{T}_{C_j} f\}$ converges in ν-measure to f. This proves that $g_1, g_2 \in \mathscr{F}$ implies that $g_1 \wedge g_2 \in F$; that is, $\mathscr{F}$ is a lattice. $\mathscr{F}$ also separates points of R^*_M. To prove this let $x \in R$ and let $f \in C(R^*_M)$ be such that $f = K(\cdot, x)$ outside a compact subset of R. For j sufficiently large and $y \in \Delta_M$

$$f(y) = K(y, x) = K\delta^K_{yC_j}(x) = \int K(z, x) \, d\delta^K_{yC_j}(z) = \int f \, d\delta^K_{yC_j} = \mathbf{T}_{C_j} f(y)$$

and $\mathbf{T}_{C_j} f(y) \to f(y)$ as $j \to \infty$. Such f belong to $\mathscr{F}$. Suppose $y_1, y_2 \in \Delta_M$ with $y_1 \neq y_2$. If $K(y_1, x) = K(y_2, x)$ for all $x \in R$, then $K(y_1, \cdot)$ and $K(y_2, \cdot)$ are proportional, a contradiction. Therefore $\mathscr{F}$ separates points of Δ_M. Suppose $y_1 \neq y_2$ and at least one of the two points is not in Δ_M. Suppose $y_1 \notin \Delta_M$ and $f \in F$. Since $\mathbf{T}_{C_j} f$ depends only upon the values of f on ∂C_j, we can choose j_0 such that $y_1 \in R_{j_0}$ and modify f on a neighborhood of y_1 without affecting $\lim_{j \to \infty} \mathbf{T}_{C_j} f$ on Δ_M. Therefore, $\mathscr{F}$ separates points of R^*_M. It follows from the Stone-Weierstrass theorem that $\mathscr{F}$ is dense in $C(R^*_M)$. Now let $\{f_i\}$ be a sequence in $\mathscr{F}$ which converges uniformly to f and let $\varepsilon > 0$. Then

$$\begin{aligned} &\nu(y \in \Delta_M : |\mathbf{T}_{C_j} f(y) - f(y)| \geq \varepsilon) \\ &\qquad \leq \nu\left(y \in \Delta_M : |\mathbf{T}_{C_j}(f - f_i)(y)| \geq \frac{\varepsilon}{3} \right) \\ &\qquad + \nu\left(y \in \Delta_M : |\mathbf{T}_{C_j} f_i(y) - f_i(y)| \geq \frac{\varepsilon}{3} \right) \\ &\qquad + \nu\left(y \in \Delta_M : |f_i(y) - f(y)| \geq \frac{\varepsilon}{3} \right). \end{aligned}$$

Choose i_0 such that $\|f - f_{i_0}\| < \varepsilon/3$. Since $|\mathbf{T}_{C_j}(f - f_{i_0})(y)| \le \|f - f_{i_0}\| < \varepsilon/3$, the first and last indicated sets are empty. Therefore

$$\lim_{j\to\infty} \nu(y \in \Delta_M : |\mathbf{T}_{C_j} f(y) - f(y)| \ge \varepsilon)$$

$$\le \lim_{j\to\infty} \nu\left(y \in \Delta_M : |\mathbf{T}_{C_j} f_{i_0}(y) - f_{i_0}(y)| \ge \frac{\varepsilon}{3}\right) = 0$$

and $f \in \mathscr{F}$; that is, $\mathscr{F}$ is closed and $\mathscr{F} = C(R_M^*)$. Now let $\mathscr{F}_0$ be a countable dense subset of $C(R_M^*)$. By passing to a subsequence of the sequence $\{C_j\}$, if necessary, we can assume that each of the sequences $\{\mathbf{T}_{C_j} f\}$, $f \in \mathscr{F}_0$, converges a.e. (ν) to f; this can be done for each f and since $\mathscr{F}_0$ is countable, a diagonal sequence of successive subsequences will satisfy the requirement. For each $f \in \mathscr{F}_0$ let

$$\Lambda_f = \{y \in \Delta_M : \lim_{j\to\infty} \mathbf{T}_{C_j} f(y) \ne f(y)\}.$$

Then $\nu(\Lambda_f) = 0$. If $y \notin \bigcup_{f \in \mathscr{F}_0} \Lambda_f$, then $\delta_{yC_j}^K \xrightarrow{w^*} \delta_y$ and $y \notin \Delta_0$. Therefore

$$\Delta_0 \subset \bigcup_{f \in \mathscr{F}_0} \Lambda_f$$

and $\nu(\Delta_0) = 0$.

THEOREM **12.17** (*Martin*) If h is a positive harmonic function on R, there is a unique measure ν on Δ_1 such that $h = \int K(z, \cdot)\, d\nu(z)$.

Proof. By Lemma 12.9 there is a measure μ on Δ_M such that $h = K\mu$. By the preceding theorem there is a unique measure ν on Δ_1 such that $K\mu = K\nu$.

THEOREM **12.18** Δ_1 is a G_δ subset of R_M^*.

Proof. Let $\mathscr{F}_0$ be a countable dense subset of $C(R_M^*)$ and let $\{R_j\}$ be an increasing sequence of open sets with compact closures $\bar{R}_j \subset R$ such that $R = \cup R_j$. For each $f \in \mathscr{F}_0$ and positive integers i, j, let

$$\Lambda(f, i, j) = \left\{ y \in \Delta_M : |\mathbf{T}_{C_j} f(y) - f(y)| < \frac{1}{i} \right\}.$$

Each $\Lambda(f, i, j)$ is open. We shall now show that

$$\Lambda_1 = \Lambda = \bigcap_{i=1}^{\infty} \bigcap_{f \in \mathscr{F}_0} \bigcap_{k=1}^{\infty} \bigcup_{j=k}^{\infty} \Lambda(f, i, j).$$

Since $\mathbf{T}_{C_j} f(y) \to f(y)$ as $j \to \infty$ for all $y \in \Delta_1$ and $f \in C(R_M^*)$, $\Delta_1 \subset \Lambda$. Suppose $y \in \Lambda$. We can then find a subsequence $\{R_{j_k}\}$ of the sequence $\{R_j\}$ such that

$\mathbf{T}_{C_{j_k}} f(y) \to f(y)$ as $k \to \infty$ for all $f \in \mathscr{F}_0$ and, consequently, for all $f \in C(R_M^*)$. By the second part of Lemma 12.13 $y \in \Delta_1$; that is, $\Lambda \subset \Delta_1$. Therefore Δ_1 is a G_δ subset of Δ_M and it is easily seen to be a G_δ subset of R_M^*.

If the boundary of the bounded open set R is sufficiently smooth, the Martin boundary of R is just the Euclidean boundary of R. We justify this statement for the case of a ball $B = B_{0,\rho} \subset E^2$, the $n \geq 3$ case differing only in computational details. For the reference point $x_0 \in B$ we take $x_0 = 0$. Now $\bar{B}$ with the relative Euclidean topology is obviously a compactification of B. If for each $x \in B$ we can show that $G_B(\cdot, x)/G_B(\cdot, x_0)$ has a continuous extension to $\bar{B}$ and that the set of such extensions separates the points of ∂B, we shall have shown that ∂B is homeomorphic to the Martin boundary of B. The matter of continuous extensions requires identifying $\lim_{z \to y} G_B(z, x)/G_B(x, x_0)$ for $y \in \partial B$. Now

$$\frac{G_B(z, x)}{G_B(z, x_0)} = \frac{\log\left(\dfrac{\|x\|}{\rho} \dfrac{\|z - x^\star\|}{\|z - x\|}\right)}{\log \dfrac{\rho}{\|z\|}}, \qquad z \in B \sim \{x_0, x\},$$

where $x^\star$ is the inverse of x relative to $\partial B_{0,\rho}$. Let $g_x(z)$ and $g_0(z)$ denote the numerator and denominator, respectively, of the quotient on the right side of the above equation. Then g_x and g_0 are defined in a neighborhood of $y \in \partial B$. These two functions have differentials at y and directional derivatives at y are known to exist from this calculus. In particular, the normal derivatives $D_{\mathbf{n}} g_x$ and $D_{\mathbf{n}} g_0$ are defined at y, where $\mathbf{n}$ is the outer unit normal to ∂B at y. If in the above equation for $G_B(z, x)/G_B(z, x_0)$ the numerator and denominator are divided by $\|z - y\|$, we obtain

$$\lim_{z \to y} \frac{G_B(z, x)}{G_B(z, x_0)} = \frac{D_{-\mathbf{n}} g_x(y)}{D_{-\mathbf{n}} g_0(y)} = \frac{D_{\mathbf{n}} g_x(y)}{D_{\mathbf{n}} g_0(y)}.$$

Using the computation of $D_{\mathbf{n}} g_x$ in Chapter 1,

$$\lim_{z \to y} \frac{G_B(z, x)}{G_B(z, x_0)} = \frac{\rho^2 - \|x\|^2}{\|y - x\|^2}, \qquad y \in B,$$

the limit being just the kernel of the Poisson integral, except for a constant factor. Since the set of functions

$$f_x(y) = \frac{\rho^2 - \|x\|^2}{\|y - x\|^2}, \qquad y \in \partial B,$$

separates the points of ∂B, the Martin compactification is homeomorphic to $\bar{B}$.

3. The Dirichlet Problem for $\mathbf{R}_M^*$

Given an extended real-valued function f on Δ_M, we might ask if there is a harmonic function h on R such that $\lim_{y\to x} h(y) = f(x)$ for $x \in \Delta_M$. Recall our blanket assumption in this chapter that R is an open connected subset of E^n. In order to construct a harmonic function corresponding to the boundary function f, the Perron-Wiener-Brelot method is extended as follows. A function u on R is hyperharmonic (hypoharmonic) if it is either identically $+\infty$ $(-\infty)$ or superharmonic (subharmonic) on R. The upper class $\mathfrak{U}_f = \mathfrak{U}_f^R$ corresponding to the boundary function f is defined in the usual way as

$$\mathfrak{U}_f = \{u : u \text{ hyperharmonic on } R,\ u \text{ bounded below on } R, \text{ and } \liminf_{y\to x} u(y) \geq f(y) \text{ for all } x \in \Delta_M\}.$$

The lower class $\mathfrak{L}_f = \mathfrak{L}_f^R$ is defined similarly. The upper and lower solutions $\overline{H}_f$ and $\underline{H}_f$ corresponding to f are defined as $\overline{H}_f = \inf\{u : u \in \mathfrak{U}_f\}$ and $\underline{H}_f = \sup\{u : u \in \mathfrak{L}_f\}$, respectively. If $\overline{H}_f = \underline{H}_f$ and both are harmonic, then f is said to be resolutive and we define $H_f = \overline{H}_f = \underline{H}_f$. With the aid of a few lemmas we shall show that each $f \in C(\Delta_M)$ is resolutive.

Among other things, we need to know that $\underline{H}_f \leq \overline{H}_f$ on R for any boundary function f. This amounts to showing that the minimum principle applies to superharmonic functions u on R relative to R_M^*; that is, if u is superharmonic on R and $\liminf_{y\to x} u(y) \geq 0$ for all $x \in \Delta_M$, then $u \geq 0$ on R. By defining a l.s.c. function u on the compact space R_M^* by the equation $u(x) = \liminf_{y\to x} u(y)$, $x \in R_M^*$, and, using the fact that $u \geq 0$ on Δ_M, the minimum principle can be seen to hold in the present context. Also note that all of the results of Lemma 8.6 are available in the present context and are used without further comment. Finally, note that $\overline{H}_f$ and $\underline{H}_f$ are either identically $+\infty$, identically $-\infty$, or harmonic on R in view of the fact that the family $\mathfrak{U}_f$ is saturated over R and that $\overline{H}_f = -\underline{H}_{-f}$.

If E is a relatively closed subset of R and u is a positive superharmonic function on R, then $\hat{\mathbf{R}}_E^u$ already has a meaning and by Lemma 9.10 there is a measure μ on $\bar{E}$ (closure relative to R_M^*) such that

$$\hat{\mathbf{R}}_E^u = \int K(z, \cdot)\, d\mu(z) = K\mu \quad \text{on} \quad R.$$

Note that $\hat{\mathbf{R}}_E^u$ is not necessarily a potential.

LEMMA **12.19** If $y \in \Delta_1$ and U is a neighborhood of y, then $\hat{\mathbf{R}}_{R\sim U}^{K(y,\cdot)}$ is a potential on R.

Proof. We need only show that the greatest harmonic minorant of $\hat{\mathbf{R}}_{R\sim U}^{K(y,\cdot)}$ is the zero function. Suppose h is the greatest harmonic minorant of $\hat{\mathbf{R}}_{R\sim U}^{K(y,\cdot)}$

and $h > 0$ on R. Then $K(y, \cdot) \geq \hat{\mathbf{R}}_{R\sim U}^{K(y,\cdot)} \geq h > 0$. We know from the Riesz decomposition theorem that $\hat{\mathbf{R}}_{R\sim U}^{K(y,\cdot)} = w + h$ where w is a potential. Since $K(y, \cdot) = h + [K(y, \cdot) - h]$ and each term on the right is a K-potential by Lemma 12.9, h is proportional to $K(y, \cdot)$ by the extremality of the latter. Therefore $\hat{\mathbf{R}}_{R\sim U}^{K(y,\cdot)} = w + \lambda K(y, \cdot)$ for some $0 < \lambda \leq 1$. We cannot have $\lambda = 1$, for then $\hat{\mathbf{R}}_{R\sim U}^{K(y,\cdot)} = K(y, \cdot)$, with the former having support in $R_M^* \sim U$, whereas the latter has support $\{y\}$. Since the balayage of $\hat{\mathbf{R}}_{R\sim U}^{K(y,\cdot)}$ relative to $R \sim U$ is just itself and $\hat{\mathbf{R}}_{R\sim U}^{w+\lambda K(y,\cdot)} = \hat{\mathbf{R}}_{R\sim U}^{w} + \lambda \hat{\mathbf{R}}_{R\sim U}^{K(y,\cdot)}$ by Corollary 9.26,

$$w + \lambda K(y, \cdot) = \hat{\mathbf{R}}_{R\sim U}^{K(y,\cdot)} = \hat{\mathbf{R}}_{R\sim U}^{w+\lambda K(y,\cdot)} = \hat{\mathbf{R}}_{R\sim U}^{w} + \lambda \hat{\mathbf{R}}_{R\sim U}^{K(y,\cdot)}$$

$$= \hat{\mathbf{R}}_{R\sim U}^{w} + \lambda w + \lambda^2 K(y, \cdot).$$

Since w is a potential, $\hat{\mathbf{R}}_{R\sim U}^{w}$ is also a potential and $\lambda = \lambda^2$ by the uniqueness of the Riesz decomposition. Therefore $\lambda = 0$ and we have a contradiction; that is, the greatest harmonic minorant of $\hat{\mathbf{R}}_{R\sim U}^{K(y,\cdot)}$ is zero and $\hat{\mathbf{R}}_{R\sim U}^{K(y,\cdot)}$ is a potential.

LEMMA **12.20** Let X be a locally compact Hausdorff space and let μ be a measure on the Borel subsets of X. If for each $x \in X$, u_x is a non-negative superharmonic function on R and $u_x(y)$ is a Borel measurable function on $X \times R$, then $u(y) = \int u_x(y)\, d\mu(x)$ is either superharmonic or identically $+\infty$ on R.

Proof. By Fatou's lemma u is l.s.c. on R. By Fubini's theorem

$$L(u : y, \delta) = \int L(u_x : y, \delta)\, d\mu(x) \leq \int u_x(y)\, d\mu(x) = u(y)$$

whenever $\bar{B}_{y,\delta} \subset R$.

THEOREM **12.21** Let X be a locally compact Hausdorff space and let μ be a measure on X. If, for each $x \in X$, μ_x is a non-negative superharmonic function on R, $u_x(y)$ is a Borel measurable function on $X \times R$, F is a relatively closed subset of R, and $\int u_x(y_0)\, d\mu(x) < +\infty$ for some $y_0 \in R$, then

$$\hat{\mathbf{R}}_F^{\int u_x\, d\mu(x)} = \int \hat{\mathbf{R}}_F^{u_x}\, d\mu(x);$$

$\hat{\mathbf{R}}_F^{\int u_x d\mu(x)}$ is a potential if each $\hat{\mathbf{R}}_F^{u_x}$ is a potential.

Proof. Let λ denote the (completed) Lebesgue measure on R. We first show that $\hat{\mathbf{R}}_F^{u_x}(y)$ is equal a.e. $(\mu \times \lambda)$ to a Borel measurable function on $X \times R$. Let Z be the set of irregular boundary points for $R \sim F$. We know from the discussion following Theorem 9.24 that Z is a polar subset of $\partial(R \sim F)$ and a subset of a G_δ polar set D by Theorem 7.1. We can therefore assume that $Z \subset D$, D is a polar set, and D is a Borel subset of $\partial(R \sim F)$. By Theorem 9.25 $\hat{\mathbf{R}}_F^{u_x}(y) = u_x(y)$ for $y \in F \sim D$ and $\hat{\mathbf{R}}_F^{u_x}(y)$ is Borel measurable on $X \times (F \sim D)$. Now let $\mu_y^{R \sim F}$ denote harmonic measure relative to $y \in R \sim F$ and let $\mathscr{F}$ be the class of non-negative Borel measurable functions f on $X \times F$ for which

$$\tilde{f}(x, y) = \int f(x, z)\, d\mu_y^{R \sim F}(z)$$

is Borel measurable on $X \times (R \sim F)$. If f is a bounded continuous function on $X \times F$, then $\tilde{f}(x, \cdot)$ is harmonic on $R \sim F$ for each $x \in X$ and $\tilde{f}(\cdot, y)$ is continuous on X for each $y \in R \sim F$ by the Lebesgue dominated convergence theorem; that is, $\tilde{f}$ is continuous in each variable separately and therefore Borel measurable on $X \times (R \sim F)$ (see the proof of Lemma 6.7). The class $\mathscr{F}$ is obviously closed under monotone limits. It is easily seen that $\mathscr{F}$ contains the indicator function of any compact subset of $X \times (R \sim F)$ and, using the results in the discussion preceding Theorem 8.13, we can show that the indicator function of every Borel subset of $X \times (R \sim F)$ belongs to $\mathscr{F}$. It follows that each non-negative Borel measurable function belongs to $\mathscr{F}$. In particular,

$$\hat{\mathbf{R}}_F^{u_x}(y) = \int u_x(z)\, d\mu_y^{R \sim F}(z), \qquad y \in R \sim F,$$

is Borel measurable on $X \times (R \sim F)$. Therefore $\hat{\mathbf{R}}_F^{u_x}(y)$ is Borel measurable on $X \times (R \sim D)$, where D has Lebesgue measure zero. It follows that $\hat{\mathbf{R}}_F^{u_x}(y)$ is equal a.e. $(\mu \times \lambda)$ to a Borel measurable function on $X \times R$ and is therefore measurable relative to the completed class of $(\mu \times \lambda)$-measurable sets. This is all that is needed to apply Fubini's theorem. By the preceding Lemma $\int u_x\, d\mu(x)$ is superharmonic on R and $\hat{\mathbf{R}}_F^{\int u_x d\mu(x)}$ is defined. Since

$$\hat{\mathbf{R}}_F^{\int u_x\, d\mu(x)}(y) = \int u_x(y)\, d\mu(x)$$

and

$$\hat{\mathbf{R}}_F^{u_x}(y) = u_x(y) \quad \text{on} \quad F,$$

except possibly for $y \in Z$,

$$\hat{\mathbf{R}}_F^{\int u_x\, d\mu(x)}(y) = \int \hat{\mathbf{R}}_F^{u_x}(y)\, d\mu(x)$$

for $y \in F \sim Z$. Now for $y \in R \sim F$

$$\begin{aligned}\int \hat{\mathbf{R}}_F^{u_x}(y)\, d\mu(x) &= \int \left(\int u_x(z)\, d\mu_y^{R \sim F}(z) \right) d\mu(x)\\ &= \int \left(\int u_x(z)\, d\mu(x) \right) d\mu_y^{R \sim F}(z)\\ &= \hat{\mathbf{R}}_F^{\int u_x\, d\mu(x)}(y).\end{aligned}$$

Therefore

$$\hat{\mathbf{R}}_F^{\int u_x\, d\mu(x)}(y) = \int \hat{\mathbf{R}}_F^{u_x}(y)\, d\mu(x)$$

for $y \in R \sim Z$ (i.e., except possibly on a polar set). We know that the left side of this equation is a superharmonic function. If we can show that the right side is also superharmonic on R, it would follow that we have equality everywhere on R, since two superharmonic functions which are equal a.e. are, in fact, equal everywhere. By Fatou's lemma the right side of the above equation is a l.s.c. function of y. By Fubini's theorem

$$\begin{aligned}A\left[\int \hat{\mathbf{R}}_F^{u_x}\, d\mu(x) : y, \delta\right] &= \int A(\hat{\mathbf{R}}_F^{u_x} : y, \delta)\, d\mu(x)\\ &\le \int \hat{\mathbf{R}}_F^{u_x}(y)\, d\mu(x)\end{aligned}$$

whenever $\bar{B}_{y,\delta} \subset R$. Therefore $\int \hat{\mathbf{R}}_F^{u_x}\, d\mu(x)$ is superharmonic on R and

$$\hat{\mathbf{R}}_F^{\int u_x\, d\mu(x)} = \int \hat{\mathbf{R}}_F^{u_x}\, d\mu(x) \quad \text{on} \quad R.$$

Assume now that each $\hat{\mathbf{R}}_F^{u_x}$ is a potential. Let $\{R_j\}$ be an increasing sequence of open sets with compact closures $\bar{R}_j \subset R$ such that $R = \cup R_j$. Then by Theorem 8.41

$$\lim_{j \to \infty} \int \hat{\mathbf{R}}_F^{u_x}\, d\mu_y^{R_j} = 0$$

for all $y \in R$. Since $\hat{\mathbf{R}}_F^{u_x}(y)$ is Borel measurable on $X \times (R \sim D)$, where D is a Borel measurable polar set, it is Borel measurable on $X \times (\partial R_j \sim D)$. Since a polar subset of ∂R_j has $\mu_y^{R_j}$-measure zero by Theorem 9.24, $\hat{\mathbf{R}}_F^{u_x}(y)$ is equal a.e. $(\mu \times \mu_y^{R_j})$ to a Borel measurable function on $X \times \partial R_j$. By Fubini's theorem

$$\begin{aligned}\lim_{j \to \infty} \int \left(\int \hat{\mathbf{R}}_F^{u_x}(z)\, d\mu(x) \right) d\mu_y^{R_j}(z) &= \lim_{j \to \infty} \int \left(\int \hat{\mathbf{R}}_F^{u_x}(z)\, d\mu_y^{R_j}(z) \right) d\mu(x)\\ &= \int \left(\lim_{j \to \infty} \int \hat{\mathbf{R}}_F^{u_x}(z)\, d\mu_y^{R_j}(z) \right) d\mu(x)\\ &= 0.\end{aligned}$$

It follows that $\int \hat{\mathbf{R}}_F^{u_x}\, d\mu(x)$ is a potential on R.

We know that the constant function 1 has a representation

$$1 = \int K(z, \cdot)\, d\chi(z),$$

where χ is the unique measure on Δ_1 in Martin's representation. We shall use χ as a reference measure.

THEOREM **12.22** (*Brelot*) Each $f \in C(\Delta_M)$ is resolutive with

$$H_f = \int f K(z, \cdot)\, d\chi(z) \quad \text{on} \quad R.$$

Proof. Let f be a continuous function on R_M^* such that $0 \leq f \leq 1$. For each positive integer j and integers i such that $0 \leq i \leq j$ let

$$E_i = \left\{y \in \Delta_1 : \frac{i}{j} - \frac{1}{2j} \leq f(y) < \frac{i}{j} + \frac{1}{2j}\right\},$$

$$F_i = \left\{y \in R : f(y) \leq \frac{i}{j} - \frac{1}{j}\right\} \cup \left\{y \in R : f(y) \geq \frac{i}{j} + \frac{1}{j}\right\},$$

$$h_i = \int_{E_i} K(z, \cdot)\, d\chi(z) = \int K(z, \cdot)\, d\chi_i(z),$$

where $\chi_i(E) = \chi(E \cap E_i)$ for all Borel sets $E \subset \Delta_M$. Since χ has support in Δ_M, each h_i is harmonic on R. Clearly $E_i \subset R_M^* \sim \bar{F}_i$ and $R_M^* \sim \bar{F}_i$ is a neighborhood of E_i. By the preceding theorem

$$\hat{\mathbf{R}}_{F_i}^{h_i} = \int \hat{\mathbf{R}}_{F_i}^{K(z,\cdot)}\, d\chi_i(z).$$

For z in the support of χ_i, $\hat{\mathbf{R}}_{F_i}^{K(z,\cdot)}$ is a potential by Lemma 12.19 and it follows from the preceding theorem that $\hat{\mathbf{R}}_{F_i}^{h_i}$ is a potential. Since $(i-1)/j < f < (i+1)/j$ on $R \sim F_i$ and $h_i \geq 0$,

$$\frac{i-1}{j}(h_i - \hat{\mathbf{R}}_{F_i}^{h_i}) \leq f h_i \leq \frac{i+1}{j} h_i + \hat{\mathbf{R}}_{F_i}^{h_i}$$

on $R \sim F_i$, $0 \leq i \leq j$. In view of the fact that $\hat{\mathbf{R}}_{F_i}^{h_i} = h_i$ on F_i, except possibly for a polar subset of F_i, this inequality holds on R except possibly for a polar set and, in fact, it is easily seen to hold everywhere on R. Therefore

$$\sum_{i=1}^{j} \frac{i-1}{j}(h_i - \hat{\mathbf{R}}_{F_i}^{h_i}) \leq \sum_{i=1}^{j} f h_i \leq f \sum_{i=0}^{j} h_i \leq \sum_{i=0}^{j} \left(\frac{i+1}{j} h_i + \hat{\mathbf{R}}_{F_i}^{h_i}\right) \quad \text{on} \quad R.$$

Since $\sum_{i=0}^{j} h_i = 1$,

$$\sum_{i=1}^{j} \frac{i-1}{j} (h_i - \hat{\mathbf{R}}_{F_i}^{h_i}) \le f \le \sum_{i=0}^{j} \left(\frac{i+1}{j} h_i + \hat{\mathbf{R}}_{F_i}^{h_i} \right) \quad \text{on} \quad R.$$

Now the left side of this inequality is a subharmonic function which belongs to the lower class defining $\underline{H}_f$ and, similarly, the right side belongs to the upper class defining $\overline{H}_f$. Therefore,

$$\sum_{i=1}^{j} \frac{i-1}{j} h_i - \sum_{i=1}^{j} \frac{i-1}{j} \hat{\mathbf{R}}_{F_i}^{h_i} \le \underline{H}_f \le \overline{H}_f \le \sum_{i=0}^{j} \frac{i+1}{j} h_i + \sum_{i=0}^{j} \hat{\mathbf{R}}_{F_i}^{h_i} \quad \text{on} \quad R.$$

As a potential, each $\hat{\mathbf{R}}_{F_i}^{h_i}$ has a greatest harmonic minorant zero. Taking the greatest harmonic minorant of the two right members of this inequality (and using a minor modification of this argument for the left members) and using Lemma 5.22,

$$\sum_{i=1}^{j} \frac{i-1}{j} h_i \le \underline{H}_f \le \overline{H}_f \le \sum_{i=0}^{j} \frac{i+1}{j} h_i \quad \text{on} \quad R.$$

Hence

$$0 \le \overline{H}_f - \underline{H}_f \le \sum_{i=0}^{j} \frac{2}{j} h_i = \frac{2}{j} \quad \text{on} \quad R$$

for every positive integer j. Therefore f is resolutive. If now f is just any continuous function on Δ_M, then it has a continuous extension to all of R_M^*. Since the addition of a constant or multiplication by a constant does not affect resolutivity, it can be assumed that $0 \le f \le 1$ and the preceding argument applies. This proves that each $f \in C(\Delta_M)$ is resolutive. For such f note that (using the above notation)

$$\sum_{i=1}^{j} \frac{i-1}{j} \int_{E_i} K(z, \cdot)\, d\chi(z) \le H_f \le \sum_{i=0}^{j} \frac{i+1}{j} \int_{E_i} K(z, \cdot)\, d\chi(z)$$

and that

$$\sum_{i=0}^{j} \frac{i-1}{j} \int_{E_i} K(z, \cdot)\, d\chi(z) \le \int f(z) K(z, \cdot)\, d\chi(z) \le \sum_{i=0}^{j} \frac{i+1}{j} \int_{E_i} K(z, \cdot)\, d\chi(z).$$

Therefore

$$\left| H_f - \int f(z) K(z, \cdot)\, d\chi(z) \right| \le \frac{2}{j}.$$

for every positive integer j. This proves the second assertion.

Notes and Comments

Chapter 1

All references to the calculus in this book concern the calculus of functions of several variables as is done in Apostol's book [1]. Many references to nineteenth-century books and papers on potential theory can be found in Kellogg [1, 3]. Lebesgue [1] has a discussion of the contributions of Gauss, Weierstrass, Riemann, Dirichlet, etc., to the early development of potential theory. Brelot [9] has an excellent historical survey commencing with the work of Frostman.

Chapter 2

The basic result of this chapter that each continuous function on the boundary of a ball has a continuous extension to the closed ball which is harmonic on the interior is due to Schwarz [1]. The fact that harmonic functions are mean-valued was known to Gauss.

Lemma 2.12 and Theorem 2.13 are due to Herglotz [1]. A discussion of the nature of the signed measure μ appearing in Herglotz's theorem has been omitted. If h is harmonic on the ball $B_{y,\rho}$ and f_r is the restriction of h to $\partial B_{y,r}$, $0 < r < \rho$, then the signed measure μ of Herglotz's theorem corresponding to h can be identified as the indefinite integral of a function f on $\partial B_{y,\rho}$ under suitable conditions on the family $\{f_r : 0 < r < \rho\}$; such results can be found in several books (cf Hoffman [1]).

Harnack's inequality is classical but Corollary 2.15 is of more recent origin. Harnack's inequality for balls in a Stolz domain is nearly obvious for positive harmonic functions on a half-space. Applying the classical Harnack inequality to a ball internally tangent to a cone having half-angle less than $\pi/2$, vertex on the boundary of the half-space, and axis normal to the boundary of the half-space, the inequality is easily seen to extend to any ball in the cone by applying a similarity transformation. Corollary 2.15 can then be obtained by using an inversion to map the half-space onto a ball. A condensation of this discussion can be found in Doob [2].

Equicontinuity of locally bounded families of harmonic functions is classical and due to Osgood [1] and Montel [1]. Lemma 2.20 is due to Choquet (cf Brelot and Choquet [2]).

The Kelvin transformation dates back to the middle of the nineteenth century. A more thorough discussion of the Kelvin transformation can be found in Kellogg [3]. Theorem 2.26 can be proved, of course, by mapping the half-space onto a ball by means of an inversion.

Chapter 3

Differentiation of a measure on the Borel subsets of E^n relative to Lebesgue measure is classical. The extension to compact metric spaces can be found in the book of Dunford and Schwartz [1]; the differentiation theory of Section 2 was adapted from that of Dunford and Schwartz. The existence of nontangential limits for positive harmonic functions was first proved by Fatou [1] for the $n = 2$ case. The history of the $n \geq 3$ case is obscure. The author is aware of at least two papers on the nontangential limit theorem for the $n \geq 3$ case (cf Bray and Evans [1] or Tsuji [1]). The use of the radial limit theorem (Theorem 3.1) and Harnack's inequality for a Stoltz domain to prove Fatou's nontangential limit theorem is due to Doob [2].

The radial limit theorem has been generalized in several directions, Littlewood [1] proved in the $n = 2$ case that the potential $G_B \mu$ of a measure on a ball B has radial limit zero at almost all boundary points of B. Littlewood's theorem was later extended to positive superharmonic functions by F. Riesz [2]. The $n \geq 3$ version of Littlewood's theorem is due to Privalov [1, 2]. The nontangential limit theorem does not extend to superharmonic functions; a discussion of this problem can be found in a paper by Tolsted [1]. Doob [2] has shown that a ratio u/h of two positive harmonic functions on a ball B has a nontangential limit at almost all points of ∂B, the "almost all" being relative to the measure μ on ∂B of Herglotz's theorem for which $h = \mathbf{PI}(\mu, B)$. Further generalizations have been obtained for harmonic functions on a half-space by Calderon [1] and Carleson [1].

Chapter 4

Superharmonic functions having continuous second partials date back to the nineteenth century, but lower semicontinuous superharmonic functions were first introduced by F. Riesz [1]. For much more about convex functions and subharmonic functions as well as numerous references on this subject see Rado [1]. The approximation theorems 4.20 and 4.21 are due to F. Riesz [2]. Convexity properties of the spherical averages of the modulus of an analytic function were first obtained by Hardy [1]; convexity properties of the spherical averages of superharmonic functions are due to F. Riesz [1].

Chapter 5

Part (i) of Theorem 5.4 is due to Schwarz [1] in the case of bounded harmonic functions; the theorem, as stated, is due to Bôcher [1]. The first fairly general existence theorem for the Green function is due to Osgood [1] in the case of simply connected regions in the plane.

The method used to define the reduction of a superharmonic function over an open set is due to Poincaré [1] who called it the "méthode de balayage" or the "sweeping out method." Greatest harmonic minorants were first considered by F. Riesz [1].

Chapter 6

The results of Section 2 are due to Cartan [1, 2, 3] as are most of the results of Section 3. Theorem 6.18 can be found in F. Riesz [2]. Theorem 6.20 is due to Evans [1] and Vasilesco [2]. Theorem 6.21, the potential theory version of Lusin's theorem, is due to Choquet [5].

Chapter 7

Polar sets were introduced by Brelot [5] to replace the notion of inner capacity zero. Theorem 7.6, if put in terms of inner capacity, is due to Frostman [1], at least for Borel sets.

Historically, Poincaré's "méthode de balayage" was devised to solve the problem of the equilibrium distribution which was known to Gauss. Physically, the problem corresponded to determining the potential at a point inside a grounded sphere due to an electrical charge on a conductor in the interior of the sphere; such a charge will distribute itself on the surface of the conductor in such a way that the potential on the conductor is constant. Since there would be no charge outside the conductor, the potential would be harmonic there, constant on the conductor, and zero on the grounded sphere. This potential was known as the equilibrium potential (i.e., the capacitary potential). That there was such a potential was accepted without question in the early part of the nineteenth century. It was eventually recognized, however, that there was a mathematical problem of proving the existence of the equilibrium potential. In the older literature this problem was known as Robin's problem. Mathematically, the problem is that of solving the Dirichlet problem for the region between the conductor and the sphere corresponding to the boundary value 1 on the surface of the conductor and the boundary value zero on the grounded sphere. Now the equilibrium potential is easily determined for a ball which is concentric with the grounded sphere and, in fact, is the potential V corresponding to a uniformly distributed charge on ∂B. By taking the reduction of V over the region between the conductor and the grounded sphere, Poincaré arrived at the equilibrium potential for the conductor under some geometrical conditions. On the one hand, Poincaré devised a method of solving the Dirichlet problem and, on the other hand, devised an operation which transformed the equilibrium potential for the ball into the equilibrium potential for the conductor. The balayage of a superharmonic function, as defined in Section 3, accomplishes the same thing. The systematic study of the réduite of a superharmonic function commenced with Brelot [8].

In physics the ratio of the total charge on the conductor to the value of the equilibrium potential on the conductor is known as the capacity of the conductor. As the capacity depends only upon the ratio, the capacity is equal to the total charge if the equilibrium potential is equal to 1 on the conductor. Inasmuch as the capacitary potential V_K, as defined at the beginning of Section 4, is 1 on K (the conductor), the definition of capacity herein agrees with the physical definition of capacity. The mathematical concept of capacity was first defined by Wiener [3] for compact sets (in the $n \geq 3$ case) in the following way. Let K be a compact set and let u be the Dirichlet solution for the open set $\sim K$ corresponding to the boundary value 1 on ∂K and 0 at infinity. Then Wiener defined the capacity $\mathscr{C}(K)$ of K by the equation

$$\mathscr{C}(K) = -\frac{1}{4\pi}\int_S D_{\mathbf{n}} u \, d\sigma,$$

where S is a smooth surface encompassing K. The notion of capacity was subsequently extended to any bounded set by De La Vallée Poussin [1]. Frostman [1] deserves much of the credit for the development of the theory of capacities; in particular, Frostman recognized the similarity of De La Vallée Poussin's capacity to an inner measure on the class of bounded sets. Brelot [2] relabeled De La Vallée Poussin's capacity as an inner capacity and also defined the outer capacity of a set. Monna [1] defined the outer capacity of a set at about the same time. After discovering that capacity is strongly subadditive and developing the theory of $\mathscr{K}$-analytic sets, Choquet [1] proved that the Borel sets are capacitable.

Theorem 7.33 is due to Cartan [2] and is stated in terms of outer capacity in much of the literature. Theorem 7.34 is due to Evans [2] and has been improved upon. Deny [1] has shown that if Z is a G_δ polar subset of an open set R having a Green function, then there is a measure μ such that $G\mu = +\infty$ on Z and $G\mu < +\infty$ on $R \sim Z$; the measure μ, however, need not be carried by Z. Choquet [6] has shown that the measure can be chosen so as to be carried by Z.

Rado [1] proved that the limit of decreasing sequence of lower bounded superharmonic functions differs from a superharmonic function on at most a set of Lebesgue measure zero; Brelot [1] showed that the exceptional set is at most of inner capacity zero. Replacing decreasing sequences by left-directed families, Cartan [2] was able to improve Brelot's result by showing that the exceptional set has outer capacity zero.

A theory of polar sets and capacity can be based upon functions other than the Green function (cf Frostman [1], M. Riesz [1], Cartan [2, 3], and Brelot [12, 13]).

Chapter 8

As pointed out in the comments on Chapter 7, the first fairly general existence theorem for Dirichlet's problem is due to Poincaré. Perron [1] introduced the upper and lower classes corresponding to a boundary function f and associated upper and lower $\overline{H}_f$ and $\underline{H}_f$. Theorem 8.11 is due to Wiener [3] and Theorem 8.17 is due to Brelot [3].

Barriers were used by Poincaré [1] to show that certain regions are regular. Barriers, in general, were defined by Lebesgue [3] but his definition included the additional condition that the barrier must have a strictly positive limit inferior at nearby points of the boundary other than the point in question. Bouligand [2] proved that the additional condition is superfluous. Theorem 8.24 is due to Poincaré [1]. Theorem 8.26 is essentially due to Lebesgue [1] who showed that if x is an irregular boundary point for a bounded open subset R of the plane, then each neighborhood of x contains a Jordan curve which lies in R and contains x in its interior. Theorem 8.27 is due to Zaremba [1]. Discussions of sufficient conditions for regularity of boundary points can be found in Wiener [2] and Lebesgue [3]. Bouligand [2] was the first to characterize irregularity of boundary points in terms of the boundary behavior of the Green function as in Theorem 8.34. Theorem 8.33 is due to Myrberg [1].

Theorem 8.39 is due to Maria [1] and Frostman [1]; the result is known as the Maria-Frostman domination principle.

Most of the results of this chapter depend only upon a few properties of harmonic functions; for example, that the harmonic functions on an open set form a linear class, that harmonicity is a local property, that the limit of an increasing or decreasing sequence of harmonic functions is again harmonic under a condition of local boundedness, and that the classical Dirichlet problem is solvable for balls. Recognizing this, Brelot [10, 11] has developed an axiomatic potential theory in which the above properties are taken as axioms. In addition to these references, see Brelot [13], Bauer [1], and Hervé [1]. Brelot and Choquet [1] have also axiomatized potential theory on a slightly less abstract level by studying spaces which are locally homeomorphic to n-dimensional Euclidean space.

Chapter 9

Most of the material of this chapter was taken directly from Brelot [7]. The method used to extend a harmonic function in the proof of Lemma 9.19 is known as Schwarz's alternating method and is classical. In the early development of potential theory the method was used to construct a solution of the Dirichlet problem on a region which is the union of two regular regions under suitable conditions. A discussion along these lines can be found in Kellogg [3]. Corollary 9.20 is due to Bouligand [1].

Chapter 10

The fine topology was introduced by Cartan. Thinness of a set originated with Brelot [2]. A proof of Fatou's nontangential limit theorem for a ball based upon the fine topology can be found in a paper by Brelot and Doob [1].

Wiener [2] proved that a boundary point x of a bounded open set R is irregular if and only if $\sim R$ satisfies the condition of Theorem 10.21. Since thinness of $\sim R$ at x and irregularity of x are equivalent by Corollary 10.12, Wiener's original result is a special case of Theorem 10.21. Theorem 10.21 can be found in Brelot [6].

Chapter 11

Although most of the basic energy concepts were known to Gauss, Frostman [1] put the subject on a sound mathematical basis. Theorem 11.3 was first proved by Frostman.

The fact that the energy of a signed measure, if defined, is always nonnegative is more apparent in the classical case. Let G be the Green function for E^n, $n \geq 3$, let μ be a measure such that $u = G\mu$ has continuous second partials, and let v be a function which has continuous second partials and vanishes outside a compact set. If ν is the signed measure having density $-\Delta v/\kappa_n$, then $G\nu = G(-\Delta v/\kappa_n) = v$. It follows from the divergence theorem that for any ball B containing the support of ν

$$\int_B G\mu \, d\nu = \int_B u(-\Delta v/\kappa_n) \, dz$$

$$= \frac{1}{\kappa_n} \int_B (\text{grad u, grad v}) \, dz.$$

In particular, if $\mu = \nu$, then

$$\|\nu\|_e^2 = \frac{1}{\kappa_n} \int |\text{grad } v|^2 \, dz \geq 0,$$

even if ν is a signed measure. The non-negativeness of the energy even for signed measures is therefore apparent in the classical case. The last equation can also be written

$$\|\nu\|_e^2 = \frac{1}{\kappa_n} \int \left[\left(\frac{\partial v}{\partial x_1} \right)^2 + \cdots + \left(\frac{\partial v}{\partial x_n} \right)^2 \right] dz;$$

the last integral on the right is known as the Dirichlet integral. For a discussion of the role of the Dirichlet integral in classical potential theory, see Kellogg [3] or Courant [1].

The theorems on the various types of convergence of elements of $\mathscr{E}$, projections of measures, and the completeness of $\mathscr{E}$ are due to Cartan [1, 2, 3]. Deny [2] has investigated the metric completion of $\mathscr{E}$ and interpreted the results in terms of the distributions of L. Schwartz.

Chapter 12

The basic facts concerning the representation of positive harmonic functions on an open connected set are due to Martin [1]. The exposition of this chapter, however, is an adaptation of the development of the Martin boundary by Constantinescu and Cornea [1]. Theorem 12.1 is due to these two authors. Commencing with the $\mathscr{Q}$-compactification of Theorem 12.1, Constantinescu and Cornea derive the Martin representation using kernels with the essential properties of the Green function.

In the original work of Martin the metric of Theorem 12.8 (restricted to $R \times R$) was used to obtain a metric completion of R which was shown by Martin to be a compactification of R. Theorem 12.22 is due to Brelot [10].

Martin's representation of a positive harmonic function can also be obtained as an application of the work of Choquet [2, 3, 4] in which elements of a convex cone (the positive harmonic functions) are represented as an integral over the extreme points (the minimal positive harmonic functions) of the cone. See also Brelot [13], Hervé [1], and Meyer [1].

Naïm [1] has extended the fine topology to the Martin boundary as well as thinness at a point of the Martin boundary. In doing so, she has established that ratios of the form $u/G(\cdot, x_0)$ and $u/K(\cdot, x_0)$, where u is a positive superharmonic function, have fine limits at minimal points of the Martin boundary; she has also established the existence of the fine limit of ratios of the form u/h, where u is a potential and h is a positive harmonic function, at almost all points of the Martin boundary, the "almost all" being relative to the measure on the Martin boundary associated with h by means of the Martin representation. She has also used the fine topology to generalize the concept of regular boundary point. Doob [1] has generalized Fatou's nontangential limit theorem in this context.

Bibliography

Apostol, T. M.

[1] *Mathematical Analysis*, Addison-Wesley, Reading, Mass. 1957.

Bauer, H.

[1] *Harmonische Räume und Ihre Potentialtheorie*, Lecture notes in Mathematics (22), Springer-Verlag, Berlin, 1966.

Bôcher, M.

[1] *Singular points of functions which satisfy partial differential equations of the elliptic type*, *Bull. Amer. Math. Soc.*, **9**, 455–465 (1903).

Bouligand, G.

[1] *Sur les fonctions bornées et harmoniques dans un domaine infini, nulles sur sa frontière*, *C.R. Acad. Sci.* Paris, **169**, 763–766 (1919).

[2] *Domaines infinis et cas d'exception du problème de Dirichlet*, *C.R. Acad. Sci.* Paris, **178**, 1054–1057 (1924).

Bray, H. E., and G. C. Evans

[1] *A class of functions harmonic within the sphere*, *Amer. J. Math.*, **49**, 153–180 (1927).

Brelot, M.

[1] *Sur le potentiel et les suites de fonctions sous-harmoniques*, *C. R. Acad. Sci.*, Paris, **207**, 836–839 (1938).

[2] *Sur la théorie moderne du potentiel*, *C.R. Acad. Sci.* Paris, **209** 828–830 (1939).

[3] *Familles de Perron et problème de Dirichlet*, *Acta Sci. Math.* (Szeged), **9**, 133–153 (1939).

[4] *Points irréguliers et transformations continues on théorie du potentiel*, *J. Math. Pures Appl.* **19**, 319–337 (1940).

[5] *Sur la théorie autonome des fonctions sous-harmoniques*, *Bull. Sci. Math.*, **65**, 72–98 (1941).

[6] *Sur les ensembles effilés*, *Bull. Soc. Math.* France, **68**, 12–36 (1944).

[7] *Sur le rôle du point à l'infini dans la théorie des fonctions harmoniques*, *Ann. Sci. École Norm. Sup.*, **61**, 301–332 (1944).

[8] *Minorantes sousharmoniques, extrémales et capacités*, *J. Math. Pures Appl.*, **24**, 1–32 (1945).

[9] *La théorie moderne du potentiel*, *Ann. Inst. Fourier* (Grenoble), **4**, 113–140 (1952).

[10] *Le problème de Dirichlet. Axiomatique et frontière de Martin*, *J. Math. Pures Appl.*, **35**, 297–335 (1956).

[11] *Axiomatique du problème de Dirichlet dans les espaces localement compacts*, Séminaire de Théorie du Potentiel, Inst. H. Poincaré, Univ. de Paris, 1957.

[12] *Éléments de la théorie classique du potentiel*, Centre de Documentation Universitaire, Paris, 1959.

[13] *Lectures on Potential Theory*, Tata Institute of Fundamental Research, Bombay, 1960.

BRELOT, M., and G. CHOQUET

[1] *Espaces et lignes de Green, Ann. Inst. Fourier* (Grenoble), **3**, 199–263 (1951).

[2] *Le théorème de convergence en théorie du potentiel, J. Madras Univ. B*, **27**, 277–286 (1957).

BRELOT, M., and J. L. DOOB

[1] *Limites angulaires et limites fines, Ann. Inst. Fourier* (Grenoble), **13**, 395–415 (1963).

CALDERON, A. P.

[1] *On the behavior of harmonic functions at the boundary, Trans. Amer. Math. Soc.*, **68**, 47–54 (1950).

CARLESON, L.

[1] *On the existence of boundary values for harmonic functions in several variables, Ark. Math.* **4**, 393–399 (1961).

CARTAN, H.

[1] *Sur les fondements de la théorie du potentiel, Bull. Soc. Math.* France, **69**, 71–96 (1941).

[2] *Théorie du potentiel newtonian: énergie, capacité, suites de potentiels, Bull Soc. Math.* France, **73**, 74–106 (1945).

[3] *Théorie générale du balayage en potentiel newtonien, Ann. Univ. Grenoble, Math. Phys.* **22**, 221–280 (1946).

CHOQUET, G.

[1] *Theory of capacities. Ann. Inst. Fourier* (Grenoble), **5**, 131–295 (1954).

[2] *Unicité des représentations intégrales au moyen de points extrémaux dans les cônes convexes réticulés, C.R. Acad. Sci.* Paris, **243**, 555–557 (1956).

[3] *Existence des représentations intégrales au moyen des points extrémal dans les cônes convexes, C.R. Acad. Sci.* Paris, **243**, 699–702 (1956).

[4] *Existence des représentations intégrale dans les cônes convexes, C.R. Acad. Sci.* Paris, **243**, 736–737 (1956).

[5] *Sur les fondements de la théorie fine du potentiel, C.R. Acad. Sci.* Paris, **244**, 1606–1609 (1957).

[6] *Potentiels sur un ensemble de capacité nulle. Suites de potentiels, C.R. Acad. Sci.* Paris, **244**, 1707–1710 (1957).

CONSTANTINESCU, C., and A. CORNEA

[1] *Ideale Ränder Riemannscher Flächen*, Ergebnisse der Mathematik und Ihrer Grenzgebiete, Springer-Verlag, Berlin, 1963.

COURANT, R.

[1] *Dirichlet's Principle, Conformal Mapping, and Minimal Surfaces*, Interscience, New York, 1950.

DE LA VALLÉE POUSSIN, C.

[1] *Extension de la méthode du balayage de Poincaré et problème de Dirichlet, Ann. Inst. H. Poincaré*, **2**, 169–232 (1932).

DENY, J.

[1] *Sur les infinis d'un potentiel, C.R. Acad. Sci.* Paris, **224**, 524–525 (1947).

[2] *Les potentiels d'énergie finie, Acta. Math.*, **82**, 107–183 (1950).

DOOB, J. L.

[1] *A non-probabilistic proof of the relative Fatou theorem, Ann. Inst. Fourier* (Grenoble), **9**, 293–300 (1959).

[2] *A relativized Fatou theorem, Proc. Nat. Acad. Sci.* USA **45**, 215–222 (1959).

DUNFORD, N., and J. T. SCHWARTZ

[1] *Linear Operators, Part I, General Theory*, Interscience, New York, 1958.

Evans, G. C.
[1] *On potentials of positive mass I, Trans. Amer. Math. Soc.*, **37**, 226–253 (1935).
[2] *Potentials and positively infinite singularities of harmonic functions, Monatsh. Math.*, **43**, 419–424 (1936).
Fatou, P.
[1] *Séries trigonométriques et séries de Taylor, Acta Math.*, **30**, 335–400 (1906).
Frostman, O.
[1] *Potentiel d'équilibre et capacité des ensembles avec quelques applications à la théorie des fonctions, Thesis, Lunds Univ. Mat. Sem.*, **3**, 1–118 (1935).
Hardy, G. H.
[1] *On the mean value of the modulus of an analytic function, Proc. London Math. Soc.*, **2**, series 14, 269–277 (1915).
Herglotz, G.
[1] *Über Potenzreihen mit positivem, reellem teil im einheitskreis*, Ber. Verhandl. *Sachs Akad. Wiss. Leipzig, Math. -Phys. Klasse*, **63** (1911).
Hervé, R. M.
[1] *Recherches axiomatiques sur la théorie des fonctions surharmoniques et du potentiel, Ann. Inst. Fourier* (Grenoble), **12**, 415–571 (1962).
Hoffman, K.
[1] *Banach Spaces of Analytic Functions*, Prentice-Hall, Englewood Cliffs, N.J., 1962.
Kellogg, O. D.
[1] *Recent progress with the Dirichlet problem, Bull. Amer. Math. Soc.*, **32**, 601–625 (1926).
[2] *On the classical Dirichlet problem for general domains, Proc. Nat. Acad. Sci.* USA, **12**, 397–406 (1926).
[3] *Foundations of Potential Theory*, Springer, Berlin, 1929 (reprinted by Dover, New York, 1953).
Lebesgue, H.
[1] *Sur le problème de Dirichlet, Rend. Circ. Mat.* Palermo, **24**, 371–402 (1907).
[2] *Sur des cas d'impossibilité du problème de Dirichlet ordinaire, C.R. Séances Soc. Math. France*, 17 (1912).
[3] *Conditions de régularité, conditions d'irrégularité, conditions d'impossibilité dans le problème de Dirichlet, C.R. Acad. Sci.* Paris, **178**, 349–354 (1924).
Lelong-Ferrand, J.
[1] *Étude au voisinage de la frontière des fonctions surharmoniques positives dans un demi-espace, Ann. Ecole Norm. Sup.*, **66**, 125–159 (1949).
Littlewood, J. E.
[1] *Mathematical Notes* (8): *On functions subharmonic in a Circle (II), Proc. London Math. Soc.*, (2) **28**, 383–394 (1928).
Maria, A. J.
[1] *The potential of a positive mass and the weight function of Wiener, Proc. Nat. Acad. Sci.* USA, **20**, 485–489 (1934).
Martin, R. S.
[1] *Minimal positive harmonic functions, Trans. Amer. Math. Soc.*, **49**, 137–172 (1941).
Meyer, P. A.
[1] *Probability and Potentials*, Blaisdell, Waltham, Mass., 1966.
Monna, A. F.
[1] *Sur la capacité des ensembles. Proc. Kon. Ned. Akad. Wetensch.* Amsterdam, **43**, 81–86 (1940).
Montel, P.
[1] *Sur les suites infinis de fonctions, Ann. Ecole Norm. Sup.*, **43**, 233–334 (1907).

MYRBERG, P. J.
[1] *Über die existenz der Greenschen Funktionen auf einer gegeben Riemannschen Fläche*, *Acta Math.* **61**, 39–79 (1933).
NAÏM, L. (Mrs. Gunter Lumer)
[1] *Sur le rôle de la frontière de R. S. Martin dans la théorie du potentiel*, *Ann. Inst. Fourier* (Grenoble), **7**, 183–281 (1957).
OSGOOD, W. F.
[1] *On the existence of the Green's function for the most general simply connected plane region*, *Trans. Amer. Math. Soc.*, **1**, 310–314 (1900).
PERRON, O.
[1] *Eine neue behandlung der ersten Randwertaufgabe für* $\Delta u = 0$, *Math. Z.*, **18**, 42–54 (1923).
POINCARÉ, H.
[1] *Sur les equations aux dérivées partielles de la physique mathématique Amer. J. Math.*, **12**, 211–294 (1890).
PRIVALOV, N.
[1] *On a boundary problem of subharmonic functions*, *Mat. Sbornik*, **41**, 3–10 (1934).
[2] *Boundary problems of the theory of harmonic and subharmonic functions in space*, *Mat. Sbornik*, **45**, 3–25 (1938),
RADO, T.
[1] *Subharmonic Functions*, Ergebnisse der Mathematik und Ihrer Grenzgebeite, Berlin, 1937 (reprinted by Chelsea, New York, 1949).
RIESZ, F.
[1] *Sur les fonctions subharmoniques et leur rapport à la théorie du potentiel*, *Acta Math.* **48**, 329–343 (1926).
[2] *Sur les fonctions subharmoniques et leur rapport à la théorie du potentiel*, *Acta Math.*, **54**, 321–360 (1930).
RIESZ, M
[1] *Integrales de Riemann-Liouville et potentiel*, *Acta Litterarum Ac Scientiarum*, **9**, 1–42 (1938–1940).
SCHWARZ, H. A.
[1] *Zur integration der partiellen differentialgleichung* $\dfrac{\partial^2 u}{\partial x^2} + \dfrac{\partial^2 u}{\partial y^2} = 0$, *J. Reine Angew. Math.*, **74**, 218–253 (1872).
TOLSTED, E.
[1] *Limiting values of subharmonic functions*, *Proc. Amer. Math. Soc.*, **1**, 636–647 (1950).
TSUJI, M.
[1] *On Fatou's theorems on Poisson integrals*, *Japan J. Math.*, **15**, 13–18 (1938–1939).
VASILESCO, F.
[1] *Sur les singularités des fonctions harmoniques*, *J. Math. Pures Appl.*, **9**, 81–111 (1930).
[2] *Sur la continuité du potentiel à travers des masses et la démonstration d'une lemme de Kellogg*, *C.R. Acad. Sci.* Paris, **200**, 1173–1174 (1935).
WIENER, N.
[1] *Discontinuous boundary conditions and the Dirichlet problem*, *Trans. Amer. Math. Soc.*, **25**, 307–314 (1923).
[2] *The Dirichlet problem*, *J. Math. Phys.*, **3**, 127–146 (1924).
[3] *Certain notions in potential theory*, *J. Math. Phys.*, **3**, 24–51 (1924).
ZAREMBA, S.
[1] *Sur le principe du minimum*, *Bull. de l'Ac. Sc. Cracovie* (1909).

Index of Notation

Subject Index